INERA Conference on Light in Nanoscience and Nanotechnology 2015 (LNN 2015)

Journal of Physics: Conference Series Volume 682

Hissar, Bulgaria
19 – 23 October 2015

Editors:

Diana Nesheva
Hassan Chamati
Julia Genova
Kostadinka Gesheva
Tatyana Ivanova
Anna Szekeres

ISBN: 978-1-5108-2092-0
ISSN: 1742-6588

Printed from e-media with permission by:

Curran Associates, Inc.
57 Morehouse Lane
Red Hook, NY 12571

CURRAN ASSOCIATES INC. proceedings.com

Some format issues inherent in the e-media version may also appear in this print version.

Printed by Curran Associates, Inc. (2016)

For permission requests, please contact the Institute of Physics
at the address below.

Institute of Physics
Dirac House, Temple Back
Bristol BS1 6BE UK

Phone: 44 1 17 929 7481
Fax: 44 1 17 920 0979

techtracking@iop.org

Additional copies of this publication are available from:

Curran Associates, Inc.
57 Morehouse Lane
Red Hook, NY 12571 USA
Phone: 845-758-0400
Fax: 845-758-2634
Email: curran@proceedings.com
Web: www.proceedings.com

Table of contents

Volume 682

INERA Conference 2015: Light in Nanoscience and Nanotechnology (LNN 2015)
19–23 October 2015, Hissar, Bulgaria

Accepted papers received: 15 December 2015
Published online: 9 February 2016

Posters

PREFACE

We are pleased to introduce the Proceedings of the Conference "Light in Nanoscience and Nanotechnology 2015" (LNN 2015) organized by the Institute of Solid State Physics, Bulgarian Academy of Sciences in the frames of the INERA Project "Research and Innovation Capacity Strengthening of ISSP-BAS in Multifunctional Nanostructures" REGPOT-2012-2013-1 NMP. The LNN 2015 Conference was dedicated to the "International year of light – 2015".

The Conference took place from 20th to 22nd of October in the beautiful spa resort Hissar, situated 140 km away from Sofia, close to the famous Valley of Roses, amidst a real abundance of curative mineral waters. The resort has an old history - even the ancient Romans knew well the curative properties of the water. Today the town has more than 4 km of Roman walls and architecture remains. During the age of the Roman Empire, the town, called Augusta, was a wealthy healing center with Emperors' palaces, wide stone streets, marble baths, statues of Roman Gods and exuberant vegetation.

Participants from 13 different countries delivered 22 invited lectures, 17 oral and 46 poster presentations, contributing in 8 different topics. Papers submitted to the Proceedings were refereed according to the standards of the Journal of Physics: Conference Series and the accepted ones illustrate the diversity and the high level of the contributions.

The Conference gave a good opportunity for interesting discussions and exchange of ideas between the participants. Not least, a significant factor for the success of the LNN 2015 was the social program, the relaxing spa facilities and the guided tour through the Roman remains of the town.

The proceedings of conferences and workshops organized in the frames of INERA Project are regularly published by the *Journal of Physics: Conference* Series. We are grateful to the Journal's staff for providing us this opportunity.

Managing editor: Diana Nesheva

Co-editors: Hassan Chamati, Julia Genova, Kostadinka Gesheva, Tatyana Ivanova, and Anna Szekeres

Journal of Physics: Conference Series **682** (2016) 011002 doi:10.1088/1742-6596/682/1/011002

Bulgarian Academy of Sciences

GEORGI NADJAKOV INSTITUTE
OF SOLID STATE PHYSICS

INERA PROJECT
Research and Innovation Capacity Strenghtening of ISSP-BAS in Multifunctional Nanostructures
REGPOT-2012-2013-1 NMP

PROCEEDINGS

INERA CONFERENCE

Light in Nanoscience and Nanotechnology 2015

Dedicated to the International year of light 2015

October 20 – 22, 2015
Hissar, Bulgaria

Journal of Physics: Conference Series **682** (2016) 011003 doi:10.1088/1742-6596/682/1/011003

Peer review statement

All papers published in this volume of *Journal of Physics: Conference Series* have been peer reviewed through processes administered by the proceedings Editors. Reviews were conducted by expert referees to the professional and scientific standards expected of a proceedings journal published by IOP Publishing.

Journal of Physics: Conference Series **682** (2016) 012001 doi:10.1088/1742-6596/682/1/012001

Chiralization and ferroelectric state induction in nanostructured liquid crystals

B Katranchev[1], M Petrov[1], P M Rafailov[1], N. Todorov[2]

[1] Institute of Solid State Physics, Bulgarian Academy of Sciences, 72 Tzarigradsko Chaussee Blvd., 1784 Sofia, Bulgaria.
[2] Faculty of Physics, University of Sofia "St. Kl. Ohridski", 5 James Bourchier blvd., 1164 Sofia, Bulgaria

E-mail: mpetrov@issp.bas.bg

Abstract. The liquid crystals (LC), due to their naturally high bulk ordering, strong birefringence and easy electrooptical driving, serve as matrix in the nanocomposites doped with non-mesogenic or mesogenic nanoparticles. The nanocomposite's structural units exhibit very complex molecular form indicating the strength and the intermolecular interaction between the matrix and dopant's molecules. Hydrogen bonds are of particular significance for the formation of the nanocomposite structural units, since the symmetry of the LC nanocomposite could be controlled and controllably decreased due to the acceptor–donor interaction between the dimeric matrix and the dopants. As a result, the LC nanocomposite can reach the lowest symmetry, known as triclinic - C_1. Using the LC p,n-alkyloxybensoic acids (nOBA) in form of hydrogen-bonded dimers as matrix and non-mesogenics - single walls carbon nanotubes (SWCNT), perfluorooctanoic acid (PFOA), 4-hydrooxypiridin (HOPY) or mesogen - cholesteryl benzoate (ChB) as dopants and choosing optimal concentrations (where the typical LC state was preserved), we obtained nanocomopsites 7OBA/SWCNT, 7OBA/PFOA, 9OBA/HOPY and 8OBA/ChB. We indicate two forms of ferroelectricity in the studied nanocomposites: developable ferroelectricity, characteristic for the 9OBA/HOPY, 7OBA/PFOA compounds and developed ferroelectricity characteristic for 8OBA/SWCNT, 8OBA/ChB.

1. Introduction

It is possible to control the optical and electro-optical properties of the liquid crystals (LC) by changing the molecular arrangement, inserting external agents (organic, in particular mesogenic, and nonorganic), usually specially functionalized nanoparticles with different structures sizes and shapes [1-5]. Since, the liquid crystalline state combines order and mobility at the molecular (nanoscale) level, molecular modification can control nano-scale assembly and the way how surface-treated nanoparticles can interact with liquid crystals, so that the macroscopic properties of the LC host can conform to specific applications.

Mixtures of liquid crystals with nonmesogenic materials (wide range of nanosize rodlike particles, such as carbon nanotubes, metallic or semiconducting nano-spheres and wires) are promising materials for the contemporary nanocomposites, able to create a new trend in the LC photonic techniques [6-18]. New family of nanoparticles, like single walled carbon nanotubes (SWCNT), CdSe quantum dots [19], hydroxypiridine [20] and perfluorooctanoic acid (PFOA) [21] or mesogen, like cholesterol benzoate ChB [22] has recently been introduced, exhibiting a great capability for homogeneous dispersions in liquid-crystal matrix and ability to alter its inherent bulk properties.

The classical p-n-alkyloxybenzoic acids (nOBA) LC are the basic components in forming supramolecular systems [23]. On the other hand, otherwise achiral nOBA LC, as an effect of the dimeric structure conformation, makes it possible to create an induced chirality [24].

In the present work we prepare mixtures of some homologues of achiral p-n-alkyloxybenzoic acids with homologues numbers 7,8,9 (7OBA,8OBA,9OBA) with non-mesogenic nanoparticle SWCNT in different concentrations. In our previous investigations [24] we found that in most of these concentrations, new chiral phases were produced and depending on the cell thickness and the anchoring energy magnitude, the new unique smectic C_G phase may or may not be induced. Furthermore, setting out for creating an adequate molecular and macroscopic model of smectic C_G phase, we are going to unify the interactions between 7,8,9OBA with mesogens, like cholesterol benzoate (ChB) or non-mesogens like SWCNT, 4-hydroxypiridine (HOPY) and perfluorooctanoic acid (PFOA) regarding the compatibility and thermal stability of the nanocomposites.

2. Experimental results and discussion

The experiments have been carried out on a nanocomposites of the LC 4, n-alkyloxybenzoic acids with n=7,8,9 (7,8,9OBA) supplied by Sigma-Aldirch (Product of Japan), with non-mesogens: i) single walled carbon nanotubes (8OBA/SWCNT) ii) 4-hydrooxypiridin (HOPY) - 7OBA/HOPY, iii) perfluorooctanoic acid (PFOA) - 7OBA/PFOA, as well as with mesogen material –iv) cholesteryl benzoate (ChB) - 8OBA/ChB.

Let us note that for the nanocomposites we choose the optimal combinations and corresponding concentrations of LC matrix (homologues of nOBA) and mesogen or non-mesogen components, for the novel achiral, chiral, or ferroelectric phases to be induced. The attention to non-ferroelectric smectic G or ferroelectric smectic C_G phases will be heightened. Thus unifying the problem, we emphasize mainly on the micropolarization, electrooptical and comparative textural analysis.

2.1. The nanocomposite 7OBA/SWCNT

The general molecular structure of the nOBA molecule is given below.

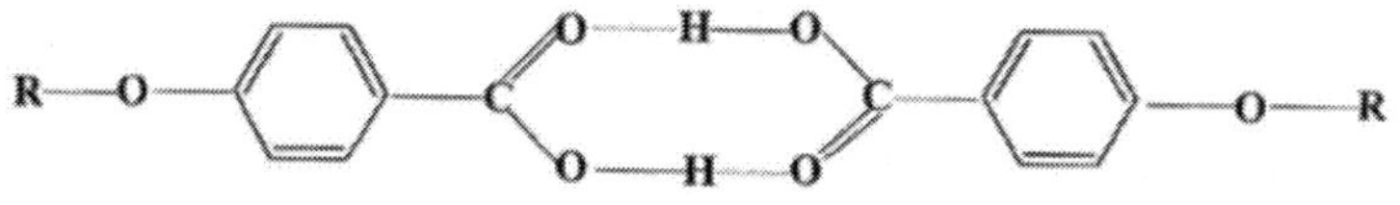

In the bulk of the nanocomposites the basic structure unit – dimer ring - may reduce its local symmetry and transform into open dimer by breaking one of the linear H-bonds. The open dimers can further undergo bending at the unbroken H-bond, thus adopting a shape similar to the bent-core banana molecules [25,26]. In this way, it creates the characteristic parameters – polar vector $\boldsymbol{P}$ and a vector perpendicular to the plane of the bent dimer $\boldsymbol{m}=\boldsymbol{n}_1\times\boldsymbol{n}_2$, where $\boldsymbol{n}_1$ and $\boldsymbol{n}_2$ are unit vectors pointing along the two wings of the dimer. $\boldsymbol{m}$ is the so called kink vector and its magnitude represents the degree of bending of the dimer. As a result, this produces the premise for the lowest possible symmetry triclinic C_1 of the nanocomposite structure.

A variety of phases are observed at SWCNT concentration of 0.01 weight % in LC cell with thickness d = 8μm. After briefly crossing the conventional N phase in an interval of ≈ 1°C, the nanocomposite forms a chiral nematic phase (N^*) at 142°C. Then the system enters at 93°C into a chiral smectic C (S_C^*) phase (on figure 1(a) the transition $N^* \rightarrow S_C^*$ is depicted). The textures of N^* and S_C^* are characterized by typical stripes indicating the helix direction, which in this case lies in the xy cell plane or is tilted with respect to it. Upon further cooling, at 90°C a reentrant chiral nematic (N_r^*) phase is reached. Figure 1(b) presents the N_r^* texture. At 86°C we observe a transition from N_r^* to a low-temperature smectic phase (figure 1(c)), which is fluid (well indicated by small colored glass particles), not smooth, and colored in red and green mosaic domains, thus closely resembling the C_G phase. At very slow cooling (0.1°C $\min^{-1}$) starting from the I phase, the domains grow as fractal nuclei and coalesce into large areas with opposite optical rotation (distinct circular dichroism). Investigating this low-temperature smectic phase we observe that upon sample rotation in either direction, the red color turns into green and vice versa. Besides this variable birefringence (red and green colors), on entering this phase from the N_r^* one we also observe a change in fluidity and the texture loses its smoothness appearing as a colored mosaic texture with equal left- and right-handed helices. This unambiguously shows that this phase is chiral and biaxial. The properties of this phase imply a drastic symmetry breaking of the constituent LC molecules. Differently from the conventional S_C phase of pristine 7OBA where the open dimers can only undergo twisting, the observed combination of chirality and biaxiality implies a loss of the linear shape of the LC molecules, i.e., formation of a bent-dimer shape.

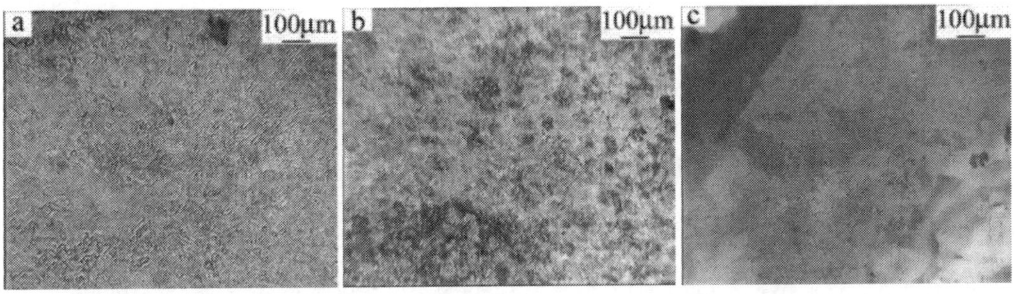

Figure 1. 7OBA/SWCNT textures (c=0.01wt%) in a nonoriented LC cell with thickness d=8μm: (a) Phase transition from N^* to S_C^* at T=93°C; (b) Reentrant N_r^* phase at T=89°C; (c) Low-temperature smectic C_G phase at T=86°C.

Reducing the cell thickness to 8 μm increases the anchoring energy and imposing unidirectional surface alignment lowers the symmetry similarly to the case when an external pressure is applied to the system. The reentrant nematics indicates that at these surface and temperature conditions monomers and open dimers dominate in the LC system and suppress the smectic layering, thus leading to a return of the long-range positional order typical for N phases. At further cooling, however, the bent dimers gain dominance and restoration of the smectic layering then leads to a fluid tilted smectic C_G state (without positional order inside the layers).

The most important problem is the compatibility of the conformed LC dimers and the SWCNT. To check the results from the microtexture analysis we performed a comparative monitoring of the different Raman spectra of 7OBA/SWCNT in parallel polarization at several temperatures (quoted in the plot) in the low-frequency region (100–1300 cm^{-1}) (figure 2). The lowest three traces are spectra of pristine 7OBA for comparison. The peak at 310 cm^{-1} appearing in the S_C phase is marked by arrows and the symbol S_C. The sharp peaks emerging upon transition to the Cr state are marked with asterisks. The inset shows at an expanded scale the RBM region of the 7OBA/SWCNT spectrum at 82°C and a spectrum of pristine SWCNT heated to 82°C before mixing to the LC.

With the radial breathing mode (RBM) centered at about 177 cm^{-1} the signature of carbon nanotubes is clearly visible in the spectra of 7OBA/SWCNT, confirming the presence of small SWCNT bundles in the nanocomposite. Except for the RBM no essential differences between 7OBA/SWCNT and pure 7OBA are found in their Raman spectra from the I phase down to 100°C, but upon further cooling a number of significant changes occur. Along with the N → S_C transition in pure 7OBA, a specific peak appears at 310 cm^{-1} as evident from the spectrum at 94°C (see figure 2). This peak can serve as a distinguishing feature of the S_C phase in 7OBA as it represents the only difference in the Raman spectra of this substance between the N and the S_C phase. The transition to a solid phase causes remarkable further changes, as can be seen from the lowest spectrum on figure 2 taken at 82°C: Additional peaks appear at 150, 178, 252, 351, 885, and 1148 cm^{-1}; the bands at 105 and 638 cm^{-1} split into two and that at 777 cm^{-1} into three sharp peaks.

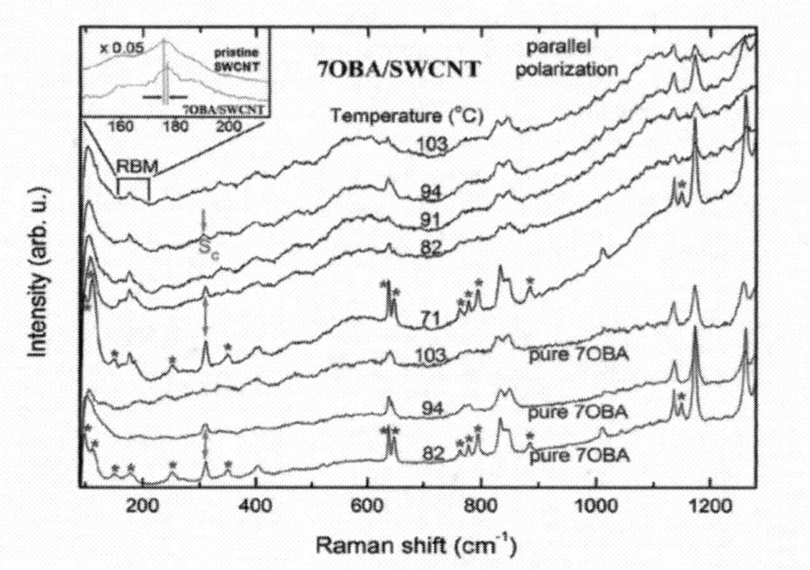

Figure 2. Raman spectra of 7OBA/SWCNT and pure 7OBA in parallel polarization at several temperatures (quoted in the plot) in the low-frequency region (100–1300 cm^{-1}).

We now analyze the Raman spectra of the 7OBA/SWCNT nanocomposite starting with the $N^* \rightarrow S_C^*$ transition expected at 94°C. The corresponding spectrum for 94°C in figure 2 indeed contains the peak at ≈310 cm^{-1} identifying the smectic phase although this peak is slightly softened and broader compared to that of pure 7OBA. In the next spectrum taken at 91°C this peak is absent which is consistent with the reentrant nematic phase from 93°C to 88°C established by the microtexture analysis. The spectrum taken at 82°C contains again a sharp and relatively intense peak at 310 cm^{-1} and resembles in its structure the pure 7OBA spectrum at 94°C thus indicating a smectic phase. Hence the behavior of the C_G phase in Raman scattering is similar to that of the conventional S_C phase. Finally, the spectrum taken at 71°C closely resembles the pure 7OBA spectrum at 82°C with all additional sharp peaks indicating solidification of the nanocomposite. The Raman spectra of 7OBA/SWCNT are thus consistent with the sequence of phase transitions found by microtexture analysis in this nanocomposite. The richer phase development in 7OBA/SWCNT is undoubtedly connected to an interaction of the LC molecules with the nanotubes and it would be interesting to find evidence for this interaction in the Raman spectra. Such interaction should stem from the intermolecular overlap of π orbitals (π stacking). The RBM band with vibrational pattern perpendicular to the nanotube-bundle axis can serve as a sensitive probe for interactions of SWCNT with molecules in contact with their surface, as well as for interactions among the SWCNT themselves, e.g., within a bundle [24]. As the 7OBA molecule also contains phenyl rings, it is logical to expect its interaction with the SWCNT to be reflected in the behavior of the RBM band of 7OBA/SWCNT. Comparing Raman spectra of the pristine SWCNT powder before mixing it into the liquid crystal with spectra of the 7OBA/SWCNT nanocomposite, we find indeed a slight RBM hardening and redistribution of its intensity in the 7OBA/SWCNT spectra. This comparison is depicted for the C_G phase in the inset of figure 2 for T = 82°C. The RBM band of the SWCNT used in the present study consists of three features: a main peak and two shoulders on both sides. For the pristine tubes their frequencies were determined as follows: 159.5, 176, and 188 cm^{-1}. In the 7OBA/SWCNT spectra the low-energy shoulder hardens by 3 cm^{-1}, the main RBM peak by 1 cm^{-1}, and the high-energy shoulder also by 1 cm^{-1} with simultaneous transfer of spectral weight from the low-energy to the high-energy shoulder. We consider this clear evidence for interaction of the LC molecules with the nanotubes and regard this interaction as the factor breaking the one H bond in the central hexagon and leading to a bent-dimer shape of the 7OBA molecules in the nanocompoosite, which is the main condition for nanoscale symmetry reduction to C_I and C_G state growth. The main feature of this phase is the out-of-layer-plane electric polarization (induced upon dimer bending), which is bulk, permanent and unlike the popular ferroelectric S_C^* closely resembles solid state ferroelectrics. The detection and evaluation of such out-of-layer-plane polarization is delicate and can be done by corresponding electro-optical excitations.

It is known that the patterns formed at phase transition LC → isotropic (I) phase require div P = 0 [27]. As already mentioned, C_G develops out of smectic layer plane's polarization ($\mathbf{P}_{out}$), unlike S_C^* where the polarization is situated in parallel to the plane of the smectic layers and directly satisfies div$\mathbf{P}$ = 0. To satisfy the requirement div $\mathbf{P}_{out}$=0 in C_G state, however, a compensation of the divergence of the out-of-smectic layer polarization, is needed. So in the presence of 10 Hz ac electric field the chiral structures (usually with ribbon or ropelike forms) shrink in correlation with the polarity inversion and cause additional elastic-deformation for compensation of the free energy of the LC system, thus ensuring the div $\mathbf{P}_{out}$=0 to be satisfied. When $\mathbf{E}$ is opposite to $\mathbf{P}_{out}$ the free energy increases by $\mathbf{P}_{out}\cdot\mathbf{E}$ inducing a transition from C_G to above-lying phases up to the I state. Especially, for an electric field of at least 1.6V/μm, we observed such a transition between developed chiral states C_G and S_C^*. We thus have electric-field induced phase transition governed by temperature-dependent threshold field. In other words, we observe an equivalent decrease of the phase transition temperature driven by the electric field [28].

Using the ferroelectrically induced decrease ΔT of the phase transition temperature, we can assess the polarization by the modified Kirkwood-Helfrich equation [27-29]: $(\Delta T/To)(\Delta H/M\rho) = 0.5\varepsilon_a\varepsilon_0|E|^2 - |P||E| \cos\alpha$, where T_o is the $C_G \rightarrow S_C^* \rightarrow I$ transition temperature at E = 0, M is the molar mass, ρ the density of the nanocomposite, and α is the angle between $\mathbf{P}$ and $\mathbf{E}$. Thus we estimate that $|\mathbf{P}_{out}| \approx 110 nC/cm^2$. Such a low-temperature smectic state we define as developed C_G.

2.2. The non-mesogen nanocomposite 7OBA/PFOA

The perfluorooctanoic acid (PFOA) was supplied by Sigma - Aldirch (product of China). The intermolecular hydrogen-bonded nanocomposites are synthesised by mixing 7OBA and PFOA in 1:1 ratio in ethanol. The molecular structure of the nanocompositen 7OBA/PFOA is presented as follows:

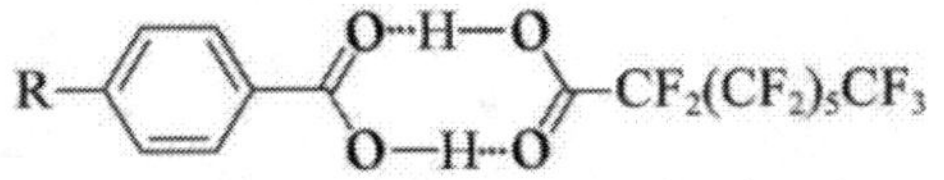

In this nanocomposite, twisted X-shaped elementary chiral units (shown in figure 3(a)) emerge 2.5°C below the isotropic phase. The left and right chiral senses are well seen (e.g., (1) and (2) in figure 3(b)). On further cooling, these chiral centers coalesce. This second stage of the chiral structure development provides large chiral monodomains indicating formation of spontaneous twofold chirality (figure 3(c)). The second twist stage yields the macroscopic effect, involving a strong texture change: left- and right-handed chiral domains grouped in regions partially separated by 2π walls (see figure 3(c)). However, here the C_G state is confined only in the domains and does not spread in the whole nanocomposite. This kind of C_G state will be referred to as developable state. The developable chiral structure is very well pronounced in this case. One notes chiral ribbons (HRs) of both helicities, indicating the presence of left and right molecular or structural twists.

To analyze the properties of the nanocomposite 7OBA/PFOA and also to detect the polarization, **P** we performed electro-optical measurements. Applying low-frequency electric field (less than 50Hz) applied across a 12μm thick LC cell, one observes the appearance of well-pronounced chiral ribbons (HR) (see figure 4(a)).

The HR coalesce into wide chiral ribbons upon increasing the field magnitude (figure 4(b)). We chose sufficiently low frequency for the polarization to be able to follow the polarity inversion. The ribbons twist into right- or left-handed spirals depending on the polarity. To compensate this twisting, an additional deformation (coiling up) of the rope- and ribbon-like texture domains arises. Following the polarity of the external field, the spirals are formed clockwise or counterclockwise. This effect indicates that the ribbons are polar and the polarization vector **P** has a component parallel to the substrate surface. Coiling of chiral structures is another compensatory mechanism to make the polarization satisfy div$\mathbf{P}_{out} = 0$.

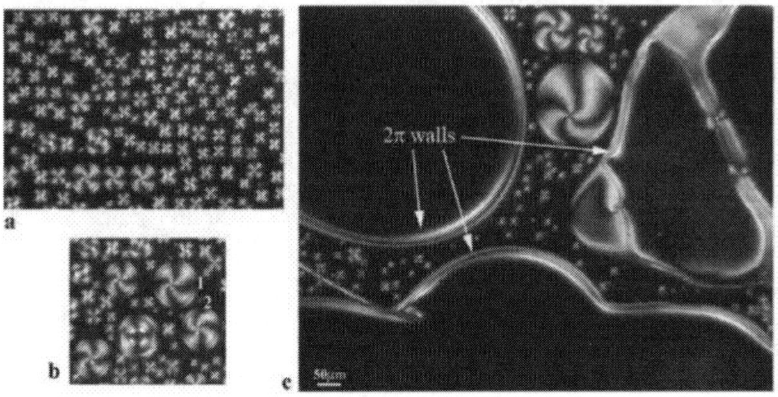

Figure 3. Microtextures of 7OBA/PFOA in LC cells with glass substrates and d=12μm: (a) The emergence of the twisted X-shaped elementary chiral units at T=118°C; (b) Left-handed (1) and right-handed (2) chiral units at T=117°C; (c) Large chiral monodomains with twofold spontaneous chirality formation at T=104°C.

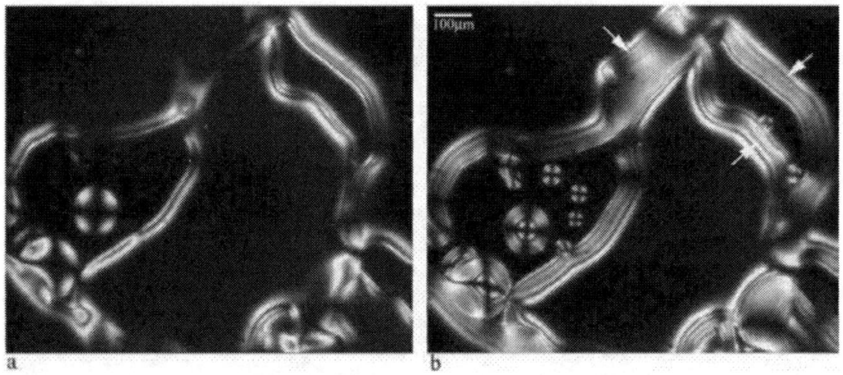

Figure 4. Electrically induced chiral ribbons (indicated by arrows) of 7OBA/PFOA in LC cell with ITO/glass substrates and d=12μm: (a) E=1.7V/μm; (b) E=7.5V/μm. Frequency 50Hz. T=111°C.

The appearance of the 2π walls in the presence of electric field and the absence of π-walls (characteristic for non-chiral state) indicates a presence of residual polarization [30-32]. The formation of 2π walls is an imprint of the spontaneous polarization, i.e., the polar symmetry of the director, unlike the π walls indicating quadrupolar symmetry as that of achiral S_C phase. As was indicated in [33], the nature of the polar response and the symmetry of the **P**-director can be determined from the director's inversion walls, since the texture reflects the response to applied electric fields by reorientation of the director.

The magnitude of the polarization can be estimated from the wall width upon application of an in-plane electric field E [33]. The 180° rotation of the director following each field reversal results in the formation of 2π walls. The ferroelectric term $f_p = -$ P.E exhibits coupling to the spontaneous polarization **P**. These contributions result in a correlation length $\xi_p=\sqrt{K/P_sE}$ where K is the mean Frank elastic constant in a one-constant approximation [29]. The inversion wall width w clearly scales as $1/\mathbf{E}^{1/2}$ (figure 5).

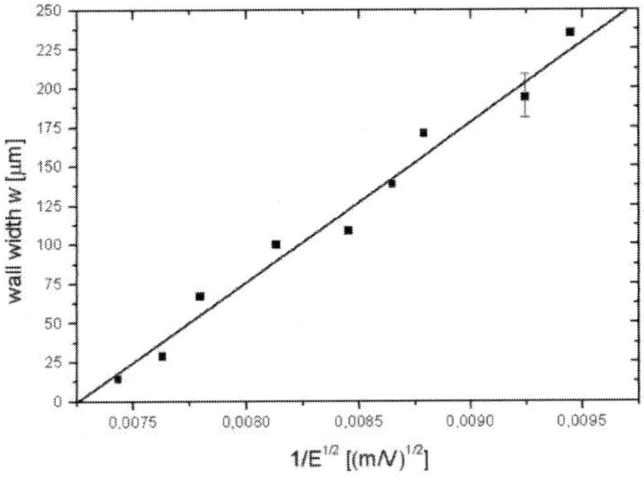

Figure 5. The 2π wall width dependence on the electric field.

From the slope of the fit, we estimated a spontaneous polarization with a magnitude of about 0.25nC/cm^2.

The developable C_G state could be indicated also by Raman spectral analysis. In figure 6 we show temperature-dependent Raman spectra of the nanocomposite 7OBA/PFOA.

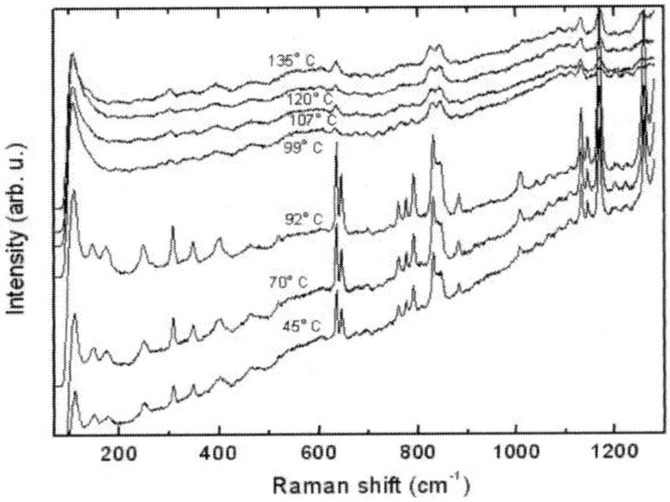

Figure 6. Raman spectra of the 7OBA/PFOA nanocomposite at different temperatures.

Between 145 and 99°C the nanocomposite 7OBA/PFOA displays the Raman signature of a nematic phase: the peak at 310 cm^{-1}, characteristic of the smectic phase, is smeared out, very faint or missing;

there is a weak single peak at 635 cm^{-1} and the doublet at 836 cm^{-1} and the lines at 1130, 1170 and 1254 cm^{-1} have relatively low intensity. Slightly above 92°C the system undergoes a transition to solid crystal state manifested by a pronounced sharpening of all spectral lines and appearance of numerous new sharp lines at 147, 176, 248, 307, 348, 400, 760, 775, 792, 882, 1009 and 1146 cm^{-1}. Additionally, the peak at 635 cm^{-1} transforms into a sharp doublet at 640 cm^{-1} and the doublet at 836 cm^{-1} undergoes a significant intensity redistribution and sharpening. Interestingly, a distinct smectic phase could not be detected with Raman spectroscopy and micropolarization analysis but at 99°C already some precursors of the additional sharp lines (e. g. of the triplet at 760 - 792 cm^{-1}) are observed. This implies that a gradual transition from the nematic state directly into the solid crystal state may be taking place and the developable smectic state can be recognized by subtle changes in the spectra between 100°C and 92°C.

The other nonmesogen dopant HOPY, similarly to the PFOA also display developable C_G state.

2.3. The non-mesogen nanocomposite 9OBA/HOPY

In a nanocomposite 9OBA/HOPY the liquid crystal matrix of 9OBA operates as a N-bonding donor and HOPY serves as a H-bonding acceptor. The building blocks of the induced 9OBA/HOPY nanocomposite could be schematically presented as is shown below

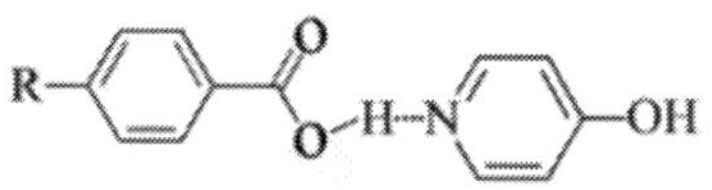

Staring with slow cooling (0.1°C/min) from the isotropic phase we obtained consecutively the well formed developable chiral textures - helical chiral ribbons (figure 7), spiral cords (manifesting the nucleation and development of the chiralization) and surrounding green and red (orange) left and right handed domains, reflecting the developed chiral domains and strongly resembling the features of the C_G phase.

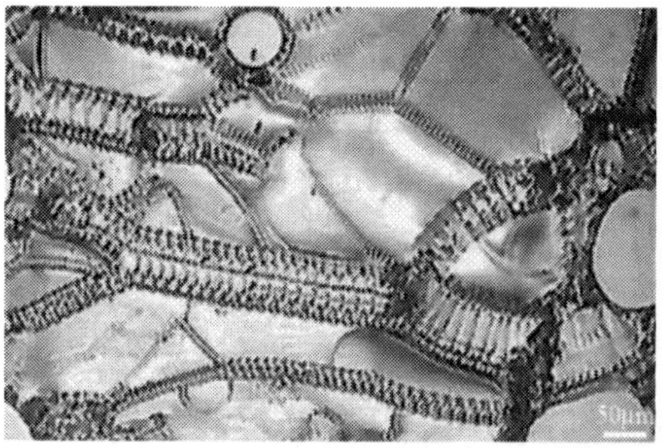

Figure 7. 9OBA/HOPY texture of well formed developable chiral and helical chiral ribbons in LC cell with glass substrates and d=12μm. T=135°C.

Applying the same procedure of the electrical field action using *dc* or low-frequency *ac* voltages, as that presented in the case of the nanocomposite 7OBA/PFOA, we estimated the polarization value in the developing chiral range to be 0.15 nC/cm^2.

Differently from the nonmesogens PFOA and HOPY and similarly to SWCNT, nanocomposites with the mesogenic dopant cholesteryl benzoate (ChB) display developed C_G state.

2.4. The mesogen nanocomposite 8OBA/ChB

In such nanocomposite a parallel molecular dipole orientation (a significant dipole moment $\mu \approx 2.4$D (Debye) perpendicular to the long molecular axis) of the dissolved chiral molecules is created. A result of this parallel orientation is the appearance of spontaneous polarization ***P*** and eventually ferroelectric state induction. The nanocomposite structural unit is indicated below:

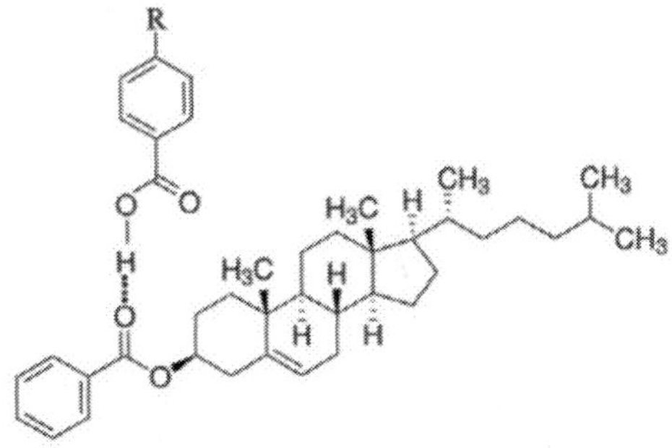

At slow (0.1°C/min) cooling from the isotropic phase we go to formation of chiral nematic N^* phase (figure 8(a)).

Applying 20V *dc* or 20V, 20 Hz *ac* electric fields one observes appearance of 2π walls (figure 8(b)). Its width linearly decreases as $E^{-1/2}$, allowing the polarization of the developable chiral structure to be evaluated (less than $0.1 nC/cm^2$). Below 114°C a chiral mosaic phase grows (figure 8(c)), considerably resembling in textural respect the developed smectic C_G state. The small values of the polarization, within the developable range, where the 2π wall exist, do not allow evaluation of the polarization value in the developed chiral state in figure 8(c).

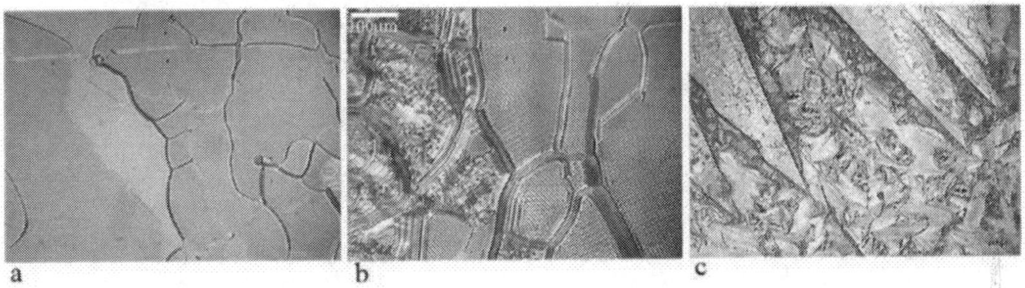

Figure 8. 8OBA/ChB textures in LCC with ITO/glass substrates and d=12μm. (a) Chiral nematic N^* phase. T=135°C; (b) Appearing of 2π walls at electric field. T=135°C; (c) Transition to chiral mosaic phase. T=105°C.

3. Conclusion

The general result is: each of the studied nanocomposites is constituted of structural units with various complex shapes, determined by the strength and specificity of the interaction between the dopant structural units and the dimer rings. This interaction combined with the surface anchoring, bulk and electrical actions, leads to drastic reduction of the LC system symmetry. As a result, transition from achiral state, characteristic for the pristine nOBA substances in the N and Sc states (with $D_{\infty h}$ and C_{2v} symmetry), to chiral low-symmetric states, including ferroelectric smectic C^* with C_2 symmetry and ferroelectric smectic C_G with the lowest C_1 triclinic one takes place. We identify the chiral nucleus generation, and the growing ribbon-like and helical structures, as developable smectic C_G states. Their response to temperature decrease or to application of *dc* and low-frequency *ac* electric fields shows the presence of a residual (spontaneous) polarization.

Acknowledgments

The work is supported by the Bulgarian National Funds for Scientific Research [grant number

DFNI-G02/18] and INERA EU project (FP7-316309-REGPOT-2012-2013-1) "Research and Innovation Capacity Strengthening of ISSP-BAS in Multifunctional Nanostructures".

References

[1] Hirst A R, Escuder B, Miravet J F and Smith D K 2008 *Angew. Chem. Int. Ed.* **47** 8002
[2] Taylor J W, Kurihara L K and Martinez-Miranda L J 2012 *Appl. Phys. Lett.* **100** 173115
[3] Qi H and Hegmann T 2008 *J. Mater. Chem.* **18** 3288
[4] Hegmann T, Qi H and Marx V M, 2007 *J. Inorg. Organomet. Polym. Mater.* **17** 483
[5] van't Zand D D, Chushkin Y, Belkoura L, Lobo C V, Strey R, Lyakova K and Clegg P S 2012 *Soft Matt.* **8** 4062
[6] Dierking I, Scalia G, and Morales P 2005 *J Appl. Phys.* **97** 044309
[7] Russell J M, Oh S, LaRue I, Zhou O and Samulski E T 2006 *Thin Solid Films* **509** 53
[8] Lynch M D and Patrick D L 2002 *Nano Lett.* **2** 1197
[9] Dierking I, Scalia G, Morales P and LeClere D. 2004 *Adv. Mater.* **16** 865
[10] Basu R and Iannacchione G S 2008 *Appl. Phys. Lett.* **93** 183105
[11] Baik I-S, Jeon S Y, Lee S H, Park K A, Jeong S H, An K H and Lee Y H 2005 *Appl. Phys. Lett.* **87** 263110
[12] Lagerwall J P F and Scalia G 2008 *J. Mater. Chem.* **18** 2890
[13] Basu R and Iannacchione G S 2009 *Phys. Rev. E.* **80** 010701
[14] Wu K-J, Chu K-C, Chao C-Y, Chen Y-F, Lai C-W, Kang C-C, Chen C-Y and Chou P-T 2007 *Nano Lett.* **7** 1908
[15] Liu Q, Cui Y, Gardiner D, Li X, He S and Smalyukh I I 2010 *Nano Lett.* **10** 1347
[16] Zamora-Ledezma C, Blanc C, Puech N, Maugey M, Zakri C, Anglaret E and Poulin P 2011 *Phys. Rev. E* **84** 062701
[17] Jeon S Y, Lee S H and Lee Y H 2006 *Appl. Phys. Lett.* **89** 056102
[18] Gommans H, Alldredge JW, Tashiro H, Park J, Magnuson J and Rinzler A G 2000 *J. Appl. Phys.* **88** 2509
[19] Thanassoulas A, Karatairi E, Cordoyiannis G, Kutnjak Z, Tzitzios V, Lelidis I and Nounesis G 2013 *Phys. Rev. E.* **88** 032504
[20] Antonova K, Katranchev B, Petrov M, Marcerou J P, Baran J and Ratajczak H J 2004 *Mol. Structure* **694** 105
[21] Rambabu M, Pisipati V G K M, Shanmukh Kumar J V and Prasad K R S 2013 *Liquid Crystals Today* **22** 4
[22] Petrov M and Simova P 1990 *Liq. Cryst.* **7** 203
[23] Podgornov F, Suvorova A, Lapanik A and Haase W 2009 *Chem. Phys. Lett.* **479** 206
[24] Petrov M, Katranchev B, Rafailov P M, Naradikian H, Dettlaff-Weglikowska U, Keskinova E and Spassov T 2013 *Phys. Rev. E* **88** 042503
[25] Niori T, Sekine T, Watanabe J, Furukawa T and Takezoe H 1996 *J. Mater. Chem.* **6** 1231
[26] Link D R, Natale G, Shao R, Maclennan J E, Clark N A, Korblova E and Walba D M 1997 *Science* **278** 1924
[27] Jakli A, Kruerke D, Sawade H and Heppke G 2001 *Phys. Rev. Lett.* **86** 5715
[28] Helfrich W 1970 *Phys. Rev. Lett.* **24** 201
[29] de Gennes P G and Prost J 1993 *The Physics of Liquid Crystals* second edition (New York: Oxford University Press)
[30] Stannarius R, Langer C and Weissflog W 2002 *Ferroelectrics* **277** 177
[31] Stannarius R, Li J-J and Weissflog W 2003 *Phys. Rev. Lett.* **90** 025502
[32] Pindak R, Young C Y, Meyer R B and Clark N A 1980 *Phys. Rev. Lett.* **45** 1193
[33] Eremin A, Floegel M, Kornek U, Stern S, Stannarius R, Nádasi H, Weissflog W, Zhu C, Shen Y, Park C, Maclennan J and Clark N 2012 *Phys. Rev. E.* **86** 051701

Light-Induced Degradation of Thin Film Silicon Solar Cells

F U Hamelmann, J A Weicht and G Behrens

Bielefeld University of Applied Sciences, Artilleriestraße 9a, Bielefeld, Germany

FHamelmann@fh-bielefeld.de

Abstract. Silicon-wafer based solar cells are still domination the market for photovoltaic energy conversion. However, most of the silicon is used only for mechanical stability, while only a small percentage of the material is needed for the light absorption. Thin film silicon technology reduces the material demand to just some hundred nanometer thickness. But even in a tandem stack (amorphous and microcrystalline silicon) the efficiencies are lower, and light-induced degradation is an important issue. The established standard tests for characterisation are not precise enough to predict the performance of thin film silicon solar cells under real conditions, since many factors do have an influence on the degradation. We will show some results of laboratory and outdoor measurements that we are going to use as a base for advanced modelling and simulation methods.

1. Introduction

Thin-film silicon in p-i-n structure is already used for commercial solar cells for decades [1]. Amorphous silicon (a-Si) uses very thin absorbing layers of just a few 100 nm thickness, the deposition processes are well known from flat panel display industry and scaleable to large areas of some square meters, which allows low production costs [2]. However, the resulting efficiencies are low compared to other solar cell technologies (5-7%). Using tandem or triple junction solar cells in combination with microcrystalline silicone offers the potential for efficiencies up to 10-15%, which makes this technology more competitive [3].

A major disadvantage of amorphous silicon is the light-induced degradation (LID) that occurs during illumination of the cells [4]. The so-called Staebler-Wronski-effect reduces the efficiency by breaking weak silicon-hydrogen bonds in the absorbing layer, leading to an increasing density of defects [5]. This effect depends very much on the quality of the intrinsic a-Si layer, the thickness of the layer, the operation temperature, the intensity of the light and other parameters [6]. The standard test for the LID uses an illumination of 1000 W/m^2 at 50°C for 1000 hours at open circuit conditions, which is different from real life operation of the solar cells [7].

In this paper we show results of a-Si cells and modules produced on industrial deposition systems, in order to find a suitable trade-off between material quality, deposition time and stable solar cell efficiency. The transparent conductive layer (fluorine doped tin oxide or boron doped zinc oxide) enables higher efficiencies with advanced light scattering, but also shows an influence on the LID of the amorphous silicon.

We also show long term measurements of industrial thin-film silicon modules under outdoor conditions. We will discuss the need for advanced models in order to get reliable simulations for the performance of thin-film silicon solar cells under different climate conditions. The LID is reduced in tandem structures due to the stable microcrystalline layer, but the series connection of the subcells makes the LID of the amorphous part still important for the cell efficiency.

Journal of Physics: Conference Series **682** (2016) 012002 doi:10.1088/1742-6596/682/1/012002

2. Method

The results presented in this paper have been measured in the lab on thin film silicon solar cells produced by production sized deposition systems (1.4 square meter glass substrates) by Applied Materials and Oerlikon. The outdoor measurements were performed on full size commercial modules. A Wacom steady state solar simulator with selfmade liquid cooled samples stage was used for IV-curve measurement under standard-test conditions (1000 W/m^2, AM 1.5, 25°C). Measurements of external quantum efficiency (EQE) were performed on a taylor made system with monochromator and additional bias illumination for tandem cells. Light induced degradation was done in a light soaking bench with controllable substrate temperature.

The outdoor characterisation of the full size modules was performed with the Sol.Connect monitoring system provided by Papendorf SE. The weather station with calibrated reference cells and pyranometers for the measurement of solar irradiation was bought from EKO Instruments. A taylor made database is logging all relevant parameters of the stations every minute: IV-curves of up to 10 modules, module temperatures, irradiation detected by pyranometer and reference cell, diffuse and direct irradiation, wind speed and direction, humidity, rain, air temperature. A detailed description of this system can be found elsewhere [8].

The modelling and simulation is not topic of this paper, the work is still in progress. First results are already published [9, 10].

3. Results

The outdoor performance of PV modules shows differences from the expected performance, usually the the modules are labelled with a peak power value (in W_p) and a temperature coefficient. With the knowledge of irradiation (in W/m^2) and module area the efficiency can be calculated. Some results are shown for amorphous (a-Si), amorphous/microcrystalline tandem (a-Si/µc-Si) and polycrystalline (c-Si) modules. In summer the irradiation is up to 1000W/m^2 at noon (figure 1), the module temperature reaches 60°C at that time. The efficiency is not constant during the day, it is lowered at high irradiation due to thermal effects, and lowered in the morning and in the evening by the weak light behaviour.

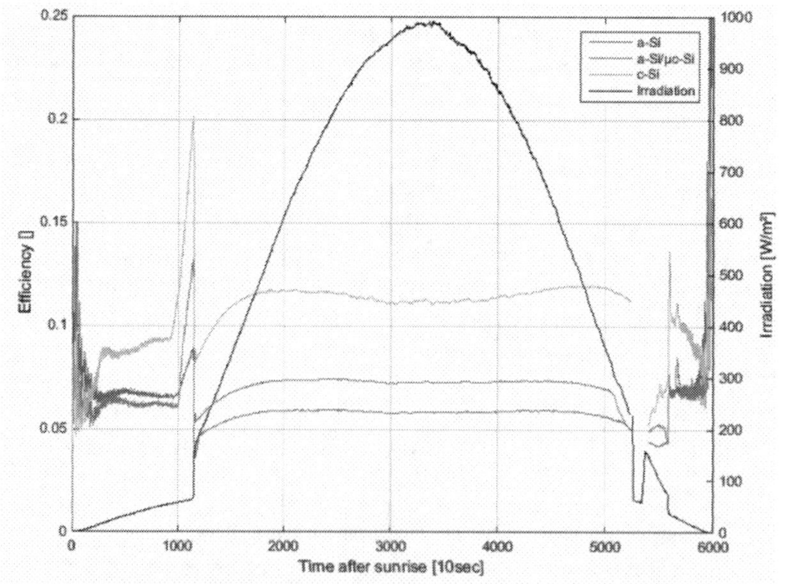

Figure 1. Efficiency of different Si-modules on a sunny summer day in Germany.

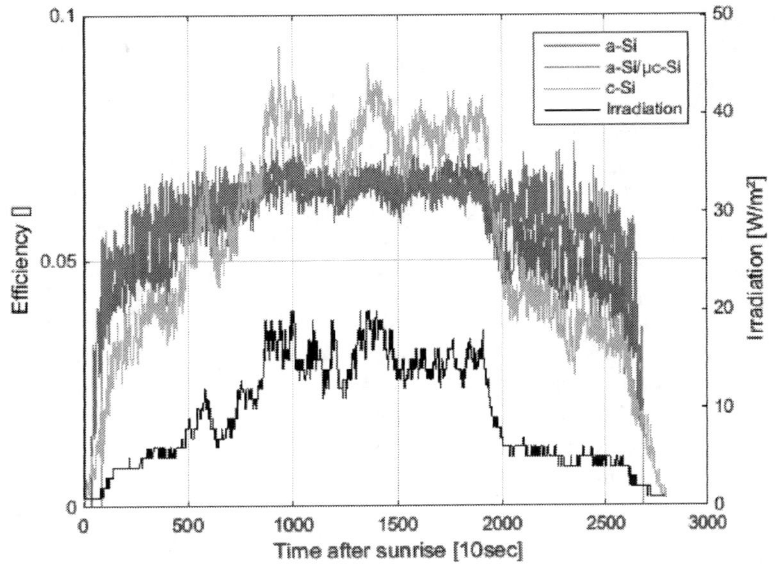

Figure 2. Efficiency of different Si-modules on a cloudy winter day in Germany.

On a dark winter day the irradiation is as low as 20 W/m^2, even with low module temperatures of below 10°C the efficiency of the c-Si module is reduced by about 50%, while the a-Si thin film module shows no significant reduction (see figure 2). This effect shows the advantage of thin film silicon at low light conditions [11]. A precise simulation of the performance of PV-modules is difficult; taking the Staebler-Wronski effect in account makes things even worse [7].

Since the amorphous part of a-Si/µc-Si tandem cells is mostly relevant for light induced degradation, single junction a-Si cells are investigated in detail about the factors influencing the light induced degradation (LID). Figure 3 shows the degradation of single junction a-Si cells on ZnO:B front contact with different i-layer thickness as a function of time. The 390 nm i-layer cell has the highest initial efficiency of 10.86 %, but also the highest degradation of 33.6% and thus the lowest stable efficiency of 7,2% (see table 1). Highest stable efficiency shows the thinnest i-layer cell (310 nm) with degradation from 10.55% to 7.34% (LID: 30.4%). Increasing i-layer thickness makes no sense the goal in production must be thinnest possible i-layer. It can be seen that the degradation does not change much after 300 h, for the following experiments the time was limited to 312 h.

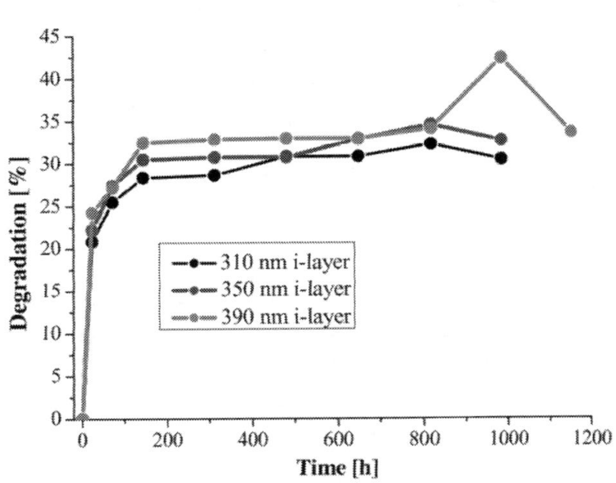

Figure 3. Degradation of a-Si cells with different i-layer thickness.

Table 1. Efficiency of cells with different i-layer thickness during LID.

i-layer	nitial	312 h	000 h	Degradation
310 nm	10.55	7.53	7.34 %	30.4%
350 nm	10.77	7.46	7.25 %	32.7%
390 nm	10.86	7.29	7.21 %	33.6%

With industrial large area deposition, homogeneity of the films becomes an important issue. With a deposition area of 1.4 m^2 thickness variations between the center and the corners of about 10% cannot be avoided. Cells from different positions of the substrate show different results (figure 4). The results from two different runs with 270 nm i-layer are almost equal, but the difference between the middle position and the corner position are significant, the LID is 25% at the corner and just 20% in the middle. Such effects are resulting in a reduced performance of full size modules compared to small cells produced in a lab.

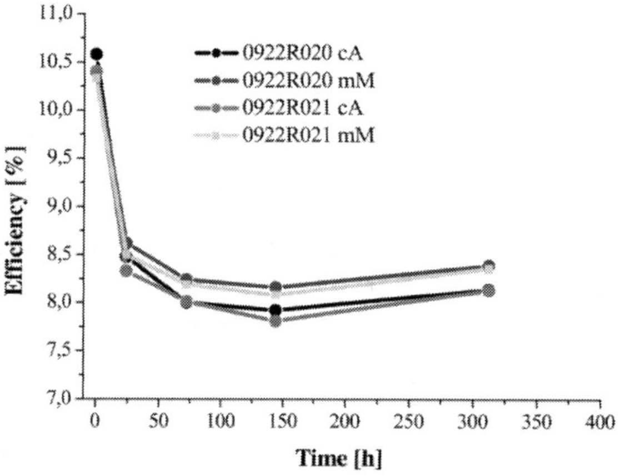

Figure 4. LID on different positions of a 1.4 m^2 substrate.

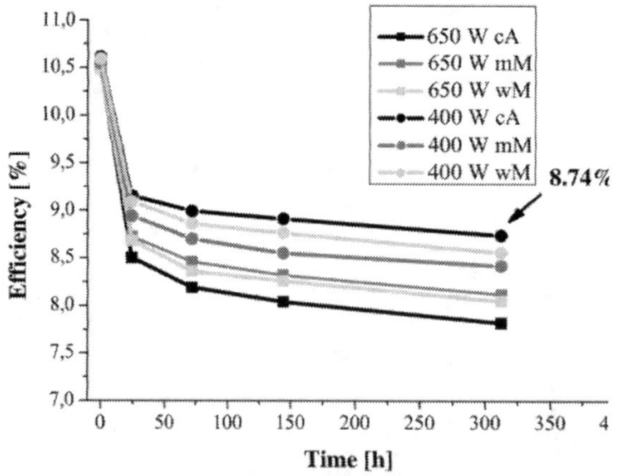

Figure 5. LID for i-layer with different deposition rates on different positions.

Another important factor is the deposition rate of the intrinsic layer. The stability to light exposure can be improved by a small growth rate, which gives the amorphous silicon a higher material quality. Unfortunately, for cost efficient production highest possible growth rates are required. For commercial use of amorphous silicon a trade off between deposition rate and LID has to be defined. In figure 5 the development of the efficiency during LID can be seen for low (400 W rf power) and high growth rate (650 W) for three different positions on the substrate. Please note that the area of highest efficiency is in the middle for high growth rate, while it is at the corner for low growth rate, with a degradation of just 15%.

Another influence is given by the front contact TCO of the substrate. The different morphology of the ZnO:B shows an influence on the LID. In figure 6 the LID results for identical a-Si layers on different ZnO:B substrates are plotted. For a similar sheet resistance of about 8 $\Omega_{sq.}$ the level of boron doping was changed with the deposition time, thin ZnO layers are highly doped (11 minutes deposition time, 2.8 sccm diborane flow), while thick layers are slightly doped (17 minutes, 1.6 sccm) with an increased surface roughness. The lowest LID can be found for the thicker, low doped TCO's.

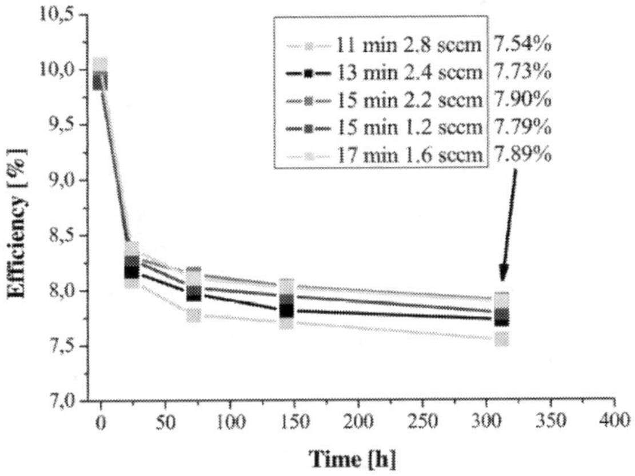

Figure 6. LID for a-Si mini modules on different ZnO:B substrates.

For this set of samples the possible regeneration by dark annealing was studied (figure 7). After the degradation for 312 h at 50°C with 1000 W/m^2 the samples were heated in an oven for 88 h at 80°C.

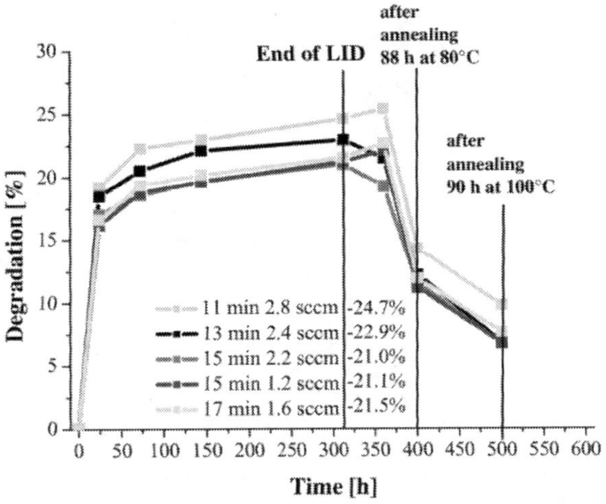

Figure 7. Regeneration by dark annealing of a-Si mini modules.

The efficiency was measured, and the samples were once more heated for 90 h at 100°C. It can be seen, that the degradation could be partly healed with this dark annealing steps. For a complete healing even larger times, or higher temperatures are needed. Some other cell layers do have an influence on the degradation, as the buffer layer between p-doped silicon and intrinsic silicon. But since this influence is small, it is not shown here. The silicon p-layer also effects the LID, as described in [12].

Finally in figure 8 the development of degradation normalised on the initial values for tandem a-Si/µc-Si mini modules produced in a factory line are shown. The degradation has been performed under different, but stable temperatures. It can be seen, that the degradation compared to the standard test at 50°C is significantly lower at 80°C, and higher at 20°C. This shows that for regions with higher temperature and solar irradiation a better performance of the modules than in central Europe can be expected.

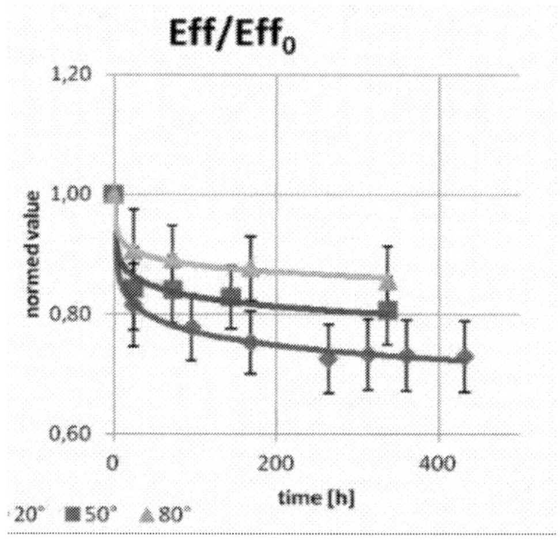

Figure 8. Degradation of a-Si/µc-Si mini modules at different temperatures.

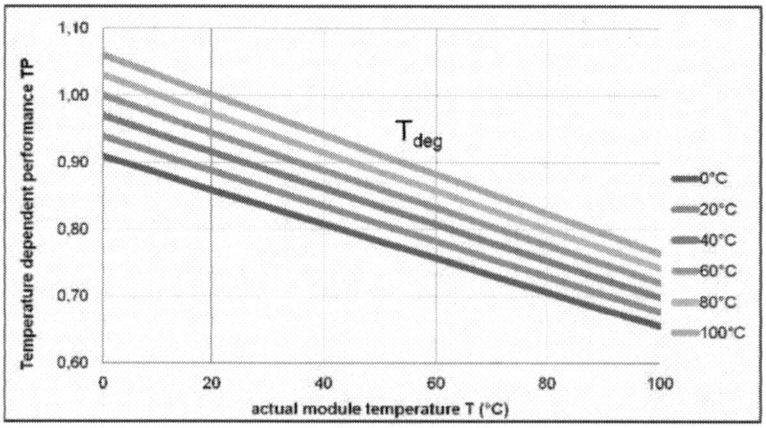

Figure 9. Temperature dependent performance of a-Si/µc-Si tandem modules for different degradation temperatures.

If the results of degradation at different temperatures with resulting different stable efficiencies are combined with the temperature dependent performance of such modules, the calculations result in the plots on figure 9. Here we have the normal temperature coefficient for the modules for several different degradation temperatures. This may give first impression about the real life performance of tandem modules. However, still the changing temperature during LID is not taken into account in this case. More work is needed for dynamic simulations.

4. Conclusions

The Staebler-Wronski-effect in a-Si solar cells and modules produced by industrial deposition equipment can be influenced by many factors during deposition. The degradation can be reduced mainly by the right choice of TCO substrate and i-layer thickness and a low deposition rate for the intrinsic silicon. Other factors such as the p-layer structure and the buffer layer also have a minor influence on the LID. All these results for a-Si can be used for the amorphous part of tandem junction thin film silicon cells. However, progress in the reduction of the LID does usually have the disadvantage of lower deposition rates, more raw material consumption and a lower production line output.

The temperature dependence of the LID and the ability for regeneration at high temperatures makes a prediction about the performance of a-Si containing solar cells difficult. In difference to standard test conditions with a stable irradiation and constant cell temperature, modules are facing changing light and temperature during degradation. The usual models are more or less good for central European climate conditions, but will differ reasonable for hotter or colder climate regions.

Acknowledgments
We acknowledge support from BMBF through Contact No. 03FH008I2 and the support from Ministry for Innovation, Science and Research of North Rhine-Westphalia.

References

[1] Carlson D E and Wronski C R 1976 Amorphous silicon solar cell *Appl. Phys. Lett.* **11**, 28
[2] Chopra K L and Das S R 1983 Thin film solar cells Springer US
[3] Green M A, Emery K, Hishikawa Y, Warta W and Dunlop E D 2015 Solar cell efficiency tables *Progress Photov.: Res. Appl.* **1** 23
[4] Staebler D L and Wronski C R 1977 Reversible conductivity changes in discharge-produced amorphous Si *Appl. Phys. Lett.* **31** 292
[5] Pankove J I and Berkeyheiser J E 1980 Light-induced radiative recombination centers in hydrogenated amorphous silicon *Appl. Phys. Lett.* **37** 705
[6] Meillaud F, Shah A, Droz C, Vallat-Sauvain E and Miazza C 2006 Efficiency limits for single-junction and tandem solar cells Solar Energy Mater. Solar Cells **90** 18
[7] Weicht J A, Hamelmann F and Behrens G 2013 Light-induced degradation of a-Si/i-Si thin film solar cells under different climate conditions and its influence on performance *Proc. 28th EU PVSEC*
[8] Behrens G, Hamelmann F, Niemann L and Weicht J 2013 Universal location based data sets for monitoring and scientific data analysis on PV-systems *Proc. 28th EU PVSEC*
[9] Weicht J, Hamelmann F and Behrens G 2014 Parameter variation of the one-diode model of a-Si and a-Si/μc-Si solar cells for modeling light-induced degradation J. Phys. Conf. Series **559** 12017
[10] Weicht J, Rasch R, Behrens G and Hamelmann F 2015 Changing of the parameters of the "three-diode model" by light-induced degradation at different degradation temperatures of a-Si/μc-Si solar cells *Proc. 31st EU PVSEC*
[11] Weicht J A, Hamelmann F U and Behrens G 2014 Low Light Behavior of a-Si and a-Si/μc-Si PV-Modules Compared to C-Si PV-Modules under Field Test Conditions *Proc. 29th EU PVSEC*
[12] Corpus-Mendoza A N, De Souza M M and Hamelmann F 2015 Design of Schottky contacts for optimum performance of thin film silicon solar cells *IEEE J. Photovoltaics* **5** 22

Journal of Physics: Conference Series **682** (2016) 012003 doi:10.1088/1742-6596/682/1/012003

Plasmonic thin films for application in improved chromogenic windows

Gunnar A Niklasson[1], Pia C Lansåker, Shu-Yi Li and Claes G Granqvist

Department of Engineering Sciences, The Ångström Laboratory, Uppsala University, P.O. Box 534, SE-75121 Uppsala, Sweden

[1]E-mail: gunnar.niklasson@angstrom.uu.se

Abstract. Nanocomposites consisting of noble metal nanoparticles in a transparent matrix exhibit plasmonic absorption in the visible wavelength range, and conducting oxide nanoparticles display a localized plasma absorption in the near infrared. The optical properties of nanocomposites are commonly modelled by effective medium theories, which describe the effective dielectric function of the composite using as input the dielectric functions of the constituents and their respective volume fractions. Plasmonic effects can be exploited in the design of energy-efficient windows in order to obtain improved performance. Electrochromic coatings that switch in the near infrared make use of the modulation of the plasma absorption of oxide nanoparticles due to charge density modulation induced by an external voltage. Plasmonic thermochromic switching in the near infrared has the potential to be significantly larger than in the case of a thin film. Very thin noble metal films are an interesting alternative to conducting oxides as transparent contacts to electrochromic devices. However, in this latter case plasmonic effects are to be avoided rather than exploited.

1. Introduction

The optical properties of nanoparticles, in particular effects related to the localized plasma resonance of metallic particles, have recently attracted major interest. Applications include sensors and biosensors [1,2], as well as plasmonic enhancement of photocatalysis [3-5]. In the field of renewable energy, applications to solar cells [6–8] and highly absorptive films for solar harvesting [9] have been proposed. The scientific study of plasmonic effects has a long pre-history, at least dating back to theoretical and experimental studies of the optical properties of noble metal nanoparticles in the early twentieth century [10,11].

Many studies have concerned noble metal nanoparticles and the dependence of the localized plasma absorption peak on particle shape and surroundings [12], substrate effects [13] and non-locality of the dielectric response [14] has been elucidated. Recently there has emerged a rising interest in alternative plasmonic materials such as TiN [15], other metal oxides and nitrides [16], alloys and intermetallics [17]. In particular oxide-based plasmonic materials exhibit their localized plasma absorption in the near-infrared (NIR) spectral range, thereby opening up a number of new applications.

In this paper we review recent work on coatings for energy-efficient windows, which take advantage of plasmonic effects. Section 2 below gives a basic theoretical background of the subject. Section 3 reviews the optical properties of ultrathin gold films and their application as transparent conductor coatings. Section 4 reviews the subject of plasmonic electrochromism, wherein the plasma resonance of highly doped oxide semiconductor nanoparticles can be tuned by an electrical signal.

Journal of Physics: Conference Series **682** (2016) 012003 doi:10.1088/1742-6596/682/1/012003

Finally, section 5 demonstrates that plasmonic effects play an essential role for the enhanced thermochromism in VO_2 nanoparticle composites.

2. Localized plasma resonance

In this section we give an introduction to the theory of plamonic absorption, but first we recall briefly some basic facts about the optical properties of metals. The optical properties of metallic materials can be approximately described by a dielectric function of the Drude form, namely [18]

$$\varepsilon_f(\omega) = 1 - \omega_p^2 / (\omega(\omega + i/\tau)), \tag{1}$$

where ω_p is the plasma frequency of the free electrons and τ is the relaxation time that describes electron scattering by phonons, defects, impurities and, in the case of nanoparticles, most importantly scattering against the particle surface. This expression only accounts for the optical properties due to the conduction electrons and, in the case of real metals, terms describing interband absorption have to be added [18]. For metal oxide based plasmonic materials, the picture becomes more complicated and the relaxation time becomes in general frequency dependent [19]. The reason is that these materials are heavily doped oxide semiconductors and scattering of conduction electrons from charged impurities, *i.e.*, the dopant ions, is very important. In this case eq. (1) has to be replaced by a Lindhard dielectric function [20] with further refinements, as described in detail by Hamberg and Granqvist [21].

The measurable optical properties are reflectance (R) and transmittance (T), from which absorption, $A = 1 - T - R$, can be obtained. The transmittance and reflectance can be computed from the dielectric function by thin film optics [22]. The absorption is related to the imaginary part of the dielectric function. A maximum in that quantity implies an absorption peak, which is visible as a dip in the transmittance of a thin film.

2.1. *Effective medium theory*

The optical properties of nanoparticles in an embedding matrix are described by the effective dielectric function, ε^{eff}, of the composite. The relation between this function and the dielectric functions of the constituents is provided by effective medium theories. Here we use to the Maxwell Garnett theory [10], which is appropriate for a topology with spherical nanoparticles dispersed in a continuous matrix. Specifically,

$$\varepsilon^{eff}(\omega) = \varepsilon_m\left(\varepsilon_p + 2\varepsilon_m + 2f\left(\varepsilon_p - \varepsilon_m\right)\right) / \left(\varepsilon_p + 2\varepsilon_m - f\left(\varepsilon_p - \varepsilon_m\right)\right), \tag{2}$$

where ε_m is the dielectric function for the matrix material and f is the volume fraction occupied by the particles. The dielectric function of the particles, ε_p, is obtained from ε_f by modifying the relaxation time to account for scattering of electrons from the surfaces of the particles. It must be understood that the dielectric functions in general are frequency dependent, although it is not written explicitly in the case of ε_p and ε_m. We now insert eq. (1) into eq. (2) and obtain the Lorentz oscillator form of the effective dielectric permittivity [23],

$$\varepsilon^{eff}(\omega) = \left(\omega_L^2 - (\omega(\omega + i/\tau)\right) / \left(\omega_T^2 - (\omega(\omega + i/\tau)\right), \tag{3}$$

where the effective longitudinal and transverse resonance frequencies are given by [23]

$$\omega_L^2 = \omega_p^2(1 + 2f) / (1 + 2f + 2\varepsilon_m(1 - f)) \tag{4}$$

and

$$\omega_T^2 = \omega_p^2(1-f)/(1-f+\varepsilon_m(2+f)). \quad (5)$$

The maximum of the optical absorption occurs at the transverse resonance frequency, and in the limit of low volume fraction and $\varepsilon_m = 1$, we obtain the well-known result $\omega_p/\sqrt{3}$. Hence Drude metallic particles dispersed in a matrix gives rise to localized plasma absorption at the transverse resonance frequency given by eq. (5) above. For noble metal particles, this frequency falls in the visible wavelength range, while for highly doped semiconductors and VO_2, which exhibit a much lower plasma frequency, the resonance frequency falls in the NIR.

2.2. Ellipsoidal particles

In the case of ellipsoidal particles the expressions above have to be generalized and it is convenient to rewrite the MG effective dielectric function in the form [24]

$$\varepsilon^{eff}(\omega) = \varepsilon_m\left(1+\frac{2}{3}f\alpha\right)/\left(1-\frac{1}{3}f\alpha\right). \quad (6)$$

For oriented ellipsoids with the incident electric field along the j-axis one obtains

$$\alpha = \frac{\varepsilon_p - \varepsilon_m}{\varepsilon_m + L_j(\varepsilon_p - \varepsilon_m)}, \quad (7)$$

where L_j is the appropriate depolarization factor. Spheres are characterized by $L = 1/3$, and in this case eq. (2) is obtained again. A random distribution of ellipsoidal particles can be described by [24]

$$\alpha = \frac{1}{3}\sum_{i=1}^{3}\frac{\varepsilon_p - \varepsilon_m}{\varepsilon_m + L_i(\varepsilon_p - \varepsilon_m)}, \quad (8)$$

where the L_is are a triplet of depolarization factors. A spheroid exhibits two localized plasma resonances, while a general ellipsoid exhibits three, one for each value of the depolarization factor L_i.

2.3. Spectral density function

We have seen that non-spherical shapes introduce more than one localized plasma resonance. More complex shapes than the ellipsoidal, together with dipolar and multipolar interactions between the electromagnetic fields scattered by particles close to one another, leads to a spectrum of resonances. By introducing a spectral density function (SDF), $g(x)$, we can write the effective dielectric function in a completely general way, which is valid for all conceivable distributions of particle shapes, local densities as well as for separated and touching particles [25,26]:

$$\varepsilon^{eff} = \varepsilon_m\left(1+\int_0^1 \frac{g(x)}{((\varepsilon_p/\varepsilon_m)-1)^{-1}+x}dx\right). \quad (9)$$

The Maxwell-Garnett theory for spheres, eq. (2), corresponds to $g(x)$ being a δ-function positioned at $(1-f)/3$. It is a very complex problem to determine the SDF by inverting experimental optical measurements. However, recent advances have made it manageable and hence it is possible to analyze the spectrum of localized plasma resonances for samples with a very complex nanostructure [27,28].

3. Ultrathin gold films as transparent contacts

Transparent conducting coatings are needed in a variety of applications including displays, solar cells, low-emittance window coatings and electrochromic devices [29]. Highly doped semiconductors, for example doped indium, tin or zinc oxides, are currently used in most of these applications [29]. Thin silver films are used in dielectric/metal/dielectric (D/M/D) stacks as low-emittance coatings. However, Ag films exhibit durability problems when used in displays and electrochromics, and here gold is a more rugged alternative.

Thin films of Au and TiO_2 were prepared by DC magnetron sputtering onto unheated glass substrates from metallic targets. The Au deposits had mass thicknesses (equivalent thicknesses in the absence of porosity) in the 1.4–9.8 nm range, as inferred from deposition rate. Spectral normal transmittance, $T(\lambda)$, was measured in the wavelength range 0.3–2.5 μm using a Perkin-Elmer Lambda 900 double-beam spectrophotometer. Figure 1 shows data for Au films with nine different thicknesses [30]. There is a consistent pattern at NIR wavelengths, namely $T(\lambda)$ decreases monotonically with increasing film thickness. The localized plasma absorption is visible as a dip in the transmittance for the two thinnest films at $\lambda \sim 0.6$–0.7 μm. The flat NIR transmittance for films of thicknesses being 4–5 nm indicates a transition from a dielectric nanoparticle film to a metallic network film and coincides with a transition to a high electrical conductivity. Au films are almost completely continuous at a thickness of 10 nm.

The effective dielectric function of the Au films was obtained from ellipsometry, as described elsewhere [31], and subsequently the SDF was obtained by advanced computational methods devised by Tuncer [27,28,31]. Figure 2 shows the SDF for three gold films. A nanoparticle film with mass thickness 1.4 nm shows a well-defined resonance that is somewhat broadened by interaction effects towards lower x. On the other hand network films exhibit a SDF extending all the way from the MG resonance to $x = 0$ [31]. Hence part of the absorption strength is for these films displaced towards long wavelengths, as evident also from figure 1. The third sample in figure 2 is an almost continuous film; $g(x)$ displays very low values, probably as a result of nano-holes in the film.

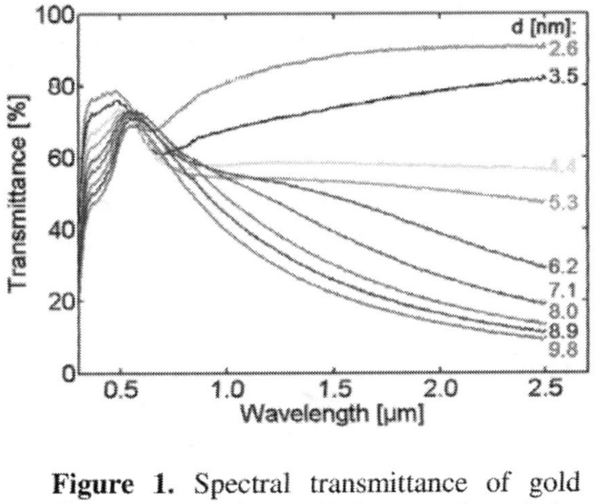

Figure 1. Spectral transmittance of gold films with the shown thicknesses d, made by sputtering onto glass. From [30].

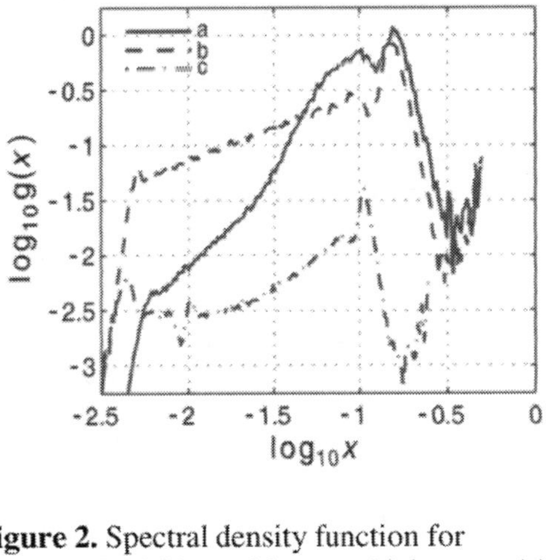

Figure 2. Spectral density function for ultrathin gold films with mass thicknesses (a) 1.4 nm, (b) 3.8 nm and (c) 9.2 nm. From [31].

The requirements for a transparent conductor are high transmittance and good electrical conductivity. In applications demanding high solar transmittance, metallic network films are advantageous but a trade-off has to be made with requirements of conductivity and durability. The visible transmittance in figure 1 is somewhat low for window applications, especially in the case of an electrochromic "smart" window which must include two transparent conductors. However, the transmittance can be increased by incorporating the Au film in a D/M/D stack. Figure 3 shows that the transmittance of a thin Au film can be considerably enhanced by having it between two TiO_2 coatings. It is seen that the transmittance reaches 80% in the mid-visible range.

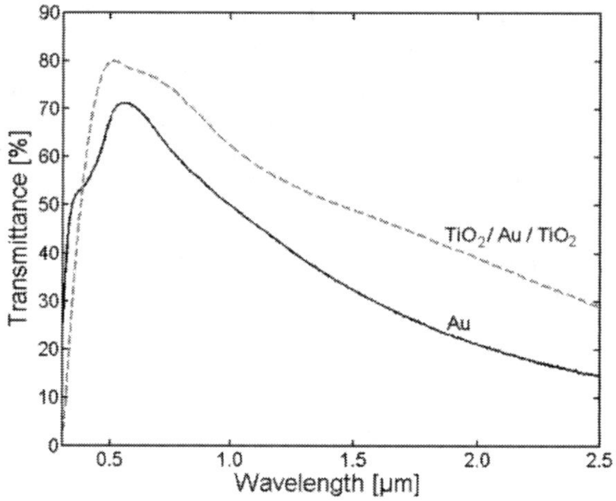

Figure 3. Spectral transmittance for an 8.0-nm-thick Au film and a $TiO_2/Au/TiO_2$ stack with 8 nm Au and 55-nm-thick TiO_2 films. From [30].

In order to evaluate D/M/D stacks for electrochromic applications, electrochemical cycling tests were performed with the optimized $TiO_2/Au/TiO_2$ film, backed by glass, immersed in an electrolyte of 0.5M $LiClO_4$ in propylene carbonate and sweeping the voltage between 3 and 5 V with regard to a Li reference electrode at a rate of 1 mV/s [30]. The Au-based film was found to remain unaffected for at least 20 cycles. For comparison, we performed an identical test with a commercial Ag-based D/M/D type transparent conductor, which revealed a prominent oxidation peak already during the first electrochemical cycle; this oxidation degrades this film severely and turns it into a non-conductor. Hence gold-based films are sufficiently stable so that they can be used as transparent conductors in demanding environments. The dielectric layers in the D/M/D stack must have a not too high resistance, although the requirements are not as strict as for an all-oxide-based transparent conductor. In real applications, doped oxide layers with moderately high conductivity would have to be used.

4. Plasmonic electrochromism

A new approach to electrochromism was developed by Milliron *et al.* [32] who demonstrated that transmittance modulation in the NIR can be achieved electrochemically in layers with nanoparticles consisting of a transparent and electrically conducting material such as In_2O_3:Sn (ITO). This opens up possibilities to use plasmonic effects, rather than polaron effects as in many electrochromic transition metal oxides [33,34], in order to produce a significant modulation of the solar energy transmittance. Furthermore, inclusion of ITO nanocrystals into a film of niobium oxide glass allowed independent modulation of the visible and near-infrared transmittance by changing the applied electrochemical potential [35].

The effect responsible for plasmonic electrochromism in transparent conducting oxide particles is that the localized plasmon frequency of, for example, ITO nanoparticles can be persistently and reversibly tuned electrochemically. The application of different potentials to a nanoparticle film modulates the charge carrier density, and hence the plasma frequency, most probably by the creation of dynamic accumulation and depletion layers at the nanoparticle surfaces [32]. Thus the localized plasma resonance absorption can be swept through the NIR so that a variable part of the solar radiation is absorbed, while the luminous absorption remains high. ITO films can display electron densities as high as ~2.5 × 10^{21} cm^{-3} [29], and similar values have been reported for ZnO-based films [36]. The corresponding localized plasma absorption lies in the NIR or in the red end of the visible spectrum.

We now present a method to calculate approximate performance limits for the luminous and solar transmittance (T_{lum} and T_{sol}, respectively) [37]. Our starting point is the dielectric function due to the free electrons in the transparent conductor. Charged impurity scattering is inevitable in a highly doped oxide and is taken into account by the theoretical framework summarized in section 2 [19–21]. The optical properties of ITO nanoparticles in a dielectric medium can be calculated by effective medium theory, eq. (2), in combination with thin film optics, and the upper panel of figure 4 shows spectral absorptance for a 5-μm-thick layer of ITO nanoparticles in a medium with ε_m=2.25; this value is characteristic for typical polymers and glasses. It is obvious that the localized plasmon peak moves from long wavelengths to ~1.3 μm when the charge carrier concentration, n_e, is increased to 2 × 10^{21} cm^{-3}. The lower panel of figure 4 shows absorptance for n_e = 1 × 10^{21} cm^{-3} and layer thicknesses in the $1 \le d \le 100$ μm range and demonstrates that high absorptance extends over a large part of the NIR for the largest values of d while the visible absorptance is still low. An electrochromic device of this type would then switch from NIR-transparent state to a NIR absorbing one as the applied potential is varied.

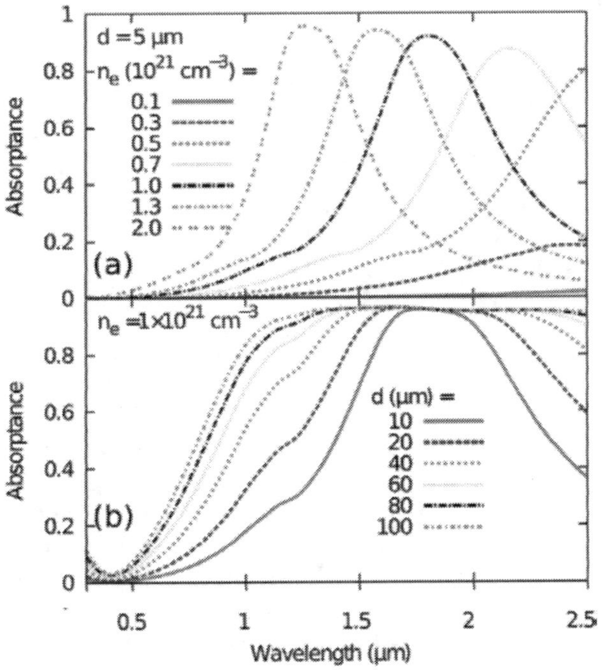

Figure 4. Spectral absorptance for a 5-μm-thick layer with electron densities of 0.1 x $10^{21} \le n_e \le 2 \times 10^{21}$ cm^{-3} (a) as well as for layers with thicknesses of $10 \le d \le 100$ μm and n_e = 1 × 10^{21} cm^{-3} (b). The layer contains 1 vol.% of dispersed spherical ITO nanocrystals. From [37].

Quantitative model calculations using a large number of values of d and n_e were carried out and the resulting data are shown as $T_{lum}(d,n_e)$ *vs.* $T_{sol}(d,n_e)$ in figure 5. It is observed that for each T_{lum} there is a well-defined minimum solar transmittance, denoted $T_{sol,min}$. This can be fitted to a purely empirical expression [37], given by the curve in figure 4, by

$$T_{sol,\min} \approx \alpha + \beta T_{lum} + \gamma T_{lum}^2 + \delta T_{lum}^3 , \tag{10}$$

where α = –1.03, β = 5.94, γ = –9.26 and δ = 5.27. This relation gives an approximate performance limit for a smart window embodying plasmonic NIR electrochromism, based on transparent conducting nanocrystals. It should be noted that eq. (10) represents the optimum solar transmittance in the colored state of an electrochromic device switching from a high T_{sol} to the value in figure 5. The

Journal of Physics: Conference Series **682** (2016) 012003 doi:10.1088/1742-6596/682/1/012003

equation also represents data pertaining to varying electron densities; for example T_{lum} being 0.6, 0.7 and 0.8 corresponds to n_es of 1.52, 1.28 and 0.98 × 10^{21} cm^{-3}, respectively, in the case d = 100 μm. The performance limits are quite impressive; for example a luminous transmittance of 0.6 and 0.8 can be combined with a solar transmittance of 0.35 and 0.5, respectively.

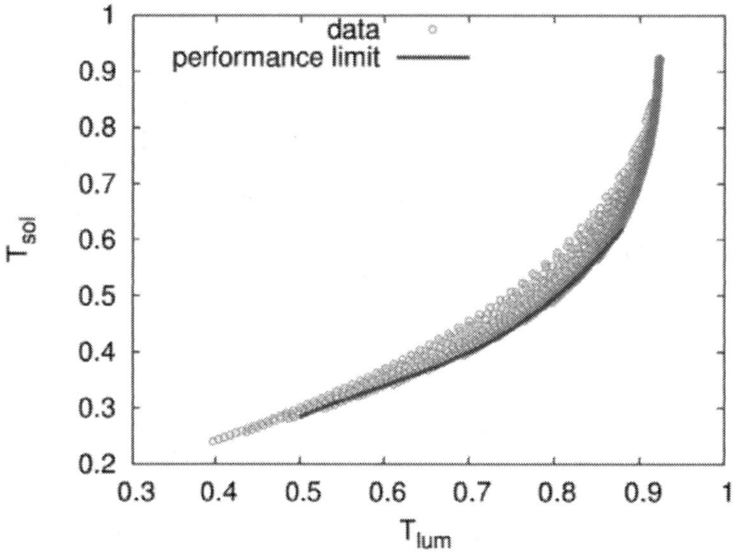

Figure 5. Luminous and solar transmittance (T_{lum} and T_{sol}, respectively) as obtained from model calculations of ITO nanocrystals in a matrix. The curve indicates an approximate performance limit for plasmonic electrochromism. From [35].

5. Thermochromic nanoparticles

Thermochromic (TC) coatings for smart windows as well as the use of TC VO_2 nanoparticles to significantly improve luminous and solar performance of TC films was reviewed in considerable detail by us at the previous INERA Conference in 2014 [38], and this material will not be repeated here. Instead we will describe the basis for, and illustrate the importance of, plasmonic effects for the optical properties of VO_2 nanoparticles. The TC switching in VO_2 occurs between a low temperature phase with semiconducting properties and a high-temperature metallic phase. In the metallic phase, the partly filled conduction band has d-character and a description of the dielectric function in terms of the Drude theory (eq. (1)) or extensions thereof appears not to be possible. Nevertheless the dielectric function of VO_2 is sufficiently metallic-like so that a well-defined localized plasma absorption peak appears in the NIR for nanoparticles and nanoparticle composites. Figure 5 shows model calculations [39] of transmittance in the semiconducting and metallic phases for randomly distributed spheroidal VO_2 nanoparticles with various aspect ratios, m. The experimental dielectric function of VO_2 in the low and high temperature phases [40,41] was used as input data. The volume fraction of particles was taken to be 1% and the thickness was put to 5 μm. The localized plasma resonance is clearly seen as a dip in the transmittance in figure 5b at a wavelength of about 1200 nm for spherical particles, which shifts to longer wavelengths for oblate spheroids. In comparison, no such transmittance dip is seen for the semiconducting low temperature phase in figure 5a. This contrast leads to a large temperature-dependent modulation of NIR transmittance and a significant TC modulation of solar transmittance [39,41]. For example, large values of T_{lum}, between 0.6 and 0.8, can be combined with a significant thermochromic modulation of T_{sol}, up to 0.25 [41].

(a) (b)

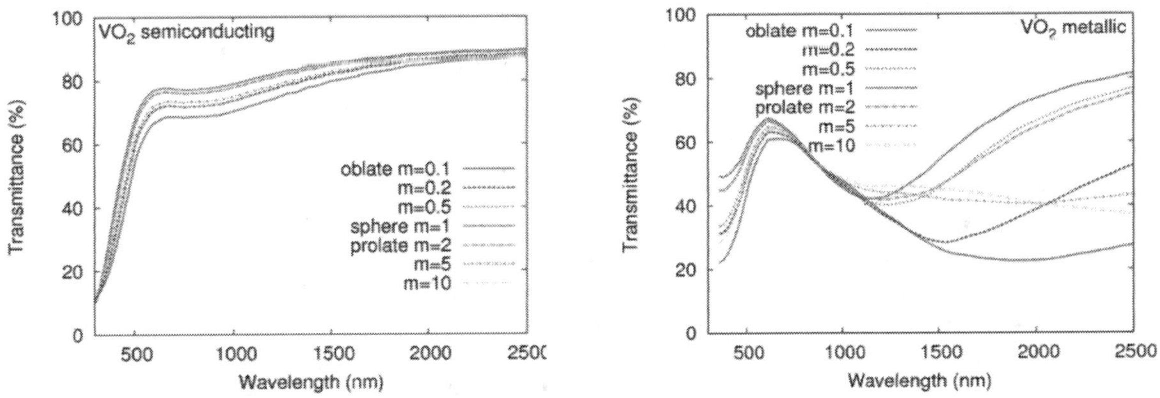

Figure 6. Spectral transmittance for spheroidal VO_2 particles, with the shown aspect ratios *m*, thickness 5 μm and a volume fraction of 0.01, dispersed in a dielectric medium. The spheroids are randomly oriented. Parts (a) and (b) refer to VO_2 in semiconducting and metallic states, respectively. From [39].

6. Conclusion

In this paper we have showed that plasmonic effects are important in a number of applications related to coatings for energy-efficient windows. An elementary treatment of the localized plasma resonance was given and it was shown that nanoparticles of a Drude metal exhibit a Lorentzian absorption peak centered at the resonance frequency. This basic concept was applied to three examples of energy-efficient window coatings. First we showed that very thin gold films with a network structure exhibit a higher solar transmittance than continuous films. When sandwiched between doped oxide layers, thin Au films can be used as transparent contacts for electrochromic devices. Secondly, the field of plasmonic electrochromism takes advantage of the NIR absorption in highly doped conducting oxide nanoparticles. The electron density in the particles comprising a particulate film can be modulated by an external applied potential thereby shifting the localized plasma absorption and giving a pronounced NIR electrochromic effect. By model computations we derived the performance limits of this concept. Thirdly, we showed that VO_2 nanoparticles in the high-temperature metallic phase exhibit a pronounced localized plasma absorption and that this effect leads to nanocomposite thermochromic coatings exhibiting an improved performance.

Acknowledgments

This work was financially supported by the Swedish Research Council for Environment, Agricultural Sciences and Spatial Planning (Formas) and by the European Research Council under the European Community's Seventh Framework Program (FP7/2007–2013)/ERC Grant Agreement No. 267234 ("GRINDOOR"). We are very grateful to Iryna Valyukh and Hans Arwin for ellipsometric measurements and to Enis Tuncer for computing the spectral density function.

This paper was presented at the INERA Conference "Light in Nanoscience and Nanotechnology", Oct. 19-22, 2015, Hissar, Bulgaria. The conference is part of the program INERA REPGOT project of the Institute of Solid State Physics, Bulgarian Academy of Sciences.

References

[1] Stewart M E, Anderton C R, Thompson L B, Maria J, Gray S K, Rogers J A and Nuzzo R G 2008 *Chem. Rev.* **108** 494-521

[2] Anker J N, Hall W P, Lyandres O, Shah N C, Zhao J and van Duyne R P 2008 *Nat. Mater.* **7** 442-53

[3] Antosiewicz T J and Apell S P 2015 *RSC Adv.* **5** 6378-84

[4] Wang P, Huang B, Dai Y and Whangbo M-H 2012 *Phys. Chem. Chem. Phys.* **14** 9813-25
[5] Zhang X, Chen Y L, Liu R-S and Tsai D P 2013 *Rep. Prog. Phys.* **76** 046401
[6] Pillai S, Catchpole K R, Trupke T and Green M A 2007 *J. Appl. Phys.* **101** 093105
[7] Catchpole K R and Polman A 2008 *Opt. Expr.* **16** 21793-800
[8] Winans J D, Hungerford C, Shome K, Rothberg L J and Fauchet P M 2015 *Opt. Expr.* **23** A92-A105
[9] Hägglund C, Zeltzer G, Ruiz R, Thomann I, Lee H-B-R, Brongersma M L and Bent S F 2013 *Nano Lett.* **13** 2252-7
[10] Maxwell Garnett J C 1904 *Phil. Trans Roy. Soc. London* A **203** 385-420
Maxwell Garnett J C 1906 *Phil. Trans Roy. Soc. London* A **205** 237-88
[11] Mie G 1908 *Ann. Phys. (Leipzig)* **330** 377-445
[12] Noguez C 2007 *J. Phys. Chem.* C **111** 3806-19
[13] Roman-Velazquez C E, Noguez, C and Barrera R G 2000 *Phys. Rev. B* **61** 10427-36
[14] Apell S P, Giraldo J and Lundqvist S 1990 *Phase Transitions* **24-26** 577-604
[15] Naik G V, Schroeder J L, Ni X, Kildishev A V, Sands T D and Boltasseva A 2012 *Opt. Mater. Expr.* **2** 478-89
[16] Naik G V, Kim J and Boltasseva A 2011 *Opt. Mater. Expr.***1** 1090-9
[17] Blaber M G, Arnold M D and Ford M J 2010 *J. Phys: Condens. Matter* **22** 143201
[18] Wooten F 1972 *Optical Properties of Solids* (New York: Academic)
[19] Gerlach E and Grosse P 1977 *Festkörperprobleme* **17** 157-93
[20] Lindhard J 1954 *Kgl. Danske Videnskab. Selskab. Mat.-Fys. Medd.* **28** 8
[21] Hamberg I and Granqvist C G 1986 *J. Appl. Phys.* **60** R123-59
[22] Azzam R M and Bashara N M 1977 *Ellipsometry and Polarized Light* (Amsterdam: North-Holland)
[23] Genzel L and Martin T P 1973 *Surf. Sci.* **34** 33-49
[24] Granqvist C G and Hunderi O 1978 *Phys. Rev. B* **18** 2897-906
[25] Bergman D J 1978 *Phys. Rep.* **43** 377-407
[26] Tuncer E 2010 *Materials* **3** 585-613
[27] Tuncer E 2005 *Phys. Rev. B* **71** 012101
[28] Tuncer E and Niklasson G A 2008 *Opt. Commun.* **281** 4374-9
[29] Granqvist C G 2007 *Sol. Energy Mater Sol. Cells* **91** 1529-98
[30] Lansåker P C, Backholm J, Niklasson G A and Granqvist C G 2009 *Thin Solid Films* **518** 1225-9
[31] Lansåker P C, Tuncer E, Valyukh I, Arwin H, Granqvist C G and Niklasson G A 2013 *Progr. Electromagn. Res. Symp. Proc.*, (Stockholm, Sweden, Aug 12–15, 2013) 443-7
[32] Garcia G, Buonsanti R, Runnerstrom E L, Mendelsberg R J, Llordes A, Andres A, Richardson T J and Milliron D J 2011 *Nano Lett.* **11** 4415-20
[33] Berggren L and Niklasson G A 2001 *J. Appl. Phys.* **90** 1860-3
[34] Triana C A, Granqvist C G and Niklasson G A 2015 *J. Appl. Phys.* **118** 024901
[35] Llordes A, Garcia G, Gazquez J and Milliron D J 2013 *Nature* **500** 323-6
[36] Ginley D S, Hosono H and Paine D C (ed.) 2010 *Handbook of Transparent Conductors*, (New York: Springer Science+Business).
[37] Li S-Y, Niklasson G A and Granqvist C G 2012 *Appl. Phys. Lett.* **101** 071903
[38] Niklasson G A, Li S-Y and Granqvist C G 2014 *J. Phys.: Conf. Ser.* **559** 012001
[39] Li S-Y, Niklasson G A and Granqvist C G 2010 *J. Appl. Phys.* **108** 063525
[40] Mlyuka N R, Niklasson G A and Granqvist C G *Phys. Stat. Sol.* A **206** 2155-60
[41] Li S-Y, Niklasson G A and Granqvist C G 2014 *J. Appl. Phys.* **115** 053513

Advanced nanostructured materials and their application for improvement of sun-light harvesting and efficiency of solar cells

D Dimova-Malinovska*

Central Laboratory of Solar Energy and New Energy Sources, Bulgarian Academy of Sciences, 72, bulv. Tzarigradsko chaussee, 1784 Sofia, Bulgaria

*doriana@phys.bas.bg, dmalinovska@yahoo.co.uk

Abstract This review describes the application of different nanostructured materials in solar cells technology for improvement of sun-light harvesting and their efficiency. Several approaches have recently been proposed to increase the efficiency of solar cells above the theoretical limit which are based on a "photon management" concept that employs such phenomena as: (i) down-conversion, and (ii) surface plasmon resonance effect (iii) decreasing of the loss due to the reflection of the radiation, (iv) increasing of the reflection from the back contact, v) increasing of the effective solar cells surface, etc. The results demonstrate the possibility for to increasing of light harvesting, short circuit current and efficiency by application of nanomaterials in thin film and hetero-junction (HJ) solar cells. The first promising results allow an expectation for application of advanced nanomaterials in the 3^{d} generation solar cells.

1. Introduction

Improvement the efficiency and decreasing the production cost of terrestrial photovoltaic devices is important for their widespread application as a power generator. Because the immense potential of solar radiation it was predicted that photovoltaic will become a very serious alternative for electricity generation in the nearest future. Additionally, the problems such as energy independence and security, stability of energy supply and variety of energy sources which become much more vital these days can be overcome applying the electricity from photovoltaic conversion in the energy mix.

The unique properties of nano-sized materials with fitted low size can improve the light harvesting by the solar cells (SCs) and thus their efficiency. Different effects due to the nano-size of the materials can be applied for improvement of the parameters of SC. Several light trapping approaches ("photon management concepts"), which include random [1] and periodic textures [2], plasmonic nanoparticles [3,4], spectral modification of the incident light applying rear earth ions [5] have been studied. In addition, application of nanostructures, such as nanorods (NRs) [6,7] and nanowires [8] in thin film solar cells technology has attracted a lot of attention, in particular because of the possibility for increasing the light harvesting and, as well as, for decreasing of the carrier collection path [9].

This review-paper presents the results of the study of the optical and structural properties of dielectric thin film SiO_2 with embedded Tb^+, applied as a down-convertor to thin film solar cells, multilayer structure ZnO:Al/Ag/ZnO:Al as a back reflector with plasmonic scattering effect and, as well as, electrochemically deposited ZnO with NRs array as antireflection coatings and substrates with increased diffused reflection. The comparison of the parameters of the solar cells without and with applied nanomaterials demonstrates improvement of the external quantum efficiency, increasing the short circuit current and efficiency and of Si thin film solar cells.

2. Experimental, results and discussion

2.1. Down conversion

Almost of the attempt to improve of the solar cell efficiency to date have aimed at modifying the structure of the solar cells itself by fitting its spectroscopic sensitivity to the fixed sunlight spectrum. A reverse concept, however, is to fit the sunlight spectrum to the solar cell sensitivity via a wavelength-shifting operation. This can be done, e.g. by the inclusion of layers containing rare earth ions on the front of the SC. Because the glass, the front contact layer and the doped layer absorb in the blue and near UV spectral range the solar cells and PV modules with glass as a cover or SCs in superstrate configuration have a poor response in this wave-length range. The efficiency of thin film solar cells can be improved with the addition of a photon down-conversion top layer which have absorption bands in the UV region and emission bands in the visible region (in the best case at the spectral sensitivity peak of the Si solar cells). An example with application of SiO_2:Tb thin films as a down convertors is presented here. The sketch of a-Si:H/c-Si (so called hetero-junction) solar cell with a down-convertor on its top is shown in figure 1.

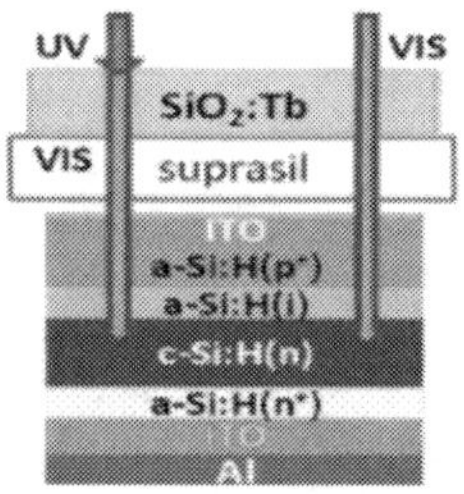

Figure 1. The sketch of a-Si:H/c-Si solar cell with a down-convertor.

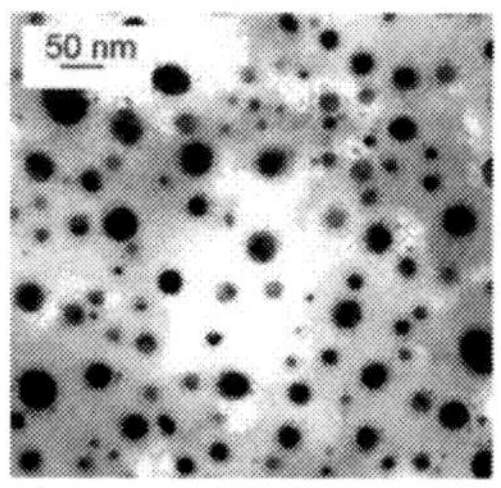

Figure 2. TEM picture of magnetron co-sputtered SiO_2:Tb thin films annealed at 600^0C.

SiO_2:Tb thin films with Tb concentration of 1% [5] are deposited by magnetron co-sputtering on suprasil substrates and are annealed at different temperature 600^0C, 700^0C, 750^0 and 850°C in atmosphere of N_2. The TEM picture of SiO_2:Tb thin films annealed at 600^0C is presented in figure 2, the size of Tb nanoclusters is in the range of 20-70 nm. The spectra of absorption (of the samples deposited on suprasil substrates) are measured before and after annealing (figure 3). The main absorption peak in the as-deposited films is at 224 nm and shifts slightly to 225 nm in the annealed samples. No significant difference is observed in the absorption spectra for the samples annealed at different temperatures. The FWHM of the main absorption peak at about 223-224 nm slightly decreases after annealing and the broad absorption bands at about 323 nm and 760 nm disappeared. The spectra of photoluminescence (PL) excited with λ = 337 nm are presented in figure 3. Peak of PL at about 550 nm is observed and its intensity is maximal for the sample annealed at 600^0C (figure 3). This emission band is due to the $^5D_4$ to $^7F_5$ transition in the Tb^+ [5 and the references there].

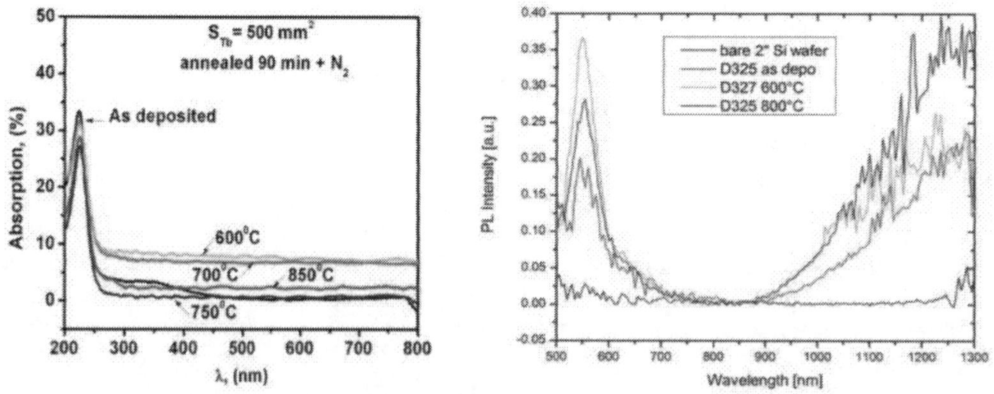

a b

Figure 3. Spectra of absorption (a) and PL (b) of SiO:Tb films.

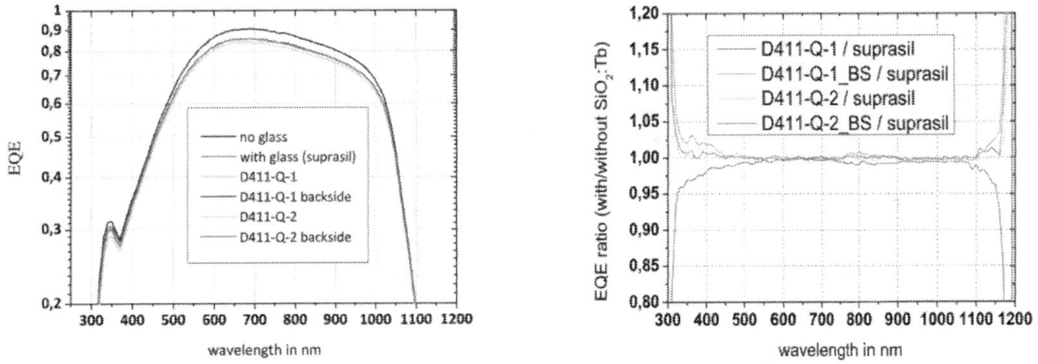

Figure 4. EQE (left) and EQE ratio (right) of Si thin films solar cells with SiO_2:Tb layers on suprasil.

SiO_2:Tb films deposited on suprasil substrates are positioned on top of an a-Si/c-Si-hetero-junction solar cell structure (figure 1). The spectra of external quantum efficiency (EQE) and EQE ratio of the solar cells measured with and without SiO_2:Tb film on the top of the cells demonstrate slight increasing of the EQE in the UV range due to the PL of the Tb^+ more pronounced for the cell D411- Q-1-BS/suprasil (figure 4).

2.2. Multilayer stacks ZnO:Al/Ag/ZnO:Al as back reflector with plasmonic scattering

It has been reported that the back reflector with plasmonic Ag nanoparticels situated between two metal oxide layers, like ZnO, TiO_2 (multilayer stacks) can provide light-trapping performance comparable to the reported for n-i-p a-Si:H solar cells with applied random texture [10]. An advantage of the multilayer structure is that using the metal oxide in the plasmonic stacks with Ag nanoparticles prevents the parasitic effect as charge recombination on metal in the case when the nanoparticles are deposited directly on Si layer [11].

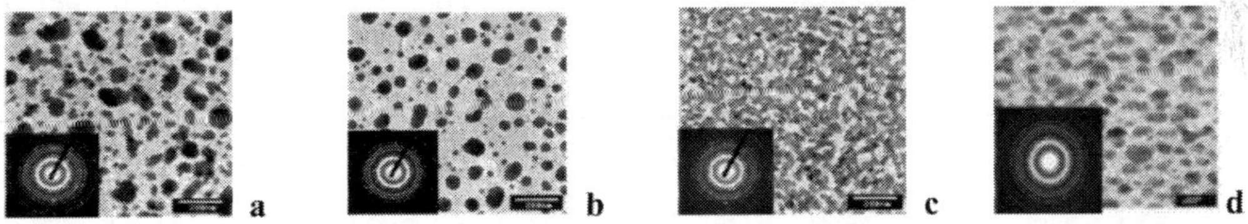

Figure 5. TEM micrographs of as-deposited (a and c) and annealed (b and d) multilayer stacks ZnO:Al(20nm)/Ag(x)/ZnO:Al(20nm) with different thickness of the Ag film for x= 16 nm (a and b) and 20 nm (c and d). The markers correspond to 200 nm [12].

It is demonstrated that the stack structure deposited by magnetron sputtering of thin films ZnO:Al (from ZnO:Al target) and Ag layer (from Ag target) between them with thickness of Ag layer $x > 12$ nm have a very low resistivity ($< 5\ 10^{-5}\ \Omega$ cm) [12]. The TEM pictures (figure 5) show that the Ag nano-particles

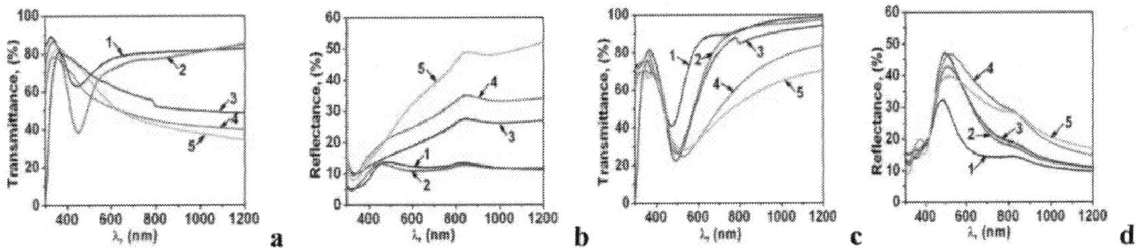

Figure 6. Spectra of reflectance of multilayer stack structures ZnO:Al(20nm)/Ag(x)/ZnO:Al(20nm): as-deposited (a and b) and annealed (c and d), for x=6 nm (curves 1), 10 nm (curves 2), 12 nm curves (3), 16 nm (curves 4) and 20 nm (curves 5) [12].

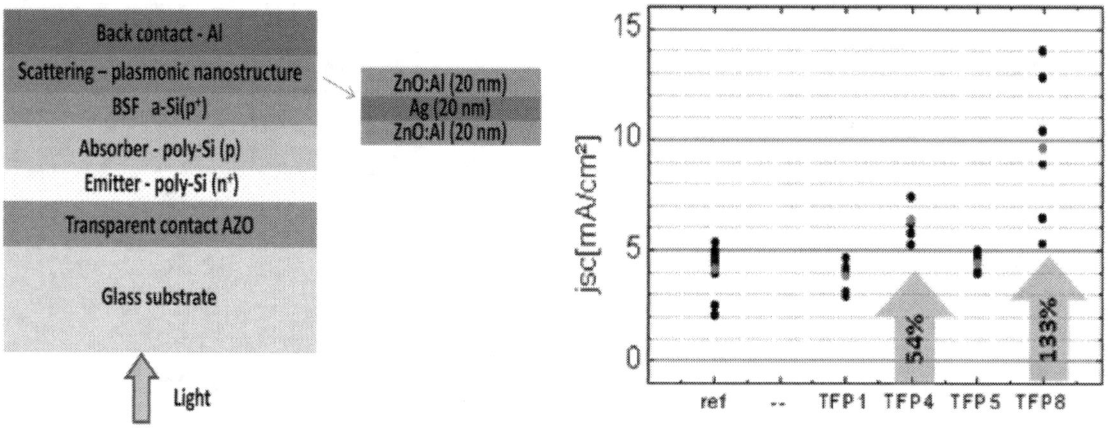

Figure 7. The scheme of the Si thin film solar cells with plasmonic structure (left) and comparison of the J_{sc} of the reference samples without and with plasmonic stack structure (right).

(NPs) form aggregates into chains in the stacks with x = 16 nm and the structure of Ag film transforms from discontinuous into semi-continuous for x = 20 nm. The annealing of the structure results in increasing of the Ag grain size due to the coalescence. The transmittance and reflectance spectra of the as-deposited and annealed films presented in figure 6 demonstrate bands at about 365 nm due to Ag electrons plasma oscillations and near lying inter-band d-shell electron transitions of the electrons [11]. The position and the width of the bands depend on the Ag NPs size and shape distribution in the middle Ag films which are determined by the thickness of the Ag film and the process of annealing. The same bands associated with the plasmonic effect and d-shell electrons inter-band transitions are observed in the spectra of haze ratio in transmission and reflection. The as-deposited stacks with thickness of the middle Ag films of 16 and 20 nm demonstrate high conductivity, high value of reflectance and plasmonic light scattering [4,12]. The application of the ZnO:Al(20nm)/Ag(20nm)/ZnO:Al(20nm) as back scattering reflector results in increasing of short circuit current of Si solar cells on the basis of poly-Si films (figure 7 and table 1). It is seen that the short circuit current, J_{sc}, relative increasing is 10% and 133% in the case of as-deposited stack with Ag thickness of 16 and 20 nm, respectively and 54% - in the case of annealed stack with x =16 nm. All presented data are average taken from measurement of 6 solar cells. The increasing of the J_{sc} is due to the increased reflection of the stacks with plasmonic properties applied as back reflector in Si thin film solar cells which is related to the optimal size, shape and distribution of the Ag nanograins [13].

Table 1. Relative average increases of the J_{sc} in thin film Si solar cells with plasmonic stack.

Samples	Plasmonic structurse	Thicknesses, (nm)	Annealing	Relative J_{sc} average increases
TFP8	ZnO:Al/Ag NPs/ZnO:Al	20/20/20	as deposited	~ 133%
TFP1	ZnO:Al/Ag NPs/ZnO:Al	20/20/20	annealed	~ 0%
TFP5	ZnO:Al/Ag NPs/ZnO:Al	20/16/20	as deposited	~ 10%
TFP4	ZnO:Al/Ag NPs/ZnO:Al	20/16/20	annealed	~ 54%

2.3. ZnO nanorods as antireflection coating in Si heterojunction solar cell

The ZnO based nano-structured (NS) layers with nanometer sized features have highly developed surface areas per unit volume. In order to fabricate solar cells with increased efficiency the ZnO films have to have such a future size on the surface that will define high light harvesting, light scattering and optical confinement [12]. It has been reported that ZnO nanorods array demonstrates antireflecting properties leading to decrease of the specular and diffused reflection of the substrate they are deposited [14,15]. One of the attractive methods for fabrication of ZnO nanostructured array is electrochemical deposition because it is a simple and low cost method and is suitable for industrial large-scale application.

ZnO nanostructured (NS) array layers are deposited as ARC on the front side of silicon heterojunction

Journal of Physics: Conference Series **682** (2016) 012004 doi:10.1088/1742-6596/682/1/012004

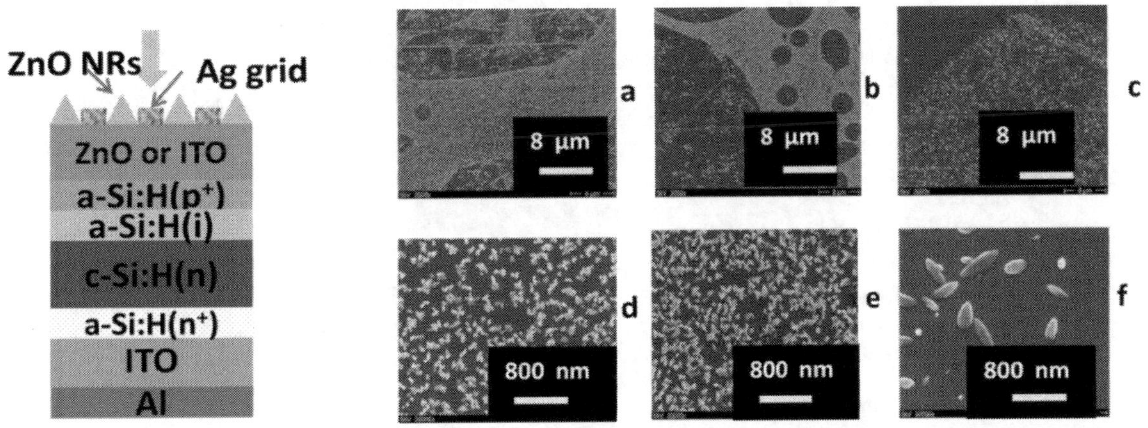

Figure 8. The scheme of the SHJ solar cells with ZnO NSs (left) and SEM of the ZnO NS array deposited on the samples 1-C6 (a and d), 1-C4 (b and e) and 1-C2(c and f) given in two different magnifications. The markers correspond to 8 µm (a, b and c) and 800 nm (d, e and f) [15].

(SHJ) solar cells with ITO as a front side transparent electrode: Ag grid/ITO/(p)a-Si:H/(i)a-Si:H/(n)c-Si/(i)a-Si:H/(n+)a-Si:H/Al [16]. ZnO nanostructured array films are grown by an electrochemical process from aqueous solution of $ZnCl_2$ (5. 10^{-4} M) and KCl (1 M) with pH ≈ 6.0 at 80^0 C using a three-electrode electrochemical cell as described in [17]. The solar cell was used as one of the electrode – the cathode. Spectrally pure graphite electrode was used as anode. The electrolyte was agitated by magnetic stirrer. The SEM micrographs of the ZnO nanostructures grown on 3 different solar cells (figure 8) demonstrate difference in the surface morphology of the grown ZnO NS array is mainly due to the different conditions

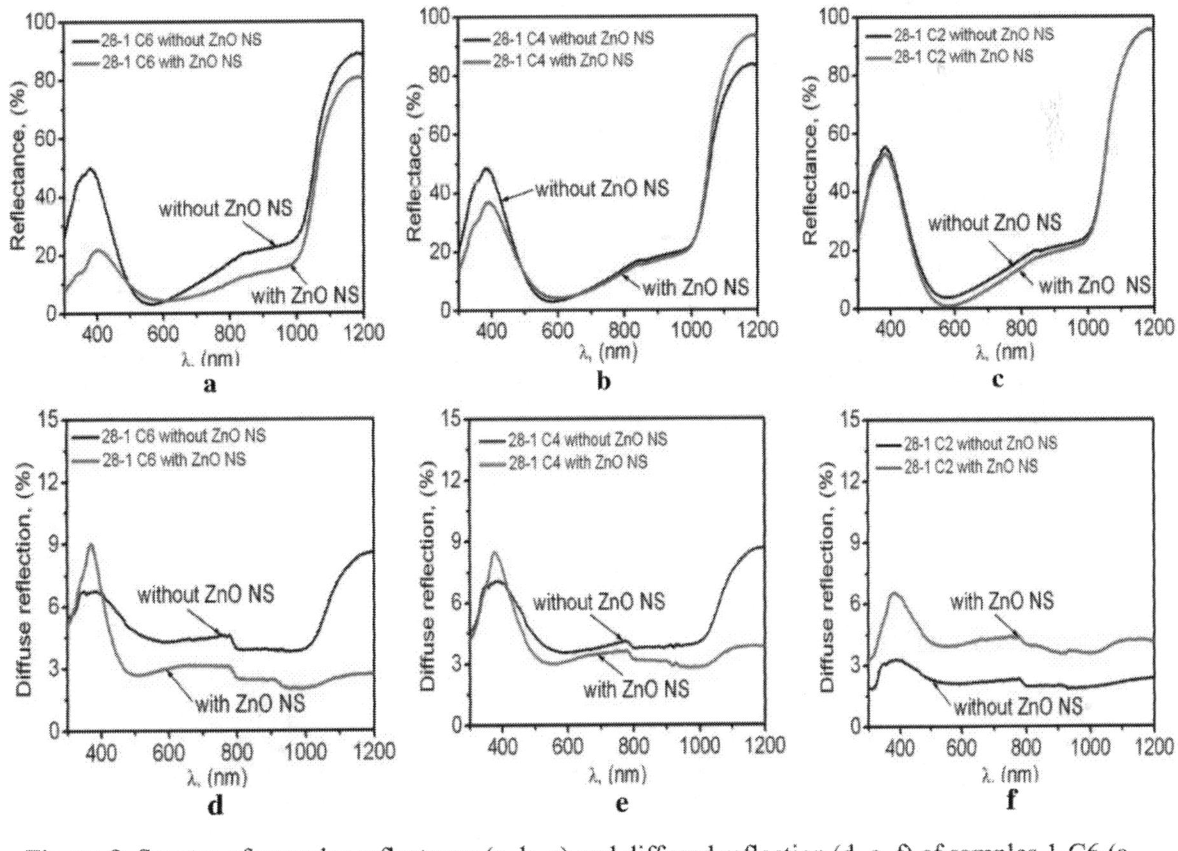

Figure 9. Spectra of specular reflectance (a, b, c) and diffused reflection (d, e, f) of samples 1-C6 (a, d), 1-C4 (b, e) and 1-C2 (e, f) without and with deposited ZnO NS array film [15].

Table 2. The deposition conditions of the ZnO NS array films and comparison of the parameter of the SHJ before after ZnO NS growth.

Solar cells parameters	**1-C2**	**1-C4**	**1-C6**
I_{sc}, (mA), with ZnO NS	30.637	30.689	30.237
I_{sc} ,(mA), without ZnO NS	30.514	29.839	28.979
ΔI_{sc} ,(mA)	0.123	0.850	1.258
ΔI_{sc} ,(%), relative increasing	0.4	2.85	4,3
V_{oc}, (mV),with ZnO NS	695	694	689
V_{oc} ,(mV), without ZnO NS	697	697	691
FF, (%), with ZnO NS	72.74	72.90	71.61
FF, (%), without ZnO NS	74.00	73.50	73.02
ΔFF	1.26	0.60	1.41
η, (%), with ZnO NS	15.492	15.535	14.917
η, (%), without ZnO NS	15.733	15.278	14.521
$\Delta\eta$, (%)	-0.241	0.257	0.396
$\Delta\eta$, (%), relative changes	-1.5	1.68	2.7
t, (min)	1.5	1.5	1.5
$I_{el.-chem.}$, (mA)	3	6	6
$U_{el-chem}$, (mV)	1000	1000	950

applied during electrochemical deposition on solar cells and to the slight difference in the thickness and resistivity of the surface ITO layer. In the case of samples 1-C6 and 1-C4 the nanorods sized about 100-200 nm are distributed in different surface area with different density. The SEM of the sample 1-C2 reviles the presence of largest nano-ZnO features about 800 nm with relatively large distance between them.

The spectra of specular reflectance and diffused reflection of the three samples of the series, before and after the deposition of ZnO nanostructured array films, presented in figure 9 show that the values of reflectance decrease which is more pronounced in the sample 1-C6. The value of diffused reflection decrease in the samples 1-C6 and 1-C4 and increase in the case of sample 1-C2. The differences in the surface morphology and microstructure of the textured ZnO films determine the difference in the optical reflection properties [18].

The study demonstrates that the ZnO nanostructured array films grown on the front side of the SHJ solar cells act as ARC leading to decreasing of the reflectance, improvement of the short circuit current (by about 4.3 %) and efficiency (by about 2.7%) of the thin film SHJ solar cells (table 2). An improvement in the short circuit current and efficiency of the SHJ solar cells is achieved in the samples where the values of reflectance and diffuse reflection decrease after ZnO NS array films deposition on the front side. The higher increase in J_{sc} and efficiency, η, is higher in the samples where larger decreasing of the values of reflectance and diffuse reflection are obtained as a result of deposition of ZnO NS array films leading to improvement of the ligh harvesting [14,15].

2.4. ZnO nanorod array as a substrate in thin film solar cells

The ZnO nanorods (NRs) prepared by an electrochemical process are implemented in the n+-i-p$^+$ a-Si:H solar cell on flexible stainless steel (SS) foil, aiming at increasing the J_{sc} without deterioration the open

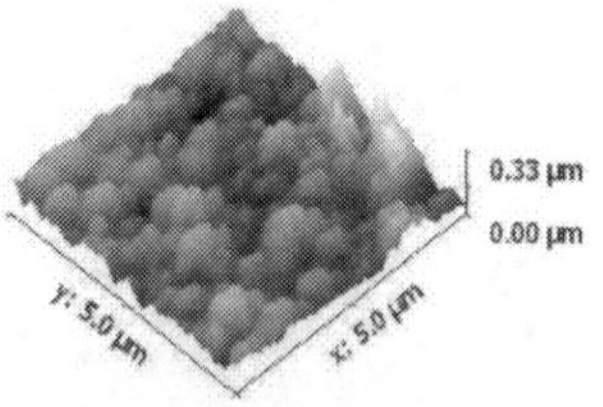

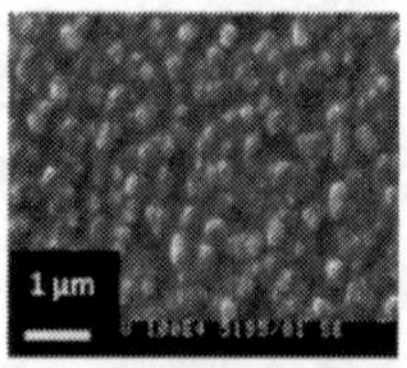

Figure 10. AFM (a) and SEM (b) pictures of ZnO NRs array on SS substrate.

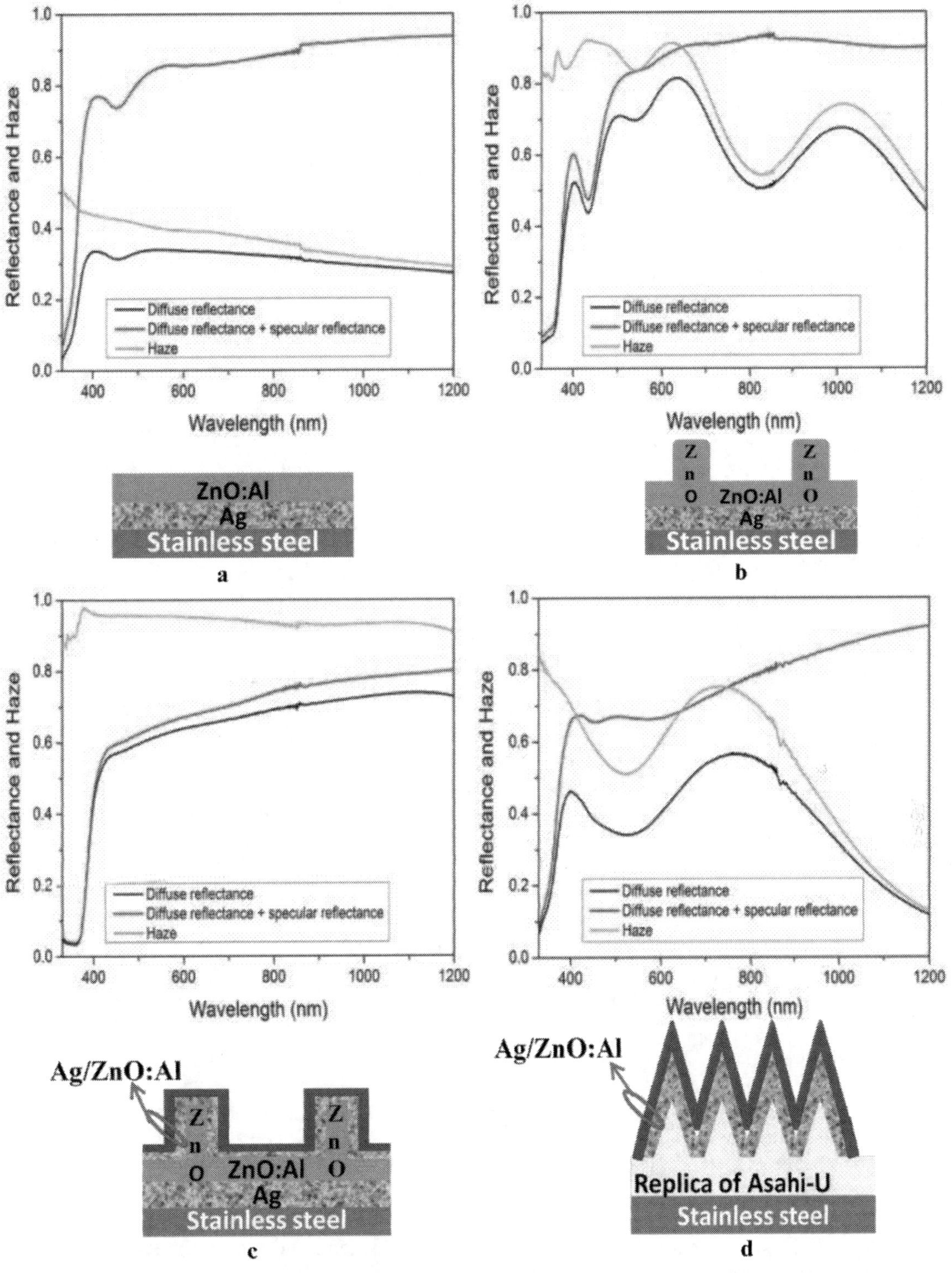

Figure 11. Spectra of diffused reflection, total reflection and haze in reflection of ZnO NRs array deposited on different structured SS substrates [7].

circuit voltage, V_{oc}, and the fill factor, *FF* [7]. Figures 10 a and b show the AFM and SEM images. The surface morphology consists of nanorods with diameter size of single NR between 200 and 500 nm [7] and high of about 200-300 nm. The thin film solar cell structure deposited on such kind of flexible substrate is

Table 3. Parameters of thin film a-Si:H solar cells with different scattering substrates [7].

N	Substrate structure	V_{oc}, mV	J_{sc}, mA/cm^2	FF	Eff.
1	SS/BR	857	11.4	0.61	6.02%
2	SS/BR/NR	803	13.4	0.47	5.11%
3	SS/BR/NR/BR	850	13.8	0.62	7.28%
4	SS/Replica of Asahi-U texture/BR	851	12.4	0.62	6.56%

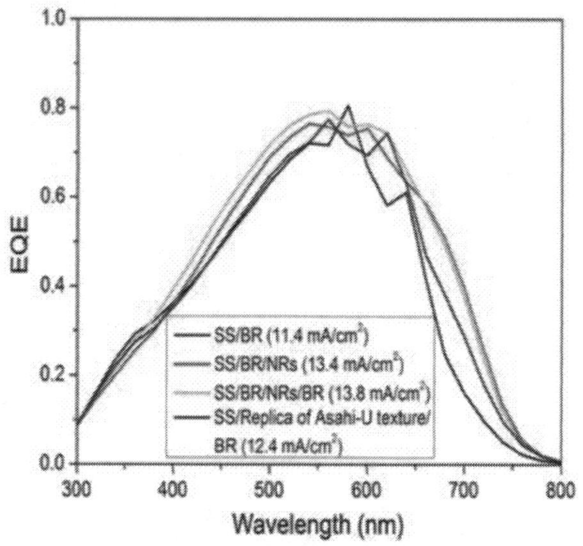

Figure 12. EQE spectra of thin film Si:H solar cells on different scattering substrate [7] .

SS/back reflector (BR)/n-type a-Si:H/i-a-Si:H/p-μc-Si:H/ITO/Ag grid. The BR consists of a stack of Ag and AZO, with thicknesses of respectively 300 nm and 80 nm. In addition to the components of the basic BR (figure 11 a), there is the scattering medium consisting of either ZnO NRs (figure 11 b) and ZnO NRs coated by a stack of Ag (300 nm) and AZO (80nm) (figure 11 c) or a nano-imprinted replica of Asahi-U texture (figure 11 d).

The spectra of reflectance, diffused reflectance and haze in reflectance of four back reflectors on SS substrate are shown in figure 11. The highest value of haze ratio in reflectance demonstrates the BR with ZnO NRs covered by two layers structure Ag (300 nm thick) and ZnO:Al (80 nm thick).

The parameters of the solar cells obtained from I-V characteristics demonstrate that when the solar cell is deposited on ZnO NRs (figure 11 b) the J_{sc} is increased by 2 mA/cm^2 (table 3). Since the ZnO NRs are not doped and thereby very resistive, the solar cell of No. 2 with a-Si:H grown directly on ZnO NRs has low FF and V_{oc}. When the ZnO NRs are coated with Ag and AZO, the solar cell shows 2.4 mA/cm^2 higher J_{sc} than the reference cell of No.1, and 1.4 mA/cm^2 higher J_{sc} than the cell of No. 4 deposited on substrate with the nano-imprinted replica of Asahi-U texture, and highest value of efficiency and no deterioration of the V_{oc} and FF. The optical enhancement of the solar cell by ZnO NRs is explicitly demonstrated in the EQE spectra in figure 12. The ZnO NRs enhance the EQE response in the wavelength range from 400 nm to 800 nm.

Conclusion

The results in this review article show that proposed novel solar cell architectures that make use of nanomaterials have demonstrated increasing light harvesting and improvement of the parameters of the solar cells as a result of the effects due to the nano-scaled structures. The third generation of solar cells with application of nanomaterials as low cost and high efficiency revolutionary technologies needs to be developed to make photovoltaic energy economically competitive. Future advances and the rate at which we can achieve large scale application of the new effects will depend entirely on the amount of resources the society chooses to invest in nanotechnology and photovoltaics.

Acknowledgements

The work has been funded by the 7 FP of EU project NanoPV № 24331.The author thanks to assist. Dr. K. Lovchinov, assist. M. Kamenova, eng. H. Nichev, eng. A. Rachkova, PhD student, eng. M. Petrov, assoc. prof. Dr. M.Ganchev and assoc. prof. Dr. O.Angelov for collaboration.

References

[1] Jäger K and Zeman M 2009 *Appl. Phys. Lett.* **95** 171108

Journal of Physics: Conference Series **682** (2016) 012004 doi:10.1088/1742-6596/682/1/012004

[2] Jovanov V, Palanchoke U, Magnus P, Stiebig H, Hüpkes J, Sichanugrist P, Konagai M, Wiesendanger S, Rockstuhl C and Knipp D 2013 *Opt. Express* A **21** 595
[3] Tan H, Santbergen R, Smets A H M and Zeman M 2012 *Nano Lett.* **12** 4070
[4] Dimova-Malinovska D, Lovchinov K, Petrov M, Karashanova D and Angelov O 2015 *Phys. Status Solidi* A **212** 189
[5] Baumgartner K, Ahren B, Sendova-Vassileva M, Dimova-Malinovska D, Hollander B, Schweizer S and Carius R 2010 *Proc. 25th EUPVSEC/5th WCPV* (Sept.2010, Valencia, Spain) pp.345-350
[6] Kuang Y, van der WerfK H, Houweling Z S, Vece M D and Schropp R E 2012 *J. Non-Cryst. Solids* **358** 2209
[7] Zhang D, Petrov M, Dimova-Malinovska D, Dörenkämper M, Bakker Kl, Schropp R and Soppe W 2014 Proc. 29th EUPVSEC (22-26 Sept.2014, Amsterdam) pp.1631-1633
[8] Kim S H, Hwak S, Hwak P and Cho J W, 2011 *Proc. 37th IEEE Photovoltaic Specialists Conference* 003008
[9] Kuang Y, Vece M D, Rath J K, van Dijk L and Schropp R E I 2013 *Rep. Prog. Phys.* **76** 106502
[10] Beck F J, Mokkapati S and Catchpole K R 2010 *Prog. Photovolt: Res. Appl.* **18** 500
[11] J. T. Guske J T, J. Brown J, A. Welsh A, and S. Franzen S 1012 *Optics Express* **20** 23215
[12] Lovchinov K, Petrov M, Angelov O, Nichev H, Karashanova D and Dimova-Malinovska D 2014 *J of Phys: Conf. Series* **514** 012012
[13] Kreibig U and Vollmer M (1995) *Optical Properties of Metal Clusters* vol 25 (Berlin: Springer)
[14] Petrov M, Lovchinov K, Ganchev M, Dimova-Malinovska D and C. Leendertz C 2013 *Scientific studies-Physics* vol 38, FASC. 4, (Plovdiv: University of Plovdiv press) pp 157-160
[15] Petrov M, Lovchinov K, Mews M, Leendertz C and Dimova-Malinovska D 2015 *Phys. Status Solidi* A **212** 166
[16] Mews M, Schulze T F, Mingirulli N and Korte L 2013 *Energy Procedia* **38** 855
[17] Dimova-Malinovska D, Lovchinov K, Ganchev M, Angelov O, Graff J S and Ulyashin A (2013) *Phys. Stat. Sol.* A **210** 737
[18] Hu Y H, Chen Y-C, Xu H-J, Gao H, Jiang W-H, Hu F and Wang Y X 2010 *Engineering* **2** 973

Journal of Physics: Conference Series **682** (2016) 012005 doi:10.1088/1742-6596/682/1/012005

Electrochromism in sputter deposited $W_{1-y}Mo_yO_3$ thin films

M A Arvizu, C G Granqvist and G A Niklasson
Department of Engineering Sciences, The Ångström Laboratory, Uppsala University, P. O. Box 534, SE-751 21 Uppsala, Sweden

E-mail: Miguel.Arvizu@angstrom.uu.se

Abstract. Electrochromic (EC) properties of tungsten–molybdenum oxide ($W_{1-y}Mo_yO_3$) thin films were investigated. The films were deposited on indium tin oxide covered glass by reactive DC sputtering from tungsten and molybdenum targets. Elemental compositions of the $W_{1-y}Mo_yO_3$ films were determined by Rutherford back scattering. Voltammetric cycling was performed in an electrolyte of 1 M $LiClO_4$ in propylene carbonate. An increase in molybdenum content in the EC films caused both a shift towards higher energies and a lowering of the maximum of the optical absorption band, as compared with WO_3 EC films. Durability under electrochemical cycling was diminished for $W_{1-y}Mo_yO_3$ EC films.

1. Introduction

Electrochromic (EC) materials are capable of undergoing reversible optical changes when an electrochemical potential is applied. The applied voltage causes double insertion/extraction of electrons from a back contact and ions from an electrolyte [1]. Among the materials exhibiting electrochromism, inorganic compounds (mainly transition metal oxides) are widely studied and have been used in industrial applications as well; for example so called "smart windows" are devices that allow the user to control the transmission of solar energy and visible light [2]. Tungsten trioxide (WO_3) is the most common inorganic EC material, and a lot of research has been carried out to investigate its properties through the years [1–4]. From the first studies on electrochromism of tungsten trioxide, it is well known that in the dark state WO_3 films acquire an intense blue color due to the formation of a wide absorption band centred around 1.3 to 1.45 eV [5,6]. Electrochromic properties in molybdenum oxide (MoO_3) were discovered a few years after the phenomenon was shown to exist in WO_3 [7]. Very soon an attempt to combine both oxides and study the effects of mixing them was carried out. Hence EC thin films were produced by co-evaporation from WO_3 and MoO_3 powders, and the outcome of those experiments was the discovery of a shift towards higher energy of the maximum of the absorption band of the mixed oxide films as compared to pure WO_3 [8]. Further studies on co-evaporated W–Mo oxides showed the existence of two types of absorption bands for molybdenum oxide; one is equivalent to the band formed in the colored state of WO_3, the so called G-band centred between 1.5 and 1.56 eV; this is the only band present in the case of small charge insertion. By increasing the amount of inserted charge an additional band, called the T-band, appears between the G-band and the interband absorption across the band gap. The T-band becomes predominant and the G-band decreases for high values of inserted charge [9]. Other synthesis techniques have also been used in order to get tungsten–molybdenum oxide EC films. Electrochemical deposition from a metal peroxide leads to enhanced dynamical optical response for small contents of molybdenum (<2%) which respect to pure WO_3 films [10]. $W_{1-y}Mo_yO_3$ thin films prepared by atmospheric pressure chemical vapor deposition showed improved EC performance with higher dark-state optical absorption, the center of the absorption closer to the peak of the eye's sensitivity and good coloration efficiency [11]. Molybdenum-doped spray pyrolysed WO_3 EC films showed a different microstructure and crystallinity, together with better EC response for nominal concentrations around 6% Mo and better electrochemical stability for small nominal concentrations (~2% Mo) as compared with pure tungsten oxide films [12]. In this present work we prepared $W_{1-y}Mo_yO_3$ EC thin films by sputtering and studied their EC performance with focus on durability issues.

2. Film deposition and composition

2.1. Sputter deposition

Thin films of W oxide, Mo oxide and W–Mo oxide were prepared by reactive DC magnetron sputtering in a deposition system based on a Balzers UTT 400 unit. Targets were 5-cm-diameter plates of W and Mo (Plasmaterials), all with 99.99% purity. The sputter system was evacuated to 2×10^{-4} mTorr. Argon and oxygen, both of 99.997% purity, were then introduced through mass-flow controlled gas inlets. Substrates for optical and electrochemical measurements were unheated 5×5 cm^2 glass plates coated with transparent and electrically conducting layers of In_2O_3:Sn having a resistance of 40 Ω/square. The target–substrate separation was 13 cm. Film thicknesses were ~300 nm as determined by surface profilometry using a Bruker DektakXT instrument. Pure WO_3 and MoO_3 films were obtained using solely the W and Mo targets, respectively, with an O_2/Ar ratio set to 0.14 and a working pressure of ~30 mTorr. The discharge power P was set to 225 W for the tungsten oxide samples and 300 W for molybdenum oxide. For the $W_{1-y}Mo_yO_3$ films, co-sputtering from W and Mo targets was performed with an O_2/Ar ratio set to 0.28 and using the same working pressure of ~30 mTorr. To change the composition of the resulting films, P was kept constant for the W target at 225 W while varying the Mo discharge power between 125 and 300 W. Films for Rutherford backscattering spectrometry (RBS) were deposited onto carbon plates. All deposited films presented X-ray diffraction patterns characteristic of amorphous materials.

2.2. Compositional characterization

Elemental characterization of the films was performed with RBS using 2 MeV ^{4}He ions backscattered at an angle of 170°. With these conditions it is possible to get up to 2 μm analysis depth with high sensitivity for heavy elements such as Mo and W. The RBS data were fitted to a model of the film–substrate system by use of the SIMNRA simulation program [13]. The resulting composition of the $W_{1-y}Mo_yO_3$ films is shown in table 1.

Table 1. Composition of $W_{1-y}Mo_yO_3$ films.

Plasma power of Mo target [W]	Mo content ($y \pm 0.01$)
125	0.02
150	0.04
175	0.08
200	0.10
225	0.14
250	0.16
275	0.22
300	0.30

3. Electrochemical data

Cyclic voltammetry (CV) measurements were performed using a Solartron 1286 Electrochemical Interface in a three-electrode electrochemical cell. The electrolyte was 1 M $LiClO_4$ in propylene carbonate, and lithium foils served as reference and counter electrodes. All measurements were performed inside a glove box with inert argon atmosphere. The voltage sweep rate was 10 mV/s.

WO_3 and $W_{1-y}Mo_yO_3$ are electrochemically stable when cycled hundreds of times in the voltage range 2.0-4.0 V *vs.* Li, and for WO_3 this is in accordance with previous findings [14]. MoO_3 is also stable, but in the range 2.5–4.5 V *vs.* Li. In order to accelerate the ageing process of the films, it is possible to increase the voltage range to 1.7–4.0 V *vs.* Li [14]. This procedure accelerates degradation by about a factor ten [14]. Figure 1 shows cyclic voltammograms for chosen $W_{1-y}Mo_yO_3$ films in that range, and in all cases fast degradation occurs within 80 CV cycles. The ageing seems to happen earlier for films containing molybdenum, but the voltammograms for these films tend to stabilize at some point and this happens earlier when more Mo is present. It is also interesting to note the appearance of a shoulder in the CV data with regard to the current density at around 3.0 V *vs.* Li in the insertion part of the cycle for samples with Mo contents $y > 0.14$. At potentials lower than 2.5 V *vs.* Li, the voltammograms follow the usual behavior for WO_3 EC films. This shoulder, although less noticeable, tends to shift to lower voltage as the number of cycles is increased.

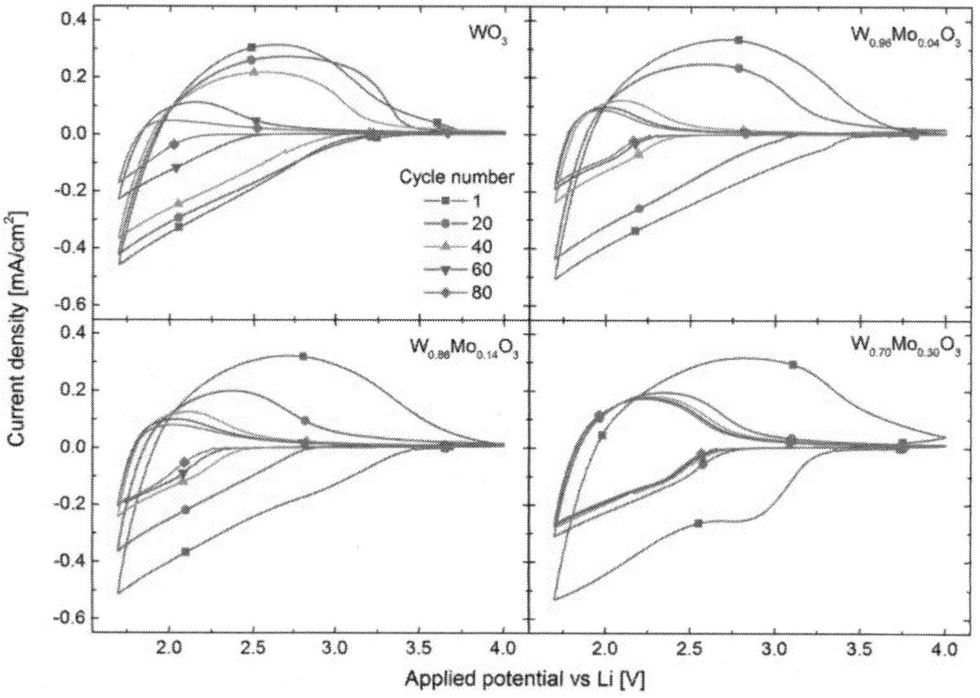

Figure 1. Cyclic voltammograms for W–Mo oxide films of the shown compositions. Data were taken after the indicated numbers of cycles and for the voltage sweep range 1.7–4.0 V *vs.* Li.

In order to further investigate the electrochemical reversibility, figure 2 shows the difference between inserted and extracted charge densities for successive voltammetric cycles defined [14] as $\delta Q = |Q_{Inserted} + Q_{Extracted}|$; the absolute value is necessary to assure that δQ is a positive quantity, since we follow the convention $Q_{Inserted} < 0$, $Q_{Extracted} > 0$ Here it is confirmed that for the $W_{1-y}Mo_yO_3$ EC films the inserted charge is larger than the extracted charge in the first cycles, and this tendency increases with increasing Mo content. Equilibrium between inserted and extracted charge is reached faster than for pure WO_3 films. This difference between inserted and extracted charge could be due to trapped lithium ions that tend to accumulate in the films, and as a consequence ion intercalation becomes more difficult and the films start to acquire a permanent yellowish coloration.

However, a depth-profiling elemental analysis of as-deposited and intercalated samples is necessary in order to confirm this supposition. In the next section we show how the optical modulation is affected when the $W_{1-y}Mo_yO_3$ EC films are cycled in the stated voltage range and the number of cycles is increased.

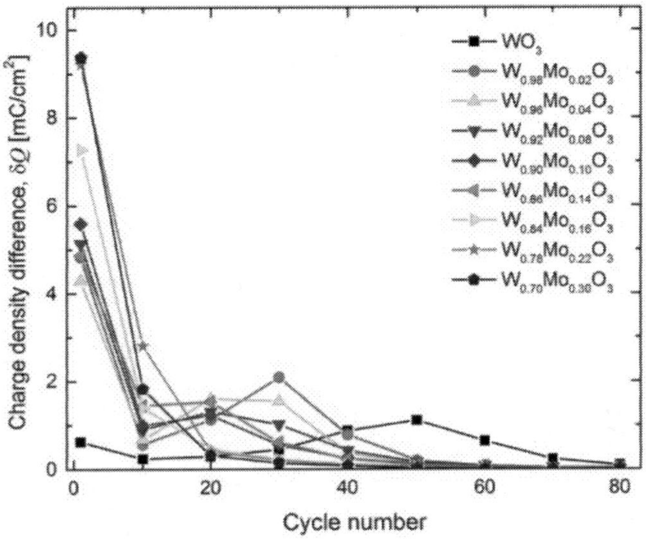

Figure 2. Evolution of the difference in charge density for charge insertion and extraction δQ, as a function of cycle number, for $W_{1-y}Mo_yO_3$ EC films. Symbols denoting data were joined by straight lines.

4. Optical measurements

4.1. In situ optical transmittance

When the samples were cycled in the voltage range 1.7–4.0 V *vs.* Li, monochromatic optical transmittance T at the mid-luminous wavelength 550 nm was registered using an Ocean Optics fiber-optic instrument. In figure 3 it is seen that the transmittance modulation $\Delta T = T_{\text{bleached}} - T_{\text{colored}}$ decreases as the number of cycles increases. It is clear that for high Mo content, $W_{1-y}Mo_yO_3$ EC films quickly lose most of their EC capacity, since the diminishing charge capacity in the films is accompanied by poor optical response. Moreover all samples eventually attained a permanent yellowish color, which is another indication of film degradation when cycling in the wide voltage window.

4.2. Optical absorption spectra

We now discuss the effect of Mo doping on the optical response of the tungsten oxide EC film in the colored state. The WO_3 and $W_{1-y}Mo_yO_3$ EC films were cycled in the voltage range 2.0–4.0 V *vs.* Li, while the MoO_3 EC film was cycled at 2.5–4.5 V *vs.* Li. After the 20th cycle, the CV data were stopped at the state of maximum coloration, the film was taken out from the glovebox, gently rinsed with isopropanol, dried with a nitrogen gun, and its spectral optical transmittance $T(\lambda)$ and reflectance $R(\lambda)$ were immediately measured in the $300 < \lambda < 2500$ nm interval using a Perkin–Elmer Lambda 900 spectrophotometer The resulting data were then used to obtain the optical absorption coefficient using the approximate expression

Journal of Physics: Conference Series **682** (2016) 012005 doi:10.1088/1742-6596/682/1/012005

$$\alpha = \frac{1}{d}\ln\left(\frac{1-R(\lambda)}{T(\lambda)}\right) \quad (1)$$

which is valid for weakly absorbing films on glass substrates, provided the refraction index of the glass is 1.5–1.7 [15]. Here d is film thickness. Figure 4 shows spectral absorption coefficient of the films as a function of photon energy for the WO_3, $W_{1-y}Mo_yO_3$ and MoO_3 EC films. The spectrum for WO_3 is typical with the maximum of the absorption band around 1.45 eV. For MoO_3, a wide absorption band around 2.19 eV is formed by the T-component and the diminished G-band. The effect of increasing Mo content in the $W_{1-y}Mo_yO_3$ films is twofold; firstly the maximum of the absorption band is shifted toward higher energies (it is clearly positioned at 1.71 eV for the $W_{0.92}Mo_{0.08}O_3$ sample) and the samples look grey instead of bluish, and secondly the maximum value diminishes and the absorption spectra became flattened. Very likely the T-component absorption in MoO_3 starts to dominate the optical behavior for the binary oxide films.

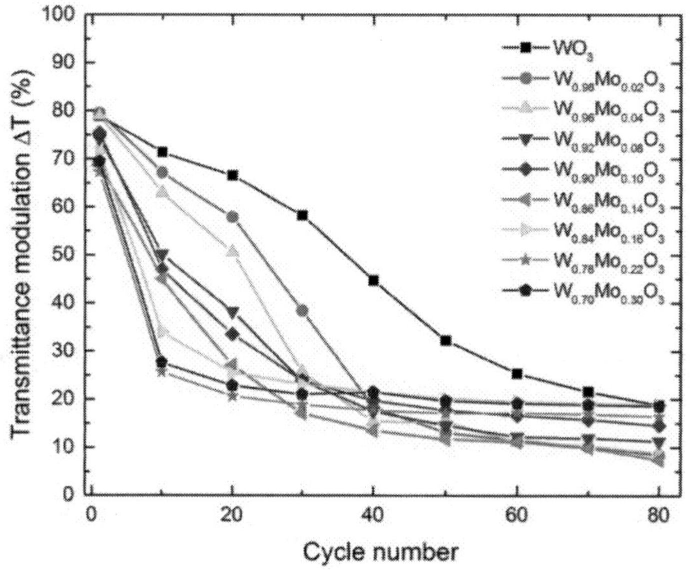

Figure 3. Evolution of the transmittance modulation ΔT at the wavelength 550 nm for films of the shown compositions. Data were taken after the indicated numbers of cycles and for the voltage sweep range 1.7–4.0 V *vs.* Li. Symbols denoting data were joined by straight lines.

5. Conclusions

Electrochromic DC sputtered thin films of WO_3, MoO_3 and $W_{1-y}Mo_yO_3$ were studied and compared as regards their optical properties and degradation. When the Mo content is increased in $W_{1-y}Mo_yO_3$ films the durability of the samples, as measured by the charge capacity and the optical transmittance modulation, decreases considerably when the potential sweep range is 1.7–4.0 V *vs.* Li. In addition, when the potential sweep range is 2.0–4.0 V *vs.* Li, the optical absorption decreases and its spectral dependence is drastically flattened. The resulting films are greyish in the colored state and there is a clear tendency of the maximum of the absorption band to shift toward higher energies, as compared to pure WO_3, at least at low Mo content. Hence, the mixed oxides can be used for “smart windows” in order to get a more neutral dark state, although at the same time somewhat compromising the durability and coloration intensity of the device.

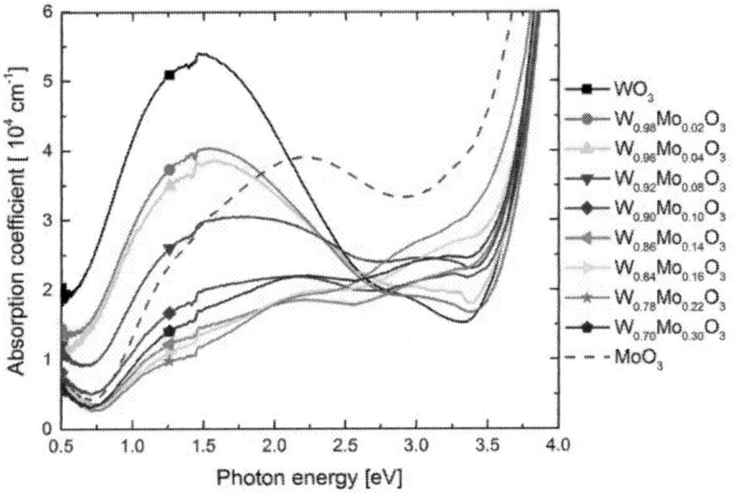

Figure 4. Optical absorption coefficient as a function of energy, in the maximum coloration (dark) state after 20 CV cycles for films of the shown compositions.

Acknowledgements
This paper was presented at the INERA Conference "Light in Nanoscience and Nanotechnology", Oct. 19-22, 2015, Hissar, Bulgaria. The conference is part of the program INERA REPGOT project of the Institute of Solid State Physics, Bulgarian Academy of Sciences. We are grateful to Daniel Primetzhofer and the staff of the Tandem Accelerator Laboratory at Uppsala University for support with RBS measurements. MAA thanks the Mexican Council for Science and Technology (CONACyT) for financial assistance to carry out postdoctoral research at Uppsala University. Our work was supported by the European Research Council under the European Community's Seventh Framework Program (FP7/2007–2013)/ERC Grant Agreement No. 267234 ("GRINDOOR").

References

[1] Granqvist C G 1995 *Handbook of Inorganic Electrochromic Materials* (Amsterdam: Elsevier)
[2] Granqvist C G 2014 *Thin Solid Films* **564** 1–38
[3] Granqvist C G 2000 Sol. Energy Mater. Sol. Cells 60 201–62
[4] Wen R-T, Arvizu M A, Niklasson G A and Granqvist C G 2015 *Surf. Coat. Tech.* **278** 121–25
[5] Deb S K 1969 *Appl. Opt. Suppl.* **3** 192–95
[6] Deb S K 1973 *Philos. Mag.* **27** 801–22
[7] Deb S K 1975 *Proc. 4th Int. Cong. Reprography and Information* (Hannover, Germany) 115–19
[8] Faughnan B W and Crandall R S 1977 *Appl. Phys. Lett.* **31** 834–36
[9] Hiruta Y, Kitao M and Yamada S 1984 *Jpn. J. Appl. Phys.* **23** 1624–27
[10] Pennisi A, Simone F and Lampert C M 1992 *Sol. Energy Mater. Sol. Cells* **28** 233–47
[11] Gesheva K, Szekeres A and Ivanova T 2003 *Sol. Energy Mater. Sol.* Cells **76** 563–76
[12] O-Rueda de León J M, Acosta D R, Pal U and Castañeda L 2011 *Electrochim. Acta* **56** 2599
[13] Mayer M 1999 *AIP Conf. Proc.* **475** 541–44
[14] Arvizu M A, Triana C A, Stefanov B I, Granqvist C G and Niklasson G A 2014 *Sol. Energy Mater. Sol. Cells* **125** 184–89
[15] Hong W Q 1989 *J. Phys.* D: *Appl. Phys.* **22** 1384–85

Preliminary study of surface modification of 3D Poly (ε - caprolactone) scaffolds by ultrashort laser irradiation

A Daskalova[1], I Bliznakova[1], E Iordanova[2], G Yankov[2], M Grozeva[2] and B Ostrowska[3]

[1] Institute of Electronics, Bulgarian Academy of Sciences, 72, Tsarigradsko Chaussee Blvd., 1784 Sofia, Bulgaria
[2] Institute of Solid State Physics, Bulgarian Academy of Sciences, 72, Tsarigradsko Chaussee Blvd., 1784 Sofia, Bulgaria
[3] Biomaterials Group, Materials Design Division, Faculty of Materials Science and Engineering, Warsaw University of Technology, 141 Woloska Str., 02-507 Warsaw, Poland

E-mail: a_daskalova@code.bg

Abstract. Three – dimensional poly (ε- caprolactone) (PCL) scaffolds as suitable biocompatible material for manufacturing tissue replacements are utilized for tissue engineering purposes. The porous structures are fabricated by rapid prototyping method (Bioscaffolder) based on hypodermic dispensing process. The consecution of experiments demonstrated the possibility on creation of surface micro formations, applying different laser fluences, at 1 kHz repetition rate for fixed time of exposure 1 sec at 800 nm central wavelength. The combination of both methods offers possibilities for successful production of 3D matrices with modified surfaces. The obtained results of laser – induced surface modifications of PCL demonstrate the potential of the method to microprocess this kind of material for possible applications in regenerative medicine.

1. Introduction

In tissue engineering, the three – dimensional (3D) scaffolds fabricated from polymers are used to provide structural support and guidance of cells for creation of new tissue generation. The created 3D templates mimic the mechanical and biological properties of native extra cellular matrix. The scaffolds have to possess specific properties for enhancing the cell adhesion and proliferation. Generally, the treatment of bone fractures using 3D scaffolds which are osteoconductive and osteoinductive are able to facilitate the osteogenesis process [1, 2]. Many synthetic and natural polymers successfully are applied as major ingredient for 3D scaffolds creation by different types of methods. One of the main techniques for creation of nanofiber meshes is the electrospining technique, where the cells tend to adhere mostly on the fibrous surface which is a serious limitation of the method. Many research groups have focused to fabricate 3D structures with smaller diameters abilities to extend their surface area to improve the cell infiltration control [3, 4]. The major limitation of the implementation of PCL scaffolds in regenerative medicine is associated mainly with the problems related to vascularisation of the artificial constructs in volume. The additional laser assisted treatment of surface topography by inducing surface roughness could provide enhanced cell infiltration [5].

In this paper, 3D – PCL scaffolds are used for producing artificial structures for hard – tissue replacements. The PCL biomaterial is bioresorbable, non-toxic and suitable for bone and cartilage engineering. The incorporation of such matrices is unsuitable for *in vitro* cell seeding without additional treatment of the surface which can improve significantly their cytocompatibility [4, 6].

The main aim of the present work is to explore the potential possibility of femtosecond laser – induced surface modifications of 3D – PCL scaffolds for further improving of cell – interface interaction. First important step to reach that purpose was achieved and it was specified that the 3D – PCL structures can be modified by femtosecond laser modification method which was obtained from the SEM morphological analysis of the surfaces.

2. Material and methods

2.1. 3D – PCL scaffolds fabrication

The material used for the fabrication of the 3D templates was Poly (ε- caprolactone) (M_n = 45000g/mol) purchased from Sigma - Aldrich. 3D – PCL scaffolds used for these measurements were printed by Bioscaffolder apparatus (SYS & ENG, Germany) based on hypodermic dispensing process. The apparatus is capable of 3D – plotting of different types of polymers for biomedical applications. It is equiped with temperature – controlled stationary table and mobile dispense head [7]. The material was heated to temperature of 56°C to melt. The samples shape was modelled by Solid Works program.

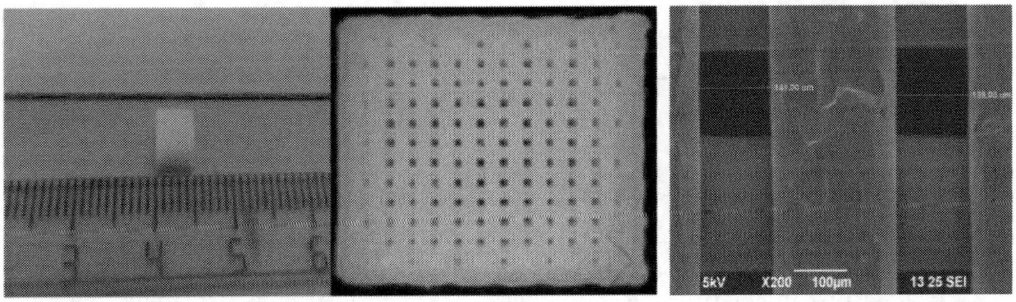

Figure 1. Images of the printed 3D – PCL wood pile scaffold: an overview of the sample (a), Dino-Lite digital microscope image of the sample (b), SEM image of non irradiated sample, x200 magnification (c).

The prepared samples have a cubic shape with a woodpile structure composed from 19 layers and 13 fibres. The obtained average fibre diameter is 209 μm and the average distance between the fibres is ~ 148 μm and porosity 40 %.

In our examinations, 3D – PCL scaffolds with dimensions 5 x 5 mm in width and 3 mm in height were printed for femtosecond laser irradiation (see figure 1). The PCL material posses hydrophobic properties (see figure 2). The obtained contact angle measurement is 90° averaged over 10 measurements).

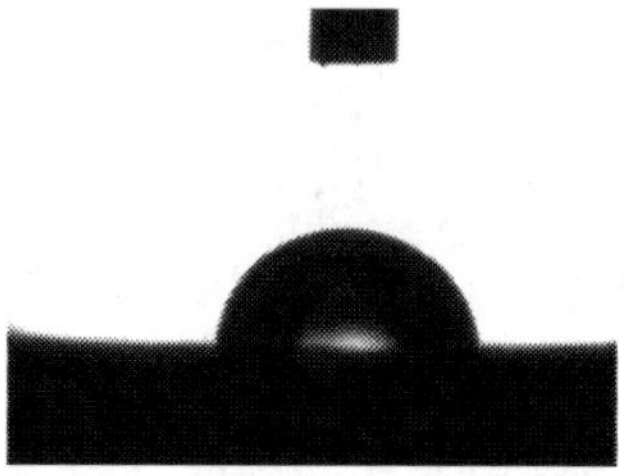

Figure 2. Wettability contact angle measurement of 3D PCL scaffold.

2.2. Experimental setup

The experimental setup depicted in figure 3, shows a schematic view of the experimental configuration. The femtosecond laser system used in this study is a Spectra – Physics product. The Spitfire Ace amplifier includes an optical stretcher, regenerative amplifier and optical compressor. The amplified femtosecond seed pulses are provided by a separate mode-locked Ti:sapphire laser (Spectra-Physics Mai Tai SP).

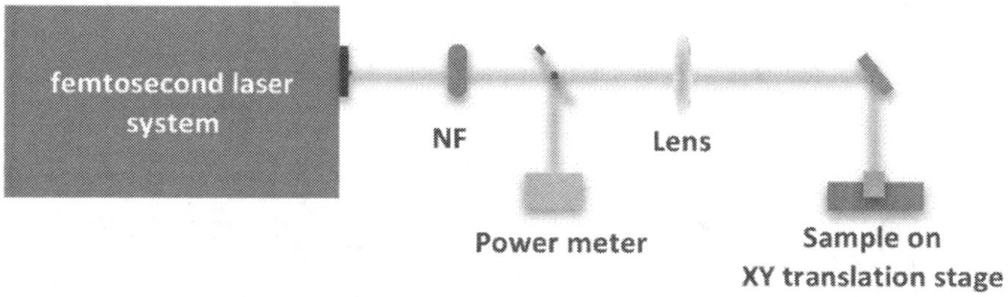

Figure 3. Schematic representative of the experimental setup for ultrashort surface modification of 3D – PCL scaffolds.

The required energy for amplifying the seed pulses is supplied by a separate pump laser (Spectra-Physics Empower 45). The Spitfire Ace amplification is provided by a titanium-doped sapphire (Ti:sapphire) crystal. The regenerative Ti:Sapphire amplified laser system emits at 800 nm central wavelength with 35 fs pulse duration and 1 kHz repetition rate. The energy per pulse is 6 mJ. The laser beam is focused onto a sample with a 20 cm focal length lens at normal incidence. The corresponding focussing spot has diameter ~ 200 μm. In order to facilitate the movement of the sample, XY translation stage has been used. A circular variable neutral density filter is used to attenuate the laser beam to desired laser power. The output power and energy are detected by power meter (Ophir Nova II). The applied laser fluences are F = 3.18 J/cm^2, F = 4.78 J/cm^2, F = 6.37 J/cm^2.

3. Results and discussion

The irradiated surfaces of the 3D – PCL scaffolds were sputter coated with gold layer and analyzed using scanning electron microscopy (Microscope JOEL ISM 6390&EDS Oxford INCA). Here, the SEM images after femtosecond laser irradiation with F = 3.18 J/cm^2, F = 4.78 J/cm^2, F = 6.37 J/cm^2 and time of exposure 1 sec for each sample are presented (figure 4). The processing of the sample was performed by scanning the laser beam over the sample surface in X, Y plane. The laser beam was focused on the surface of the PCL samples.

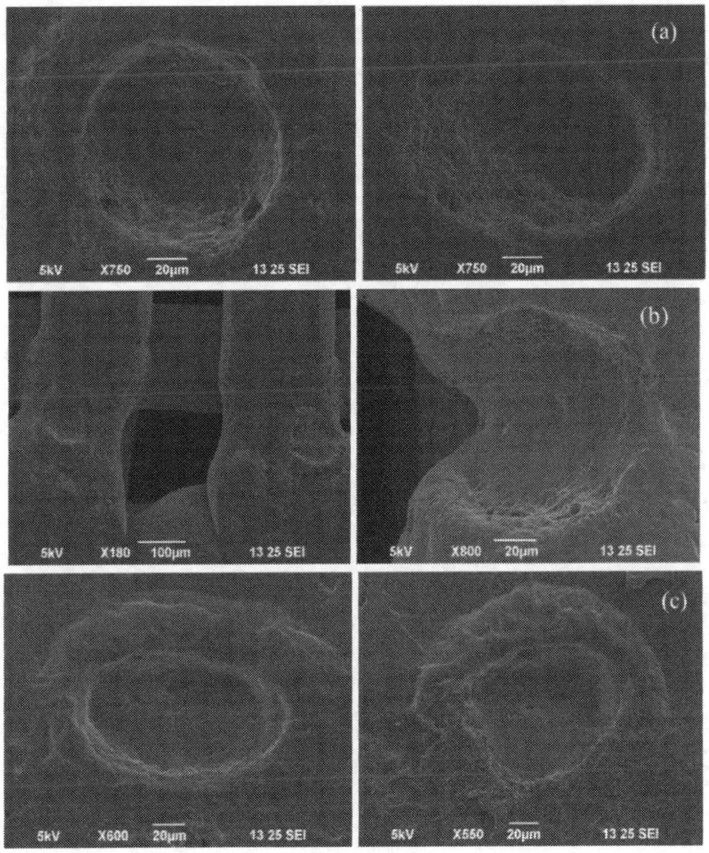

Figure 4. SEM images of 3D – PCL scaffolds irradiated with F = 3.18 J/cm^2 (a); F = 4.78 J/cm^2 (b); F = 6.37 J/cm^2 (c) and time of exposure 1 sec at 800 nm. Each row represents two positions on the sample under same condition.

Figure 4 clearly shows the potential of femtosecond laser for inducing surface modifications of 3D PCL as scaffolds. The negligible heat distribution at the surrounding areas at low laser intensities (figure 4a) were observed. The acquired SEM microscope images reveal that by increasing the laser energy and at constant exposure time (figure 4c) is observed distinct formation of rims created around the laser modified zones which is not that pronounced for lower laser intensities (figure 4a). Figure 5 shows more detailed representation of the laser irradiated area on the laser modified surface of the sample. The three examples of the SEM images on figure 5a, 5b and 5c correspond to x43, x150 and x650 magnifications, respectively. The samples are irradiated with laser fluence F = 6.37 J/cm^2, exposure time of 1 sec at 800 nm central wavelength. The femtosecond laser interaction with PCL results in clean ablation with minimal heat affected zones when working in low fluence regime. Femtosecond processing regime proceeds with plasma – induced ablation, which leads to rapid formation of plasma and vapor which leads to negligible heat affected zones, absence of collateral damage, very clean ablation. The main reason of this phenomenon is the nature of ultra – short interaction processes where the photon

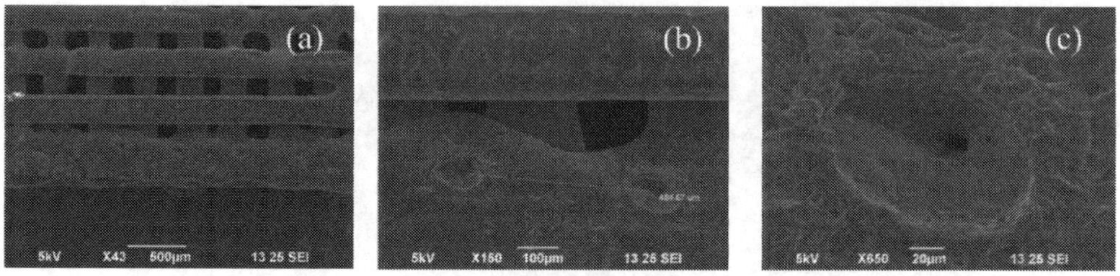

Figure 5. SEM images of 3D – PCL scaffolds with different magnifications – x43 (a), x150 (b), x650 (c). Laser fluence F = 6.37 J/cm^2 and exposure time - 1 sec at 800 nm.

energy is less than the band gap of the material. The laser photons at 800 nm have 1.55 eV energy while the most polymers have from 4 to 7 eV photon energy. In the case of femtosecond ablation of polymers the near IR photons are nonlinearly absorbed through multiphoton absorption leading to multiphoton ionization, which enables precise microstructuring [8].

The combination of rapid prototyping technique with laser assisted ablation leads to clean ablation craters at the material surface and could be applied for improving the surface properties like roughness and wettability thus enhancing the cell adhesion and differentiation. The obtained experimental results indicated that to increase the level of precision of laser – material microprocessing and obtaining well – defined surface modifications is required precise control over laser beam position on the material surface. In fact, the output laser system parameters such as energy and exposure time enable to expand the laser system capabilities for more detailed studies of the optimal structure for medical applications.

4. Conclusions

In this study, we report first attempts for performing micro modifications on 3D – PCL scaffolds surfaces using femtosecond laser technique. The preliminary results of the present study are utilized to explore the potential of femtosecond laser radiation to create micro topographical features by improving the surface porosity which is a crucial factor for cell – surface interaction. The creation of additional sample modification will assess the mechanical properties of the produced matrices and will provide a path to enhance cell's infiltration in volume for improved vascularisation. The preliminary SEM investigations confirmed that the obtained laser created modifications on 3D – PCL rapid prototyped samples result in minimum heat affected zones and collateral damage in the low fluence regime. The next step is to perform extended study on the cell cultivation on the fs laser modified surface. Furthermore, investigating the cell adhesion, proliferation and differentiation on 3D laser irradiated PCL scaffolds.

Acknowledgments

This work was performed and financially supported in the frame of COST Action MP1301 NEWGEN, "New generation biomimetic and customized implants for bone engineering", 2013 – 2017, and partly supported and financed by the INERA project – "Research and Innovation Capacity Strengthening of Institute of Solid State Physics - Bulgarian Academy of Sciences in Multifunctional Nanostructures", Grant agreement no: 316309.

References

[1] Manjunath K, Shiek A, Dhanasekaran M and Winkins S 2014 *Int. J. Nanomed.* **9** 183–195

[2] Wong H, Chu P, Leung F, Cheung K, Luk K and Yeung K 2014 *Prog. Nat. Sci.: Mat. Int.* **24** 561–567

[3] Guimaraes A, Martins A, D Pinho E, Faria S, L Reis R and M Neves N 2010 *Nanomed.* **5(4)** 539 – 554

[4] Rebollar E, Cordero D, Martins A, Chiussia S, L Reis R, M Neves N and Leóna B 2011 *Appl. Surf. Sci.* **257** 4091–4095

[5] Blakeney B, Tambralli A, Anderson J, Andukuri A, Lim Dong-Jin, Dean D and Jun Ho-Wook 2011 *Biomat.* **32** 1583 – 1590

[6] Billiet T, Vandenhaute M, Schelfhout J, Vlierberghe S V and Dubruel P 2012 *Biomat.* **33** 6020 – 6041

[7] Ragaert K 2011 *Micro-Extrusion of Thermoplastics for 3D Plotting of Scaffolds for Tissue Engineering Gent:* University of Gent

[8] Deepak K L N, Venugopal Rao S and Narayana Rao D 2011 *Proc. of SPIE* **8190** 81900R-1

Application of laser light scattering for determination of the border aerosol-air in a specialized physical laboratory setup

K S Damov[1, 2] and M T Iliev[2]

[1]South-West University "Neofit Rilski", 2700 Blagoevgrad, Bulgaria
[2]Sofia University "St. Kliment Ohridski", 5 James Bourchier Blvd., 1164 Sofia

E-mail: damov@swu.bg

Abstract. The current article examines the application of laser light scattering in a specialized laboratory setup. It is used for determination of the kinematic viscosity and mass density of Aerodispersed Systems formed in Limited Volume (High Concentration Aerosols) by the method of free flow out. The measurement chamber is first filled with the investigated aerosol. After a predetermined delay time the aerosol is allowed to flow out through a calibrated pipe with fixed size located few centimetres above the chamber's bottom. The lowering of the upper border aerosol-air is continuously scanned using a laser beam directed along the axis of the cylindrical chamber. The kinematic viscosity and mass density of the investigated aerosol phase are calculated by formulas obtained by the authors. The suggested application of laser light scattering led to higher accuracy of the determination the position of aerosol-air border, thence the certainty of this method. This improvement allowed the use of computer controlled optoelectronic setting. The use of laser light scattering significantly improves the method for determination of the kinematic viscosity and mass density of Aerodispersed Systems formed in Limited Volume.

1. Introduction

The aerosol phase of the Aerodispersed Systems formed in Limited Volume (ASLV) exerts additional hydrostatic pressure which causes aerosol flow through the horizontal tube above the bottom of the container (figure 1). In our previous papers [1-3], after applying the continuity equation and Poiseuille's law we come to the following formula for the kinematic viscosity of the aerosol:

$$\nu = \frac{\eta}{\rho} = -\mathrm{b}\frac{\bar{H}}{\Delta H}\Delta t;\ \mathrm{b} = \frac{\pi r^4 g}{8\ell S} \tag{1}$$

where r is the radius of the horizontal tube, l – its length, g – gravitational acceleration, S – the cross-section of the aerosol container, η – the dynamic viscosity, ρ – the density of the aerosol phase. The interval ΔH crossed by the lowering of the upper borderline for time Δt, $\bar{H}$ is its average height with respect to the bottom of the container, The density ρ can be considered as the average density in the aerosol volume $V = S\bar{H}$, b – calculated equipment constant.

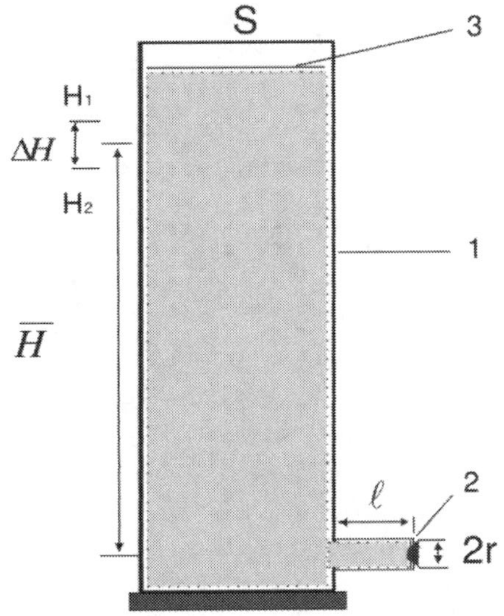

Figure 1. Schematic diagram showing (illustrating) the basis of the method for determination of the viscosity and the mass density of ASLV: 1– aerosol container, 2– calibrated exit tube, 3– aerosol-air border.

The equation (1) gives the possibility to obtain from a single experiment an array of values of the kinematic viscosity of the aerosol under investigation for different heights from the container bottom. The viscosity is specific for each type of aerosol and it can serve as its "individual characteristic" [4]. With the help of equation (1) we cannot determine separately the dynamic viscosity and the density of the aerosol, which change in the process of evolution of the system. This requires an assessment of the aerosol parameters, which will enable us to get additional information for the mass density ρ of the aerosol phase.

The most frequently used aerosol was cigarette smoke; it contains well-defined particles with a spherical shape, and it has been researched by authors [5, 6].

Einstein formula [7] for the dynamic viscosity of a dispersion system is well known:

$$\eta = \eta_0(1+2.5\varphi) \tag{2}$$

where η is the dynamic viscosity of the dispersion system, η_0 - of the dispersion medium (air) and φ is the ratio of the volume of the dispersion phase to the total volume.

The following assessment of parameters was made for the aerosol we used. The countable concentration of particles is $N \approx 10^{15}$ m^{-3} [6]. The mean diameter of the aerosol particles measured with an electronic microscope is $D \approx 0.16\times10^{-6}$ m [6]. For the total volume of aerosol particles in 1m^3 we have $\approx 10^{-6}$ m^3, $\varphi \approx 10^{-6}$.

This allows us to disregard the second term in formula (2), i.e. $\eta \approx \eta_0 = 1.80\times10^{-5}$ N.cm^{-2}. Then the formula (1) transforms to [8–11].

$$\rho = \frac{\eta}{\nu} = -\frac{1}{b}\frac{\Delta H}{\bar{H}}\frac{\eta_0}{\Delta t} \tag{3}$$

Equation (3), makes it possible to determine the density of the aerosol phase ρ, after measuring Δt and $\bar{H}$.

Let's make a short theoretical analysis of the error we are making in our measurements. According to equation (1) after we apply the well-known method we come to the following equation for the relative error

$$\frac{\Delta v}{v} = 2\frac{\Delta d}{d} + \frac{\Delta \bar{H}}{\bar{H}} + \frac{\Delta t}{t} + \frac{\Delta \ell}{\ell} + 2\frac{\Delta D}{D} + \frac{\Delta(\Delta H)}{\Delta H} \tag{4}$$

In formula (4): $\Delta d / d$ is the relative error in the measurement of the exit tube diameter ($d = 2r$), $\Delta \bar{H} / \bar{H}$ – of the average height, $\Delta t / t$ – of the measured time interval, $\Delta \ell / \ell$ – of the measured length of the exit tube, $\Delta D / D$ – of the diameter of the measurement chamber ($S = \pi D^2/4$), $\Delta(\Delta H) / \Delta H$ – of the measured distance between level H_1 and H_2.

The analysis of the individual relative errors shows, that the biggest part in the final error is due to the error of the measured time interval Δt. In practice it ranges from a second to several tens of seconds. The value of Δt depends on several factors – the viscosity of the aerosol, the size of the experimental setup (measurement chamber and the exit tube), the meteorological factors etc.

We have to keep in mind, that the aerosol-air border is not clearly outlined. Under day light (or the light of an electric bulb) it looks blurred due to the diffusion.

The scanning is usually made from one side (left or right) where the fluctuations are largest. This along with the blurred border leads to a significant error in the end result. Thus the accuracy of the measured time interval Δt significantly depends on the precise determination of the position on the horizontal aerosol-air border.

In order to resolve this problem, we referred to laser scattering. It has already been used for determination of the horizontal aerosol-air border by the authors of [12]. The scanning of the aerosol-air border was done visually. They have examined the natural lowering of the border caused by sedimentation. Its duration is in the order of several hours. The long period of this process assured them high accuracy of the end results, even in the case of visual scanning.

The scattering, reflection and absorption of light by the aerosols depend on the size, shape and nature of their particles and also on the wavelength of the incident light. Because of its large practical application the theory of light scattering has already been highly developed.

The theories of light scattering treat the particles of the dispersed aerosol phase (in approximation) as spheres of transparent isotropic dielectric. It is assumed that they scatter light incoherently, i.e. independently of each other. D. Sinclair [13] supposes, without evidence that in order to fulfil this assumption, the distance between the individual particles has to be at least 10 times (or even better 100 times) larger than their radius. Later it has been experimentally proven that with highly concentrated hydrosols [14] for incoherent scattering this ratio has to be larger than 3–4 times. As an exception this condition applies also to the case of extremely high concentrations of larger particles, because there the interference of light scattered by adjacent particles is usually negligible.

In this work by means of laser scattering we determine very precise the location of liquid-air border in containers filled with aerosol systems. As result of this the viscosity and density of the given aerosol systems could be characterised or compared better than the other visual methods.

2. Experimental equipment

To determine the position of the upper aerosol borderline in the course of time, we used the method of scattering of laser light under an angle of 90° with respect to the beam direction and detection of light with phototransistors [4, 8].

We assume that the scattering of the laser beam by aerosol particles of the size indicated above is described with the theory of Mie [6]. Since we used He-Ne laser with unpolarised light and power 5 mW, from the indicatrice of the dispersed light at an angle of 90° [6], there follows that the intensity of dispersed light is enough to be registered with phototransistors.

Figure 2 shows the scheme of the experimental equipment, where 1 is an aerosol container, 2 – container for initial formation of aerosol, 3 – peristaltic pump, 4 – entry to chamber, 5 – exit calibrated tube, 6 – laser, 7 – mirror, 8 – photosensors, 9 – electronic block, 10 – computer.

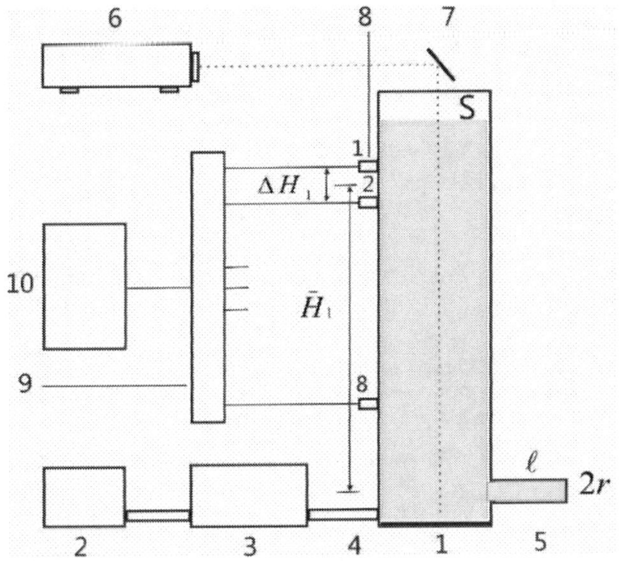

Figure 2. Scheme of the experimental equipment.

In our setup we used a transperant Polymethyl methacrylate (PMMA) cylinder, 30 mm in diameter as an aerosol chamber. The exit tube's (also made of PMMA) length is $\ell = 30$ mm and has a diameter $2r = 8$ mm. The calculated (1) installation constant is $b = 4.65\times10^{-5}$ m^2s^{-2}. The Stokes number is $Stk \approx 10^{-6} << 1$, therefore the particles are accurately following the change in direction of the flow.

We use 8 sensors (phototransistors), spaced of 30 mm from each other, the first one being 30 cm above the exit tube. After filling the chamber, we let the aerosol stay for 1 min before the outflow.

3. Results and Discussion

The results of the kinematic viscosity of several types of aerosols obtained are shown in table 1. Each of the results listed is an average of 7 experimentally obtained viscosity values. The experiments were carried out under standard laboratory conditions and atmospheric pressure of 970–980 hPa. On line 10 in table 1 we have shown, for comparison, the kinematic viscosity of air at 20°C and normal atmospheric pressure [15].

From the results shown, it is clear that the kinematic viscosity of the limited volume aerosols is one or two orders of magnitude higher than the one of air.

Table 1. Values of the kinematic viscosity of several types of aerosols obtained by the described method.

	Aerosol type	**ν, 10^{-4} m^2s^{-1}**
1	Arda Cigarette smoke hard pack	6.0 ± 0.3
2	Arda Cigarette smoke soft pack	5.0 ± 0.3
3	Phoenics Cigarette smoke soft pack	4.7 ± 0.2
4	Paraffin vapour	3.9 ± 0.2
5	Glycerol vapour	10 ± 2
6	Ammonium chloride smoke	13.5 ± 0.6
7	Colophony resin smoke	9.6 ± 1.5
8	Screening smoke white	5.3 ± 0.2
9	Colored screening smoke red	13 ± 2
10	**Air at 20°C**	**0.15**

Table 2 shows the data we obtained for the mass density of the aerosol phase of several types of aerosols using the described method. Each of the listed values is an average of 7 measurements. On line 10 in table 2 we have shown, for comparison, the kinematic viscosity of air at 20°C and normal atmospheric pressure [15].

It is seen from the results shown that the mass density of the aerosols with limited volume is one or two orders lower than that of air. This fact is confirmed by the data for the kinematic viscosity of aerosols and air shown in table 1.

Table 2. Values of the mass density of aerosol's phase of several types of aerosols obtained by the described method.

№	**Aerosol type**	**ρ, 10^{-3} kg.m^{-3}**
1	Cigarette smoke – Arda hard pack	30.6 ± 1.8
2	Cigarette smoke – Arda soft pack	37 ± 2
3	Cigarette smoke – Phoenics soft pack	38.5 ± 1.5
4	Paraffin vapour	46.6 ± 1.4
5	Glycerol vapour	22 ± 3
6	Ammonium Chloride smoke	13.5 ± 0.6
7	Colophony resin smoke	21 ± 2
8	Screening smoke – white	33.9 ± 1.4
9	Colored screening smoke – red	15 ± 2
10	**Air at 20°C**	**1200**

4. Conclusions

An important fact is that on the basis of laser light scattering an experimental setup was constructed for investigation of ASLV. Although the aerosol phase of ASLV has a much smaller density than the density of air at normal pressure and temperature, they flow out of the aerosol container like liquids a result of the additional hydrostatic pressure they possess.

The presented method based on laser light scattering and Poiseuille's law makes it possible to examine ASLV in the process of their evolution even in the case of application of an additional external impact on them – like electric, magnetic and/or acoustic fields.

In result of the processes of coagulation and sedimentation within them, the density of the aerosol phase is constantly decreasing. The preliminary results show that this method is applicable to a period of time of the evolution process in relation to the moment of their production in order of 10 to 15 min. The lowest density of the aerosol phase that can be measured using our method is in the order of 2.8×10^{-3} kg m^{-3}.

The preliminary results have also shown that this method can be used for the development of a new type of characterization of the final product of the cigarette industry (cigarettes, cigars, cigarillos, pipe tobacco, filters), by determining and analyzing the parameters of their smoke.

Acknowledgements

INERA EU project Research Potential (FP7-316309-REGPOT-2012-2013-1) "Research and Innovation Capacity Strengthening of ISSP-BAS in Multifunctional Nanostructures" is acknowledged.

5. References

[1] Antonov A and Damov K 1989 Method for changing the viscosity of aerosol systems, *XV AU Union Conference*, **2**, p.16, Odessa /in Russian/
[2] Damov K and Antonov A 1995 Investigating the kinematic viscosity of aerosol systems in quasi-liquid state, *Compt. Rend. Acad. Bulg. Sci.*, **48**, №2, 29-33

[3] Damov K and Antonov A 1997 On the measurement of kinematic viscosity of verosol vystems, 9th *International Conference on Surface and Colloid Science*, p.79, Sofia

[4] Damov K 1991 Method and structure for determining, the dynamic viscosity of aerosols, *Miscellany of Investigations by Young Researchers*, SWU Blagoevgrad, 132-136 /in Bulgarian/

[5] Lee S, Schneider T, Grant L and Verkerk P May 1985 *AEROSOLS : Research, risk assessment, and control strategies, Proc of the Sec. U.S.-Dutch Int. Symp, (Williamsburg: Virginia)* (Michigan: Lewis Publishers, INC)

[6] Green H and Lane W 1964 *Particulate Clouds: Dust, Smokes and Mists* (London:Princeton)

[7] Bibbik E 1981 *Reology of Dispersion Systems*, (St Petersburg: St. Petersburg University press) /in Russian/

[8] Damov K and Antonov A 2007 Determining the mass density of aerosol system in limited volume via laser light scattering, *Journal of Applied Electromagnetism*, **9** - № 1, June, 57-64

[9] Damov K and Antonov A 2008 Investigation of aerosol system evolution in limited volume via laser light scattering, *Journal of Applied Electromagnetism*, **10** - №1, June, 45-54

[10] Damov K and Antonov A 2005 Method of measuring density distribution in model aerosol system, *Proc. of the International Scientific Conference 8-11.06.2005*, South-West Univ. "Neofit Rilsky", Blagoevgrad, **1**, 346-351 /in Bulgarian/

[11] Damov K and Antonov A 2007 Investigation of aerosol system evolution in limited volume via laser light scattering, *Proc. of the International Scientific Conference 6-10.06.2007*, South-West Univ. "Neofit Rilsky", Blagoevgrad, **2**, 117-121

[12] Antonov A, Petrova A and Uskeselieva L 1977 *Compt. Rend. Acad. Bulg. Sci.*, **30**, 10-14

[13] Sinclair D 1950 *Handbook on Aerosols*, (Washington: U.S. Government printing office)

[14] Reist P 1984 *Introduction to Aerosol Science* (New York: Macmillan Publishing Company)

[15] White F 1998 *Fluid Mechanics, 4th edition*, (London: McGraw-Hill)

Monitoring of respiratory volumes by an long period grating sensor of bending

N Raicevic[1], M D Ivanovic, P Belicev and J Petrovic
Vinca Institute of Nuclear Sciences, University of Belgrade
Mike Petrovica Alasa 12-14, 11000 Belgrade, Serbia

E-mail: nevenar@vin.bg.ac.rs

Abstract. Here, we present a method of respiratory volumes monitoring using a single fiber-grating sensor of bending. Measurements are conducted using simple monochromatic interrogation scheme that relies on a photodiode measurement of the power transmitted through a long period grating (LPG) sensor at fixed wavelength. Good sensor accuracy in measurements of tidal and minute respiratory volumes for different types of breathing is achieved.

1. Introduction

Long-period fibre gratings have been used as bend-rejection filters, gain-equalizers, and sensors [1, 2]. Their sensing properties stem from sensitivity of the fibre cladding modes to changes in environmental parameters such as temperature and the refractive index of the surrounding medium, and on forces applied to the fibre (pressure, strain, torque, bending). The sensing applications of LPGs are in biomedicine [3, 4, 5, 6, 7], medicine [8, 9, 10], environmental studies [11], radiation dosimetry [12], etc. LPGs can be sensitive to a number of parameters at the same time, which is known as cross-sensitivity and is detrimental in all applications except in multi-parameter sensing schemes [13].

Here we investigate application of LPG sensors of bending in volumetric measurement of air inhaled and exhaled during breathing. The idea to use the sensor of bending is motivated by the discovery of Konno and Mead [14] that the volume of the inhaled air is correlated to the thoracic and abdominal movements caused by breathing. Further incentives for the use of LPGs for this purpose are the non-invasiveness of the measurement method and immunity to electrical noises. As the end application we target the non-invasive ventilation (NIV) during which a mechanical ventilator provides respiratory support to a patient. The current clinical standard relies on air-flow sensors interfaced with the patient though an oronasal mask or a mouthpiece. Such an interface promotes air leaks that introduce errors in the measurement of respiratory volumes, making NIV success dependent on the capability of the ventilator to deal with air leaks or on clinician's observation of the patient's chest wall movement [15, 16]. Here we review our investigations of the LPG bending measurement scheme as one possible solution fully independent of the mask.

Content of this paper is as follows: in Section 2 we describe the working principle of LPG sensor and support this by the results of numerical modelling. We further present the full experimental characterization of the grating and describe a simple monochromatic scheme that was used for

[1] Nevena Raicevic: nevenar@vin.bg.ac.rs

respiratory monitoring. In Section 3 we review the results of volumetric measurements of breathing performed on a set of 15 healthy volunteers. Conclusions are given in Section 4.

2. Long period grating sensors of bending

2.1. Principles of operation

The LPG is one-dimensional periodic perturbation of the fibre refractive index profile or geometry along the fibre axis. Period of LPG varies from several tens to several hundreds of micrometers which, for NIR light, can induce out coupling of fundamental core mode and formation of resonant co-propagating cladding modes [1]. The formed cladding modes represent loss of light from the core, and can be recognized as attenuation bands (AB) in the transmission spectrum. Attenuation bands occur at discrete, resonant wavelengths (λ) that satisfy phase-matching condition between core (n_{co}^{eff}) and ν-th cladding mode (${}^{\nu}n_{clad}^{eff}$):

$$\lambda_{\nu} = \left(n_{co}^{eff} - {}^{\nu}n_{clad}^{eff}\right)\Lambda \tag{1}$$

Here Λ represents the grating period. In order to calculate the mode distribution of core and cladding modes fulfilling Eq. (1), finite element method [17] has been performed due to its flexibility in meshing different geometries and shapes. Theoretically, mode analysis was done for LPGs inscribed in the core of standard SMF-28 with the following parameters: n_{co}=1.44922, n_{cl}= 1.444024, r_{co}=4.1μm (core radius), and r_{cl}=62.5μm (cladding radius). The refractive index modulation representing the grating is set to Δn_{co}=5.75·10^{-4}. Intensity distributions of core mode and four lowest cladding modes formed in single mode fibre are presented in figure 1a).

The transmission spectrum of LPG, with ABs marked, is given in figure 1b). Beside resonant wavelength, AB is characterized with magnitude and width. Magnitude of each AB is influenced by overlap of core and cladding mode as well as by LPG length. Width of AB is related to fibre dispersion and the grating dispersion.

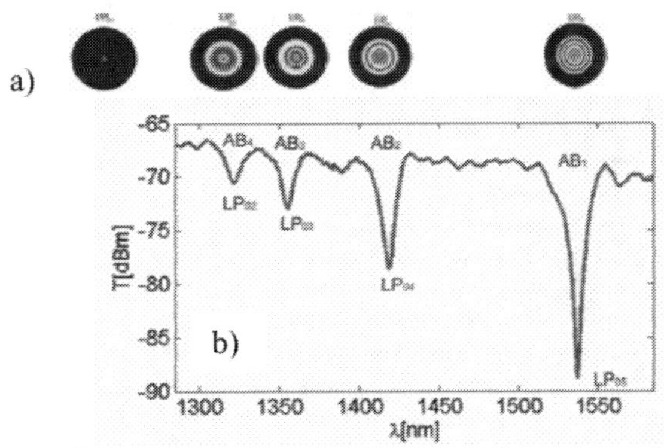

Figure 2. a) Intensity distribution of core mode (LP_{01}) and cladding modes LP_{02} , LP_{03} , LP_{04} , LP_{05} and LP_{06}; b)Transmission spectrum detected using Optical Spectral Analyzer and broad band optical source. Four ABs are occurring.

To model the bending of the grating, conformal transformation has been used. With this procedure it is possible to model curved fibre with equivalent straight one whose refractive index is modified [18] by the expression:

$$n(x) = n_{core}\left(1 - \frac{x}{R}\right) \tag{3}$$

where n_{core} is the core refractive index, R is the curvature radius and x is the distance from fibre center. Strictly speaking, the dependence on x is exponential, but since $x/R \ll 1$, it can be approximated by the linear term. The change in refractive index profile comes from the strain-optic effect and results in cladding mode redistribution. In figure 2 we presented intensity distribution of modes LP_{05} for curvatures $K = 0\ m^{-1}$, $K = 0.5\ m^{-1}$ and $K = 1\ m^{-1}$ obtained for a single mode fibre with LPG of period Λ=340μm at appropriate wavelengths. Mode redistribution results in changes in the resonant wavelength, magnitude and shape of AB, which all can be detected by the spectrum analyzer. In the case of LP_{05} mode, bending induces reduction in the core-cladding mode overlap and hence a reduction in the magnitude of the attenuation band, while the asymmetrisation of the grating refractive index profile changes its shape.

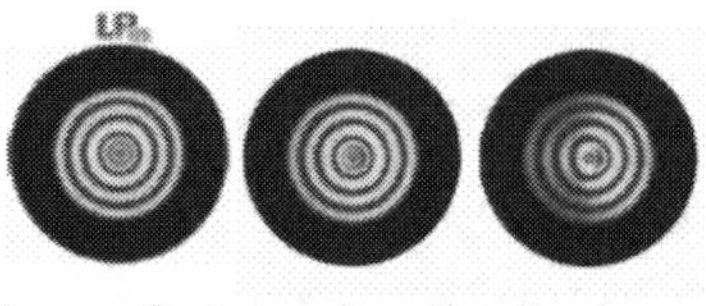

Figure 2. Intensity distribution of mode LP_{05} before and after LPG bending.

2.2. Sensor characterization

Grating response to bending is analyzed utilizing gauge specially designed for this purpose, figure 3a). Measurement of the transmission spectra are conducted using and broadband optical source (Spectral Products ASB-W-005) and Optical Spectral Analyzer (OSA) (YOKOGAWA AQ6370C). Response to bending of an LPG sensor is presented in Fig 3 b). It can be seen that transmission spectrum of this sensor exhibits a red shift with increase of the curvature. The working range in which the sensor has linear response is marked by gray in figure 4 a).

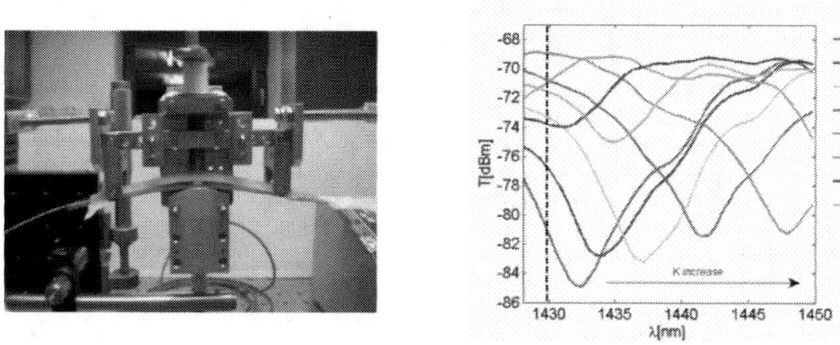

Figure 3. (a) Gauge for characterization of the fibre sensors of bending. (b) Spectral response of the LPG to bending.

Changes in the transmitted power at a fixed wavelength (dashed line in figure 3b)) are also presented as a function of the grating curvature, figure 4 b) (K-T plot). It can be seen from figure 3 b) that the slope and the width of the linear grating response depend on the choice of the laser wavelength.

The above results indicate that in curvature recognition, both the resonant wavelength and transmitted power can be used as measurement observables. As the high cost of the OSA limits its widespread use in medical applications, we have investigated a simple monochromatic interrogation

Journal of Physics: Conference Series **682** (2016) 012008 doi:10.1088/1742-6596/682/1/012008

scheme that relies on a photodiode measurement of the power transmitted through an LPG sensor at fixed wavelength. For this purpose a distributed feedback laser with a central wavelength 1470.73 nm is used, with narrow tunable range of 5nm, which is controlled via temperature or current.

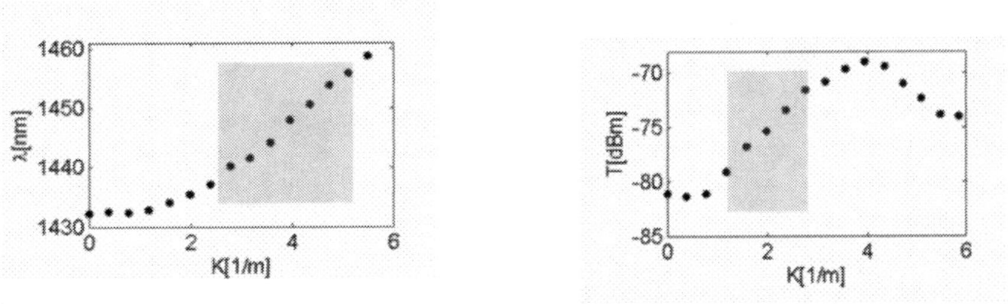

Figure 4. (a) K-λ dependence; (b) K-T dependence; Shaded area represents linear response of LPG to bending.

3. Application in respiratory monitoring

We have applied LPGs to continuous monitoring of the respiratory volumes based on the measurement of chest wall movements. The proposed solution uses LPG curvature sensor attached to the patient's thorax and a simple and cost-effective monochromatic interrogation scheme described above [19, 20].

The measurement protocol consists of two steps, calibration and test measurements. As a reference instrument we used a spirometer (SpiroTube, Thor Medical, Budapest). After eliminating the drift by zeroing the end of each breath, the calibration function was determined by different order polynomial fitting of a sensor to a simultaneously recorded spirometer signal. Finally, the linear scaling was chosen as optimal regarding robustness and accuracy. The calibration function was determined for each subject and used for calculation of the respiratory volumes in all test measurements performed on the same subject.

A number of measurements on 15 healthy volunteers, aged 32.6±8.0 years, were used as a statistical set for the data analysis. During all measurements the subjects were in supine position equivalent to the position of the patients during mechanical ventilation. LPG sensor was attached to a surface of the thorax in an area with stiff underlying tissues and fixed by an elastic bandage that goes around subject's torso and over the sensor. Since LPGs can sense the heart beats, sensor was positioned at the right side of the thorax.

4. Results and discussion

A typical set of the calibration and test results is shown in figure 5. Scatter plots a) and b) show linear correlation between the change of the local torso curvature and change in lung volume measured by spirometer. The calibration curve is represented by an inner solid line. The outer lines designate the Δ region that contains 68% of all measured values, and corresponds to standard deviation. For the calibration measurements, subjects were asked to breathe naturally with typical tidal volumes of 400-800 ml, and for the tests to perform one minute of each natural, shallow (approximately half the tidal volume of natural breathing) and a combination of deep, natural and shallow breathing (hereinafter called mix type of breathing). The calibration uncertainty Δ_{calib} of the whole group of (6.8±2.2)% and test uncertainty Δ_{test} of (7.9±2.4)%, (11.4±2.9)% and (10.2±3.8)% for natural, shallow and mix breathing, respectively, validates the correlation fit.

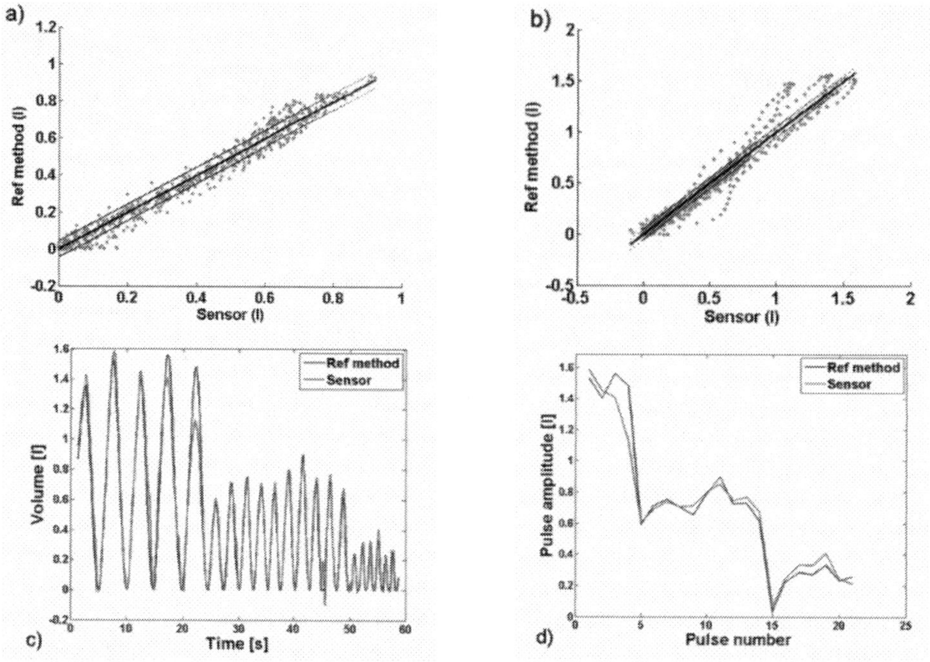

Figure 5. a), b) Scatter plot (dots) and the calibration function (inner line) and Δ region (outer lines) in calibration and test step, respectively; c) calibrated sensor output; d) tidal volume extracted from c).

Tidal volumes shown in figure 5 d) were obtained as a difference between maximum (end-inspiratory) and minimum (end-expiratory) of the same breathing cycle. Sum of the tidal volumes over one minute yields the minute volume and represents a typical average measure used by doctors. We calculated the measurement error of the LPG sensor with respect to the spirometer in the following way: the mean tidal volume error was defined as the mean relative error of tidal volumes over a number of respiratory cycles, and the minute volume error as the relative error of the minute volumes. These errors were used as the performance metric [19]. Results obtained for the whole group show the tidal volume error of (10.8±3.8)% with the minimum error of 0.4% and the maximum error of 16.6%, and the minute volume error of (8.7 ± 4.6)% with the minimum error of 0.2% and the maximum error of 18.2%, all calculated for normal breathing.

When compared to other measurement methods based on the discovery of Konno and Mead, the proposed method of respiratory volume monitoring using a single LPG curvature sensor is simpler, easier to implement and less susceptible to signal distortion. Additionally, it does not require subject cooperation for calibration, does not suffer from large baseline drift, and exhibits good calibration and/or measurement accuracy for clinical use [19]. Its main disadvantage is susceptibility to the body movements not originating from respiration, which can be largely corrected by the advanced signal postprocessing and better sensor fixation to the body.

4. Conclusion

We have presented the application of an LPG sensor of bending in monitoring of respiratory volumes. The method is based on the observation of a linear correlation between the changes in the local torso curvature and the lung volume. A good agreement of tidal and minute volumes measured simultaneously by the LPG sensor and spirometer, proves the sensor accuracy and consistency in time. The measurements were performed by a single LPG sensor, analyzed here in detail, and a monochromatic measurement scheme that requires only one narrowband laser diode and a photodiode.

Therefore, besides the mandatory satisfactory accuracy, the proposed scheme is characterized by simplicity and a potential low-cost, which indicates its suitability for application in non-invasive mechanical ventilation.

Acknowledgment
The authors acknowledge financial support from the Ministry of Education, Science and Technological Development of Serbia (Project No. III45010) and thank Lj. Hadzievski, B, Bojovic, A. Danicic, M. Vukcevic, T. Allsop and D. J. Webb for collaboration and their contribution to this work.

References

[1] Othonos A and Kalli K 1999 *Fibre Bragg gratings: fundamentals and application in telecomunications and sensing* (Artech house)
[2] Kashyap Kashyap R 1999 *Fibre Bragg Gratings* (New York: Academic)
[3] Rego G 2013 *The Scientific World Journal* **2013**
[4] Wolfbeis O S 2000 *Anal. Chem* **72** 81R
[5] OMishra V, Singh N, Tiwari U and Kapur P 2011 *Sens. Actuat.* A **167** 279
[6] Al-Fakih E, Azuan N, Osman A, Rafiq F and Adikan M 2012 *Sensors* **12** 12890
[7] Kalinowski H J, Abe I and Simões J A 2010 *Trends in Photonics* 315
[8] Dziuda L, Skibniewski F W, Krej M and Baran P M *2013 J. Biomed. Opt.* **18** 057006
[9] Grillet A, Kinet D, Witt J, Schukar M, Krebber K, Pirote F and Depre A 2008 *IEEE Sens. J.* **8** 1215
[10] Dziuda L, Skibniewski F W, Krej M and Baran P M 2013 *J. Biomed. Opt.* **18** 057006
[11] Melo L, Burton G, Davies B, Risk D and Wild P 2014 *Sens. Actuators B* **202** 294
[12] Kher S, Chaubey S, Kashyap R and Oak S M 2012 *Photonics Techn. Lett. IEEE* **24** 742
[13] James S W and Tatam R P 2003 *Meas. Sci. Technol.* **14** R49
[14] Konno K and Mead J *1967 J. Appl. Phys*. **22** 407
[15] Folke M, Cernerud L, Ekström M and Hök B 2003 *Med. Biol. Eng. Comput.* **41** 377
[16] Mehta S and Hill N S 2001 Noninvasive Ventilation, *Am. J. Respir. Crit. Care Med* **163(2)** 540
[17] Kawano K and Kitoh T 2001 *Introduction to optical waveguide analysis* (John Wiley and sons)
[18] Schermer J and Schermer R 2007 *IEEE J. Quantum Electr.* **43** 899
[19] Petrović M D, Petrović J, Daničić A, Vukčević M, Bojović B, Hadžievski L, Allsop T, Lloyd G and Webb D J 2014 *Biomed. Opt. Express* **5** 1136
[20] Petrović M D, Petrovic J, Simić G, Ilić I, Daničić A, Vukčević M, Bojović B, Hadžievski L, Allsop T and Webb D J 2013 *Conf. Proc. IEEE Eng. Med. Biol. Soc.* Osaka, Japan

Journal of Physics: Conference Series **682** (2016) 012009 doi:10.1088/1742-6596/682/1/012009

Chemical vapour deposition growth and Raman characterization of graphene layers and carbon nanotubes

Y-C Lai[1], P M Rafailov[2,5], E Vlaikova[2], V Marinova[1,3], S H Lin[4], P Yu[1], S-C Yu[1], G C Chi[1], D Dimitrov[2,3], P Sveshtarov[2] V Mehandjiev[2] and M M Gospodinov[2]
[1]Department of Photonics, National Chiao Tung University, Hsinchu 30010, Taiwan
[2] Institute of Solid State Physics, BAS, 72 Tzarigradsko Chaussee blvd., 1784 Sofia, Bulgaria
[3]Institute of Optical Materials and Technologies, BAS, Sofia 1113, Bulgaria
[4]Department of Electrophysics, National Chiao Tung University, Hsinchu 30010, Taiwan

E-mail: rafailov@issp.bas.bg

Abstract. Single-layer graphene films were grown by chemical vapour deposition (CVD) on Cu foil. The CVD process was complemented by plasma enhancement to grow also vertically aligned multiwalled carbon nanotubes using Ni nanoparticles as catalyst. The obtained samples were characterized by Raman spectroscopy analysis. Nature of defects in the samples and optimal growth conditions leading to achieve high quality of graphene and carbon nanotubes are discussed.

1. Introduction

Graphene, due to its outstanding properties as high electrical mobility [1], high thermal conductivity [2] and exceptionally low absorbance [3] play a key role in scientific and technological development. Up-to-date application of graphene in integrated devices requires wafer scale deposition technology with controllable number of graphene layers. On the other hand, the implementation of graphene as transparent conductor needs reproducible deposition of single or few layers. Especially, synthesis of bilayer graphene becomes increasingly desirable as it was reported to develop a tuneable band gap upon mechanical stress and electrical gating [4]. Among the well-known methods of graphene growth as mechanical exfoliation (characterized with very low concentration of structural defects, however with limited industrial applications due to the lack of thickness, size and location control), chemical vapor deposition (CVD) turns out to be the most prospective, inexpensive and readily accessible approach to satisfy the complex requirements of optoelectronics and microelectronics for large-area, high-quality graphene with low-density of structural defects.

During the CVD process, transition metals as Cu, Ni, Fe, play a fundamental role in reducing the decomposition temperature of methane [5]. One essential limitation of utilizing Ni as deposition substrate is that single- and few-layered graphene is obtained only over few-micron regions and the film is not homogeneous and with no uniform thickness over the entire substrate. The impossibility to control the number of layers is partially due to the fact that the segregation of carbon from the metal carbide upon cooling occurs rapidly within the Ni grains and heterogeneously at the grain boundaries

[5] To whom correspondence should be addressed.

[6]. In contrast, the peculiar combination of very low affinity between carbon and copper combined with the ability to form transient soft bonds facilitates formation of nucleation sites on the copper foil surface that form preferentially single-layer graphene through lateral growth [5]. In that aspect Cu is the perfect candidate for making large area, single layer graphene films with uniform thickness due to the low solubility of carbon in Cu.

In this work we study CVD growth of graphene on copper foil as well as the plasma-enhanced CVD growth of carbon nanotubes. We characterize the obtained samples with Raman spectroscopy and discuss some consequences of the transfer of graphene layers on Si wafers.

2. Experimental

The graphene samples were grown in quartz tube furnace by a CVD method involving high purity methane (99.9995%) and hydrogen gases. The growth procedure is displayed in figure 1. Under vacuum of 270 mTorr, the furnace is heated with 50 sccm flow of H_2. When temperature is stabilized at 1000°C, the hydrogen flux is decreased to 20 sccm and methane is introduced into the tube while keeping the pressure around 630 mTorr to grow graphene for 20 min. Rapid cooling is then applied after growth, and methane and hydrogen gas flows are continued throughout the cooling process until the temperature falls below 200°C.

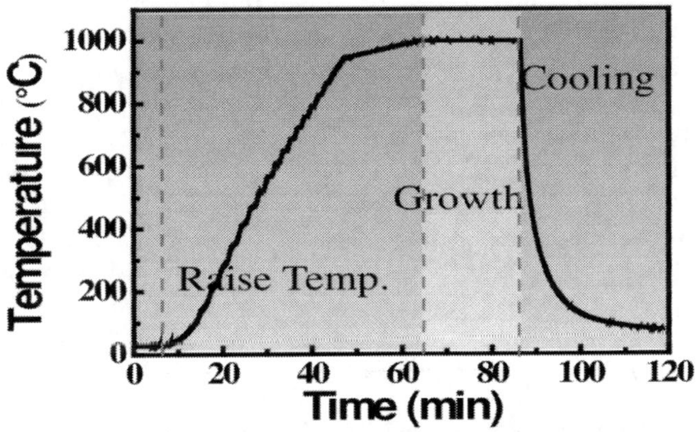

Figure 1. Stages of the graphene growth procedure.

To transfer the obtained graphene we used a spin coater to deposit photoresist PMMA (A4) on the top of the graphene layer to mechanically support graphene after copper is etched away. $Fe(NO_3)3.9H_2O$ was used as copper etchant. The PMMA/graphene/Cu foil was put to float in the etchant solution to selectively etch out the Cu-foil. Afterwards the obtained the PMMA/graphene layer was rinsed several times with doubly ionized water and transferred to a SiO_2/Si substrate, followed by removal of the PMMA using hot acetone.

Carbon nanotubes were grown in an Oxford Instruments Plasmalab 100 reactor from C_2H_2 gas precursor in a mix of Ar and H_2 as carrier and NH3 as additional radical source (gas purity 99.999 %) at a temperature in the range 800–950 C at chamber pressure 1.5 Torr. The NH_3:CH_4:H_2 ratio was maintained constant at 10:3:20 by mass-flow controllers regardless of the chamber total pressure. The chamber features a hot table consisting of a meandered sheet metallic heater and graphite susceptor Ceramic heat shields around the hot plate were used to reflect emitted radiation thus reducing heating power loss. The gas mixture entered the reaction space through a showerhead. Capacitively coupled RF (13.56 MHz) plasma was ignited between the showerhead and the grounded hot table both in the pretreatment and the growth step. Some of the chamber construction details, the substrate holder and the high temperature furnace arrangement are shown in the figures. The substrates were semiconductor

Journal of Physics: Conference Series **682** (2016) 012009 doi:10.1088/1742-6596/682/1/012009

silicon wafers coated with TiN diffusion barrier and 5 nm Ni catalyst film. The latter was activated by plasma treatment at the growth temperature in a predeposition step.

The Raman spectra were measured on a Jobin Yvon Labram spectrometer with a CCD detector and entrance slits set to 1.5 cm^{-1} spectral width. The 632.8 nm line of a He/Ne-laser and the 488 nm line of an Ar^+ laser were used for excitation, the absolute accuracy being 1 - 2 cm^{-1}. The laser beam with power not surpassing 0.5 mW was focused on a spot of about 1 - 2 μm in diameter on the sample surface using microscope optics with an objective of x50 magnification.

Electrical measurements for determination of the carrier mobility were performed by means of silver-paste contacts.

3. Results and Discussion

3.1. Growth and characterization of graphene

The Raman spectrum of graphene and carbon nanotubes has several unique features [7,8]. The G-band at around 1580 cm^{-1} arises from an E_{2g} zone-center vibration of the sp^2 -hybridized carbon network. The D-band is a double-resonant (DR) defect-induced feature arising from a breathing-like vibration at the Brillouin zone boundary (K point) [9]. Its frequency is excitation-energy dependent and normally varies in the interval 1320 – 1360 cm^{-1} for visible laser energies. The 2D band is the second order of the D band and is also excitation-energy dependent. Since momentum conservation is provided by two phonons with opposite wave vectors, no defects are required for the DR activation process of the 2D band. Thus the intensity ratio I(D)/I(2D) of the D band and the 2D band is proportional to the amount of structural defects. Another DR defect-induced feature is the D' mode ($\approx$1620 cm^{-1}) involving vibrations near the Γ point. Combination modes of graphene in the intermediate frequency range are the so-called M-band at $\approx$1750 cm^{-1} (absent in single-layer and twisted (non-AB-stacked) double-layer graphene) and a doublet between 1800 and 2100 cm^{-1} [10]. The defect-induced combination mode of the D and D' band appears at $\approx$ 2920 cm^{-1}.[8] Around 2450 cm^{-1} there is another characteristic band for graphene-based materials denoted as G* which arises from the combination mode of iTO phonons and longitudinal acoustical (LA) phonons [8]. Finally, non-dispersive bands activated by the mutual perturbation of adjacent layers in twisted double-layer graphene, called the I-band were reported recently to lie on the higher-energetic side of the D-band [11], and may facilitate distinguishing the AB-stacked bilayer graphene from the twisted one [12].

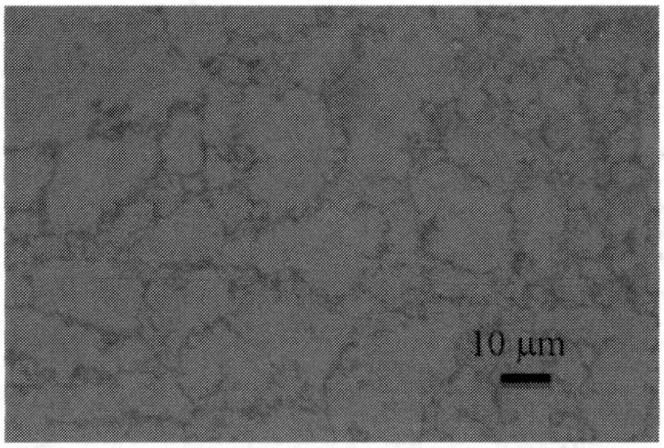

Figure 2. Optical micrograph of the obtained graphene film after transfer on Si/SiO_2.

In the Raman spectrum of graphene, the ratio between the intensity of the G band and the 2D band (I(G)/I(2D)), the lineshape of the 2D band and its width (FWHM) can be used as a measure for the number of single-atomic layers as the 2D band has characteristic lineshapes for mono- and bi-layer graphene, due to differences in their electronic structure [7,8]. In characterising single-layer graphene

(SLG) and discriminating between it and the various kinds of bilayer graphene (BLG) we consider the 2D peak frequency the most important parameter as it blue-shifted [13] with respect to that of SLG for any arbitrary twisting angle between the two layers of BLG [14]. On the other hand, a 2D peak with single-Lorentzian shape of BLG was found to be in some cases even narrower than that of SLG [14]. In the following analysis we will use also the second-order modes for additional confirmation of our conclusions.

Figure 2 shows an optical microscope image of the obtained graphene transferred from the Cu foil to Si/SiO_2. Obviously, there are highly uniform regions with size of several tens of μm^2 without any visible change in contrast. To investigate the quality and proof the visible homogeneity of the synthesized graphene films spot-resolved Raman measurements were performed over those regions. Figure 3 shows representative Raman spectra of the obtained graphene both as grown on Cu foil and transferred onto SiO_2/Si substrate excited at 2.54 eV.

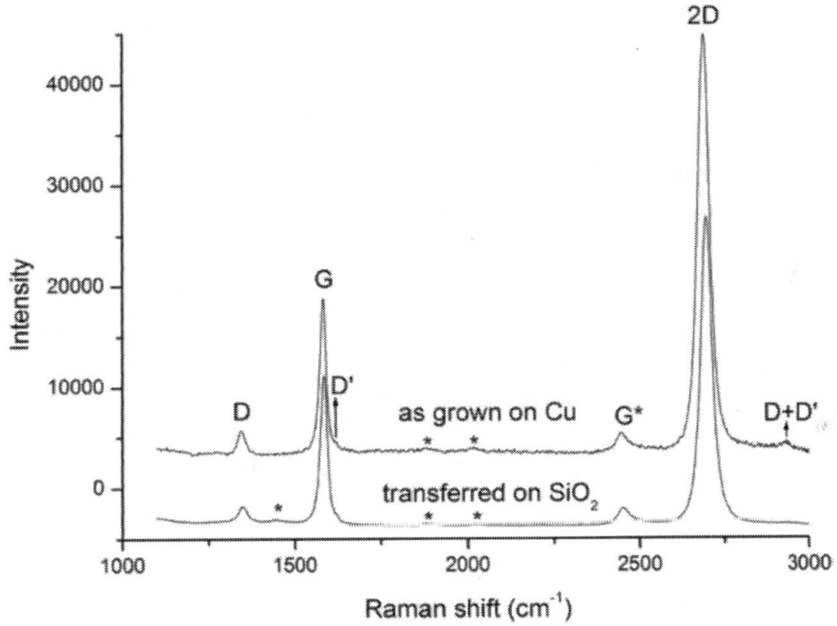

Figure 3. Raman spectra of the obtained graphene film recorded "as grown" and after transfer onto Si/SiO_2 as marked in the figure. Asterisks mark important intermediate bands: the doublet between 1800 and 2100 cm^{-1} and the probable superlattice feature at ≈ 1375 cm−1. To enable reliable comparison the spectra are normalized to the same G-band intensity.

The obtained graphene exhibits FWHM (2D) of 37 cm^{-1} and I(2D)/I(G) ~ 2.5. Low-intense but well-formed defect-induced bands (I(D)/I(G) < 0.05) are observed. The 2D lineshape is fitted with a single Lorentzian centered around 2690 cm^{-1}. These findings indicate SLG with few defects [15]. Surprisingly, the transfer onto SiO2/Si seems to diminish the amount of structural defects as the defect-induced modes D, D' and D+D' are weaker after transfer (see figure 3). On the other hand, transfer onto SiO_2/Si probably leads to partial overlap and creation of folded BLG. This is indicated by the blue-shift (5 – 10 cm^{-1}) of the 2D band and decrease of the I(G)/I(2D) ratio after transfer as well as by the appearance of a new band around 1450 cm^{-1}, which falls in the region of appearance of perturbation-induced bands, indicative of bilayer superlattice structures [12]. Examination of the combination modes in figure 3 confirms the presence of SLG. The asymmetric shape of the excitation-energy dependent D* band is typical for single-layer graphene [16]. The M-band at ≈1750 cm^{-1} is completely absent in all spectra and shape and frequency of the doublet between 1800 and 2100 cm^{-1} are characteristic for SLG Raman spectra excited at 488 nm [10].

The 2D band blue shift could also be due to PMMA-residues from the transfer process that usually turns graphene into p-type doping [17]. The lack of large doping-induced shift of the G-band above 1586 and the sufficiently high I(2D)/I(G) ratio [18,19] in all spectra, however, give us confidence that the doping level of our samples, if any, is sufficiently low. Likewise, the lack of strain-induced changes in the shape of the G- and 2D band shows that possible transfer-induced wrinkles are below detectable limit. These additional arguments are important as both charged impurities and strain do not contribute to the D band [8] and their influence cannot be assessed from the D band intensity.

The sheet resistance of the obtained graphene was established to be 1810 Ω/sq with a carrier mobility slightly over 1000 cm^2/Vs. The high resistance value is not surprising if one takes into account the contact resistance to the silver paste and the polycrystallinity of the graphene layer. Sheet resistance of similar-quality graphene measured via thermally evaporated gold electrodes [20] was found to barely surpass 1000 Ω/sq. This gives us confidence that the intrinsic sheet resistance of the obtained graphene is of the order of several hundreds Ω/sq which makes it suitable for application as transparent conductor in optoelectronics.

3.2. Growth and characterization of carbon nanotubes

Vertically aligned carbon nanotubes were grown on Ni/TiN/Si substrates from acetylene precursor after catalyst nanoparticles activation in hydrogen plasma [21]. Growth occurs via carbon species adsorption, surface migration and trapping at atomic steps at the Ni nanoparticles' surface. The growing nanotube detaches from the substrate the catalytic particle which stays at the nanotube tip and to a large extent defines the nanotube diameter. Tube growth terminates when all the catalytic particle surface gets completely covered with a crust of fixed non-mobile carbon species. NH_3 provides cyano–containing radicals in the plasma which combine with acetylene–derived species to produce the carbon source at the catalytic seed surface thus lowering the needed free energy change. The RF electric field provides orienting force that keeps top-seeded growth morphologically stable. This produces straight nanotubes aligned perpendicular to the substrate. Scanning electron microscopy (not shown) identified all nanotubes to be multi-walled and therefore of metallic nature with a length of about 1.2 μm.

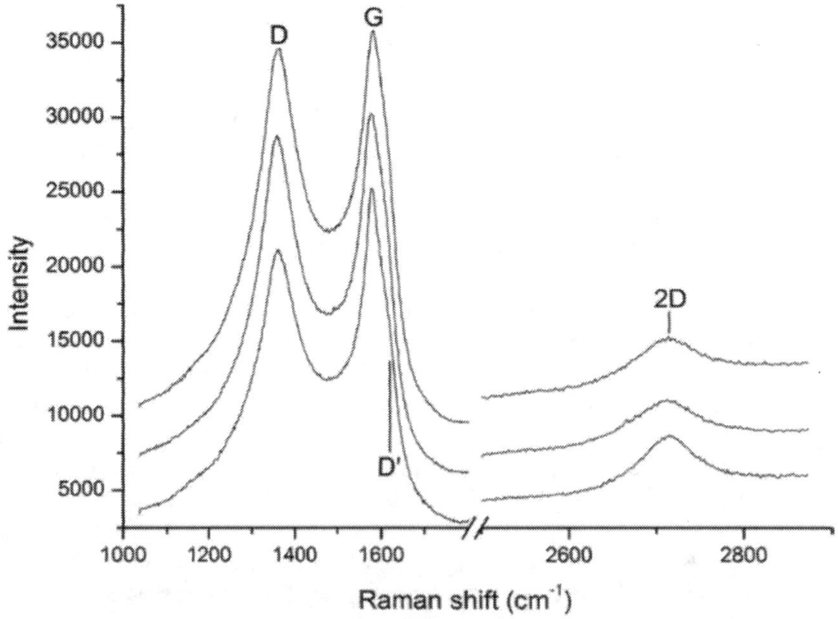

Figure 4. Raman spectra of CVD grown carbon nanotubes.

Journal of Physics: Conference Series **682** (2016) 012009 doi:10.1088/1742-6596/682/1/012009

Figure 4 displays representative Raman spectra of the obtained carbon nanotubes. The spectra exhibit the typical Raman features of multiwalled carbon nanotubes. The high intensity of the double resonant defect-dependent bands D and D' points to a considerable amount of defects. This is, however, typical for multiwalled carbon nanotubes which contain a lot of incommensurate structures. Furthermore, the intensity ratio I(D)/I(2D), though varying over the sample, remains below unity indicating that the honeycomb lattice of the distinct walls is not deteriorated by the defects.

4. Conclusion

Single-layer graphene films were grown on Cu foil substrates by low-pressure chemical vapour deposition. Raman characterization of the obtained Cu layers reveals single-crystal graphene with little amount of defects and good homogeneity. We observe spectroscopic indication for small fraction of bilayer graphene on the sample transferred onto silicon and explain this with a possible partial folding during the transfer process. The spectroscopic characterization and the assessment of the electrical properties reveal good-quality graphene suitable for application as transparent conductor in hybrid optoelectronic structures. Straight aligned perpendicular to the substrate carbon nanotubes were grown by plasma-enhanced chemical vapour deposition and characterized with Raman spectroscopy. The majority of them are multi-walled and therefore of metallic nature.

5. Acknowledgements

The authors acknowledge support from the FP7 Project INERA 316309 with the EU. This work was funded by the Bulgarian Science Fund under the project DFNI-T02/26 and the Ministry of Science and Technology (MOST), Taiwan under the contracts: MOST 104-2221-E-009-164; and 103-2911-I-009-516.

6. References

[1] Neto A H C, Guinea F, Peres N M R, Novoselov K S and Geim A K 2009 *Rev. Mod. Phys.* **81** 109

[2] Lin Y-M, Dimitrakopoulos C, Jenkins K A, Farmer D B, Chiu H-Y and Grill A 2010 *Science* **327** 662

[3] Nair R R, Blake P, Grigorenko A N, Novoselov K S, Booth T J, Stauber T, Peres N M R and Geim A K 2008 *Science* **320** 1308

[4] Zhang Y, Tang Ts-T, Girit C, Hao Zh, Martin M C, Zettl A, Crommie M F, Shen Y R and Wang, F 2009 *Nature* **459** 820

[5] Mattevi C, Kima H and Chhowalla M 2011 *J. Mater. Chem.* **21** 3324

[6] Obraztsov A N, Obraztsova E A, Tyurnina A V and Zolotukhin A A 2007 *Carbon* **45** 2017

[7] Ferrari A C, Meyer J C, Scardaci V, Casiraghi C, Lazzeri M, Mauri F, Piscanec S, Jiang D, Novoselov K S, Roth S and Geim A K 2007 *Phys. Rev. Lett.* **97** 187401

[8] Ferrari A C and Basko D M 2013 *Nature Nanotechnology* **8** 235

[9] Thomsen C and Reich S 2000 *Phys. Rev. Lett.* **85** 5214

[10] Cong C, Yu T, Saito R, Dresselhaus G F, and Dresselhaus M S 2011 *ACS NANO* **5** 1600.

[11] Gupta A K, Tang Y, Crespi V H, and Eklund P C 2010 *Phys. Rev. B* **82**, 241406(R) (2010)

[12] Wang Y, Su Z, Wu W, Nie S, Xie N, Gong H, Guo Y, Lee J H, Xing S, Lu X, Wang H, Lu X, McCarty K, Pei S, Robles-Hernandez F, Hadjiev V G and Bao J 2013 *Appl. Phys Lett.* **103** 123101

[13] Poncharal P, Ayari A, Michel T and Sauvajol J-L 2008 *Phys. Rev. B* **78** 113407

[14] Kim K, Coh S, Tan L Z, Regan W, Yuk J M, Chatterjee E, Crommie M F, Cohen M L, Louie S G, and Zettl A 2012 *Phys. Rev. Lett.* **108**, 246103

[15] Cancado L G, Jorio A, Martins Ferreira E H, Stavale F, Achete C A, Capaz R B, Moutinho M V O, Lombardo A, Kulmala T S and Ferrari A C 2011 *Nano Lett.* **11**, 3190–3196.

[16] May P, Lazzeri M, Venezuela P, Herziger F, Callsen G, Reparaz J S, Hoffmann A, Mauri F and

Maultzsch J, 2009 *Phys. Rev. B* **87** 075402
[17] Pirkle A, Chan J, Venugopal A, Hinojos D, Magnuson C W, McDonnell S, Colombo L, Vogel E M, Ruoff R S, Wallace, R. M. 2011 *Appl. Phys. Lett.* **99**, 122108
[18] Casiragi C, Pisana S, Novoselov K S, Geim A K and Ferrari A C 2007 *Appl. Phys. Lett.* **91**, 233108
[19] Hao Y, Wang Y, Wang L, Ni Z, Wang Z, Wang R, Koo C K, Shen Z, and Thong J T L 2010 Small **6**, 195–200
[20] Kholmanov I N, Magnuson C W, Aliev A E, Li H, Zhang B, Suk J W, Zhang L L, Peng E, Mousavi S H, Khanikaev A B , Piner R, Shvets G and Ruoff R S 2012 *Nano Lett.* **12** 5679
[21] Bell M S, Teo K B K, Lacerda R G, Milne W I, Hash D B and Meyyappan M 2006 *Pure Appl. Chem.* **78** 1117-1125

From superresolution to nanodetection: overview of far field optical nanoscopy techniques for nanostructures

P C Montgomery[1], A Leong-Hoi[1], F Anstotz[1], D Mitev[2], L Pramatarova[2] and O Haeberlé[3]

[1]Laboratoire des Sciences de l'Ingénieur, de l'Informatique et de l'Imagerie (ICube), University of Strasbourg-CNRS,23 rue du Loess, 67037 Strasbourg, France
[2]Georgi Nadjakov Institute of Solid State Physics, Bulgarian Academy of Sciences, 72 Tzarigradsko Chaussee blvd., 1784 Sofia, Bulgaria
[3] Laboratoire MIPS, Universite de Haute-Alsace, Ecole Nationale Supérieure d'Ingénieurs Sud Alsace (ENSISA), 12 rue des Frères Lumière, 68093 Mulhouse, France

E-mail : paul.montgomery@unistra.fr

Abstract. Far field optical nanoscopy has been brought to the forefront with the 2014 Nobel Prize for chemistry in fluorescent nanoscopy for revealing intra-cellular details of tens of nm. In this review, we present an improved classification scheme that summarizes the many optical nanoscopy techniques that exist. We place particular emphasis on unlabelledsuperresolution techniques that provide real improved resolving power and unlabellednanodetection techniques for characterizing unresolved nanostructures. Superresolution is illustrated with sub-100 nm imaging of diatoms with tomographic diffractive microscopyand adenoviruseswith submerged microsphere optical nanoscopy. Three sub-categories of nanodetectionare then presented. Contrast enhancement is illustrated with surface enhanced ellipsometric contrast microscopy for the study of bacterial motility and strobed phase contrast microscopy for measuring the mechanical properties of vesicle membranes. High sensitivity phase measurement using interference microscopy demonstrates how nanostructured surfaces and structures can be characterized in biomaterials, laser textured stainless steel and defects within thin polymer films. Finally, deconvolution is illustrated with the use of through-focus scanning optical microscopy in critical dimension measurement and characterization of 40 nm linewidths in microelectronic devices. In this way we show how new far field optical nanoscopy techniques are being developed for unlabelled characterization of nano and biomaterials.

1. Introduction

To help in the understanding and development of nano and biostructures, new imaging techniques are required that are capable of sub-cellular resolution and nanodetection over wide fields. The use of light in far field optical nanoscopy has been brought to the forefront with the 2014 Nobel Prize for chemistry [1], with fluorescent stimulated emission depletion microscopy (STED) [2] achieving a real optical lateral resolution of 30 nm and single molecule microscopy [3, 4] capable of revealing intra-cellular details of tens of nm using the superlocalization of fluorescent marker molecules. Several far field nanoscopy techniques and principles without using labelling had already been developed in materials science at the beginning of the 1990's [5, 6] and in [7] we presented a first classification scheme of the main optical nanoscopy techniques in order to help to better understand their different approaches and performances that we improved in [8]. In the present review, we further extend the classification scheme (figure 1) with additional techniquesand categories which differentiatethe unlabelled techniques(grey area) from those requiring the use of fluorescent or gold nanomarkers.

As highlighted in the work of S. Hell with STED [9], overcoming the classical resolution limit due to diffraction in fluorescence microscopy came from the realization that resolution was limited by the principle of controllable on/off states in fluorescence molecules, followed by contrast limits and superlocalization. Structured illumination microscopy (SIM) [10] also allows superresolution.

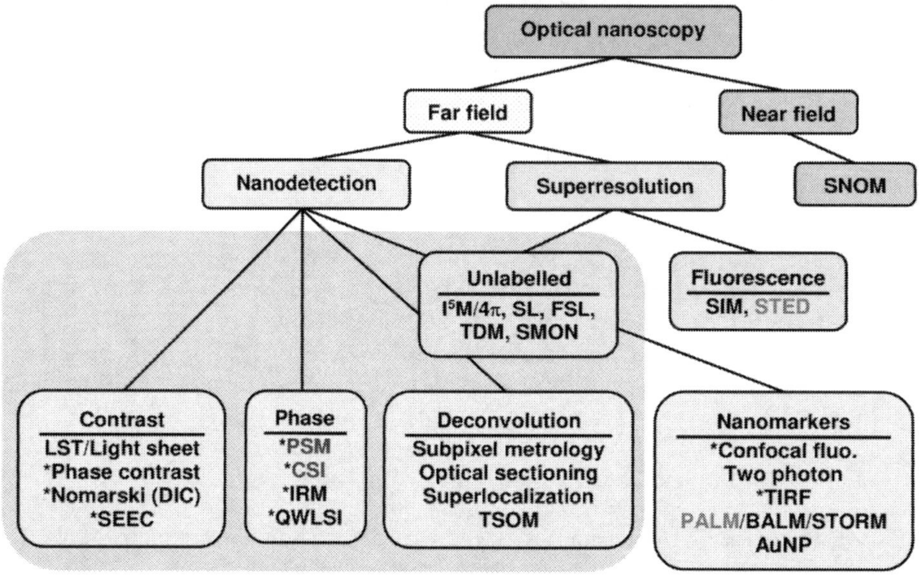

Figure 1. Improved classification scheme of optical nanoscopy techniques.
Notes: *indicates the possibility of real time imaging, red words show the two techniques honored by the 2014 Nobel Prize for chemistry, blue words show techniques being developed further at ICube and the grey area shows unlabelled far field nanoscopy techniques.

Abbreviations: 4π, 4π microscopy; AuNP, gold nanoparticles; BALM, bleaching (or blinking) assisted localization microscopy; Confocal fluo, confocal fluorescence microscopy; CSI, coherence scanning interferometry; FSL, far field superlens; I^5M, combination of I^2M (interference illumination microscopy) and I^3M (incoherent interference illumination microscopy); IRM, interference reflection microscopy; LST, laser scanning tomography; Nomarski, differential interference contrast (DIC) microscopy; PALM, photo-activated localization microscopy; PSM, phase-stepping microscopy; QWLSI, quadriwave lateral shearing interferometry; SEEC, surface enhanced ellipsometric contrast microscopy; SIM, structured illumination microscopy; SL, scattering lens microscopy; SMON, submerged microsphere optical nanoscopy; SNOM, scanning near-field optical microscopy; STED, stimulated emission depletion; STORM, stochastic optical reconstruction microscopy; TDM, tomographic diffractive microscopy; TIRF, total internal reflection; TSOM, through-focus scanning optical microscopy.

An important distinction is made between superresolution techniques that provide real improved resolving power compared with the classical λ/2 limit of diffraction and nanodetection techniques that are used to characterize unresolved nanostructures [11]. Concerning the category ofsuperresolution techniques, we illustrate first among the unlabelled approaches results from tomographic diffractive microscopy (TDM), giving a resolution of 97 nm in the imaging of diatoms [12]. A second technique is described by submerged microsphere optical nanoscopy (SMON) [13] that uses the unusual properties of photonic nano-jets [14] to observe adenoviruses that are less than 100 nm in size.

The second category concerns nanodetection techniques and is divided into four sub-categories that make use of the contrast, phase, deconvolution or nanomarkers to achieve nanodetection. Unlabelled contrast enhancement is illustrated with surface enhanced ellipsometric contrast microscopy (SEEC) [15], using polarized light and a dedicated thin film coated substrate to reveal nano-thin slime trails of bacteria for studying their motility. A second contrast enhancing technique is also presented, real time phase contrast microscopy to reveal the vibration modes of the 5 nm thick lipid walls of giant vesicles [16]. Very high sensitivity phase measurement is demonstrated by our own results using improved modes of coherence scanning interferometry (CSI, marked in blue in figure 1) of the study of biomaterial nanostructures [17, 18], nanotextured stainless steel surfaces [19] and μm defects in thin Mylar polymer films[20] by means of carefully controlled image processing. Finally, a technique using deconvolution, through-focus scanning optical microscopy (TSOM)[21], is presented to demonstrate how nanostructures on semiconductor surfaces can be measured to within nm. The well-controlled use of the optical response of ordinary optical microscopes demonstrates not only the measurement of 40 nm linewidths in microelectronic devices to within 1 nm, but the availability of a wealth of other useful information such as sidewall angle and the detection of nanoparticles and the distinction between different numbers of clumps of nanoparticles.Nanodetection techniques using nanomarkers [1, 3, 4] such as photo-activated localization microscopy PALM, bleaching (or blinking) assisted localization microscopy (BALM) and stochastic optical reconstruction microscopy STORM are not described here, having been covered in [7, 9]. In this review, we highlight in particular nanoscopy techniques that are capable of superresolution or nanodetection and do not require labelling of the sample but nonetheless reveal details or information about nanostructures.

2. Unlabelled superresolution nanoscopy

The lateral resolution in standard optical microscopy is generally accepted to be limited by the effects of diffraction, given by the Rayleigh criterion, $R_l = 0.61\times\lambda/NA$, where λ is the wavelength of the light from the sample, and NA is the numerical aperture of the imaging lens, given by $NA = n\times\sin\alpha$, where n is the refractive index of the imaging medium and α is the half-angle of the light cone that can enter the objective. Using visible light and a high NA objective in air, this gives a typical resolution limit of about 0.4 μm and 0.2 μm with an oil immersion lens in blue light [22]. This value can be improved further by up to 30% in confocal microscopy by means of an optimized imaging system consisting of point illumination and a pinhole in front of the detector to eliminate out-of-focus signals. Further selectivity is achieved with total internal reflection microscopy (TIRF) [23] and two photon microscopy [24].

In recent decades, several superresolution techniques have been developed that give a real increase in resolving power of the imaging system, so leading to the observation of finer details, the most successful being in fluorescence microscopy. Well known techniques are STED microscopy [2, 9], and SIM microscopy [10].While fluorescence microscopy techniques lead to the advantages of improved resolution and the targeting of specific proteins in living cells, they are dependent on the use of labelling with fluorescent molecules which is intrusive and limits the time of light exposure to tens of seconds due to photobleaching. Unlabelled nanoscopy techniques have the advantages of being less intrusive and open to a wider number of objects that can be imaged. Several new techniques that do not require labelling exist, amongst themI^5M, 4πillumination [25],SL microscopy [26, 27] and FSL microscopy[28]. In this section we highlight just two new techniques, namely TDM and SMON.

2.1. Tomographic diffractive microscopy (TDM)

Much of the information concerning the shape of a microscopic sample, such as its surface and 3D structure can be saved in a digital hologram using the technique of digital holographic microscopy (DHM). However, reconstructed images from digital holograms generally have low lateral resolution due to limited spatial frequencies in the Fourier domain which is degraded further by speckle noise.

In TDM, the tomographic mode of DHM, one solution to improve the resolution is to vary the illumination angle so as to increase the synthetic aperture size and increase the spatial frequencies available [12].In addition, the use of many different illumination angles averages the speckle noise, improving further the details observable. Using phase shifting interferometry, the incident wave diffracted by the specimen is measured in the Fourier domain at different illumination angles. A three-dimensional representation of the object is then reconstructed using the series of images recorded at each angle in the spatial frequency domain. By means of the first approximation of Born [22] and a simple Fourier transform, the distribution of the complex refractive index within the sample is then obtained.TDM thus allows high resolution imaging of transparent unlabelled specimens, up to twice the resolution possible in classical microholography, or $\lambda/(4\times NA)$ in the transmission mode.

An example of some results using TDM is shown in figure 2, consisting of a series of reconstructed images viewed from 6 different angles of a diatom which is 22 µm wide and 40 µm long [12]. Each image is made from 400 different illumination angles. Using an illumination wavelength of 475 nm and a high numerical aperture oil immersion objective (NA = 1.4) a lateral resolution of 97 nm has been achieved, thus demonstrating an experimental value of about $\lambda/(3.5\times NA)$. A further reduction in the illumination wavelength to 405 nm allowed this value to be improved to a resolution of 90 nm [29]. Both of these results clearly show the possibility of imaging below 100 nm in a transmission non-fluorescent microscope. While the axial resolution is not as good, being 300 nm, this can nonetheless be improved to 200 nm by using TDM in the reflection mode [30]. The techniques developed here will find many applications for high resolution, non-invasive imaging of live cells.

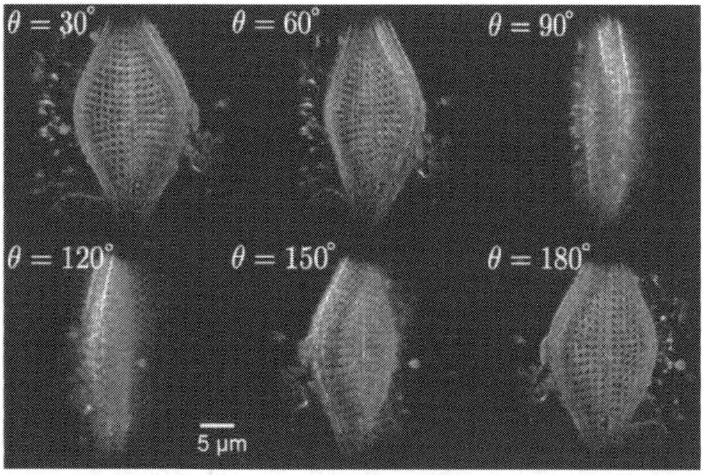

Figure 2. 6 views at different angles of a diatom measured with TDM in the transmission mode, demonstrating a lateral resolution of 97 nm.

2.2. Submerged microsphere optical nanoscopy (SMON)

A second emerging technique for achieving superresolution without labelling is the SMON technique, in which a glass microsphere is placed on the sample, in front of the objective, resulting in a lateral resolution of 50–120 nm in water [13]. Two effects are produced by the glass bead. The first one is the formation of a high intensity subdiffraction limited sized photonic-jet below the microsphere. Photonic nano-jets are better known for their interesting properties ofsuperresolution focusing of laser beams [14] and use in two photon fluorescence enhancement [14, 31] and laser surface nanostructuring [32].By reversing this optical path, superresolution imaging can be achieved.

The second effect is the conversion of the evanescent waves close to the sample into propagating waves that can be collected by the microscope objective to perform far-field imaging. In standard illumination conditions, these waves are very weak but it is the particular focusing properties of the nano-jet by the microsphere that plays an important role in enhancing the signal contrast of the propagating waves outside the microsphere. The resulting image thus contains fine subdiffraction limited detail of the sample that cannot normally be retrieved by classical imaging.

The results illustrated in figure 3(a) show a SMON image of anodic aluminum oxide (AAO) nanopores that are approximately 50 nm in size. A 50x water immersion objective (NA = 0.75) was used in the transmission modewith a 100 μm microsphere of $BaTiO_3$. A resolution of well below 100 nm is clearly demonstrated. The results in figure 3(b) and (c) are of clusters of 75 nm adenoviruses, shown with SEM in (b) and SMON in (c) using an 80x water immersion objective (NA = 0.9). The samples were prepared by first coating the microscope cover slide with a 5 nm thick layer of gold to enable both optical and SEM imaging. The results show that individual adenoviruses can be identified and thus the superresolution capability of SMON. The TDM and SMON techniques are good illustrations of unlabelled superresolution techniques that are capable of lateral resolutions below the classical limit of 200 nm without using labelling.

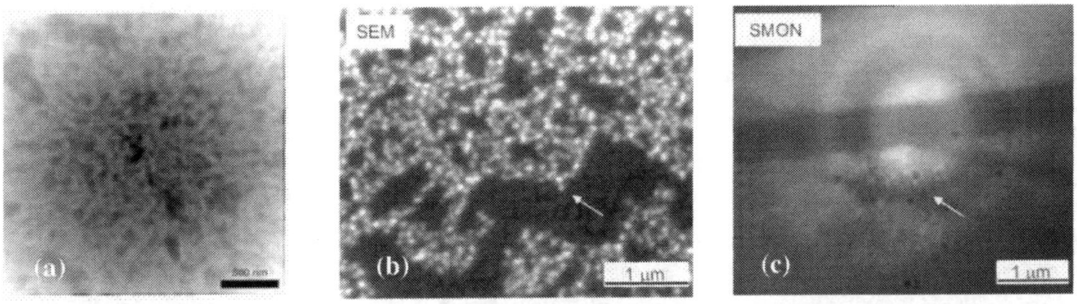

Figure 3. Results of SMON imaging in water with 100 μm $BaTiO_3$ microspheres; (a) SMON image of 50 nm sized AAO nanopores and imaging of clusters of 75 nm adenoviruses by (b) SEM and (c) SMON.

3. Unlabelled nanodetection nanoscopy

As shown in figure 1, the second family of far field nanoscopy techniques concerns those that use nanodetection by which nanostructures are made visible or some parameter connected with them is made observable or measurable without necessarily resolving all the details. This can be achieved without labelling by increasing the contrast, measuring the phase, or using deconvolution techniques. In [7] we described high contrast techniques such as Nomarski and laser scanning tomography (LST) [6] which is more recently known as light sheet microscopy [24, 33]. These different methods are now illustrated with the SEEC technique and strobed phase contrast microscopy for increasing the contrast, high resolution CSI by measuring the phase and TSOM that uses deconvolution techniques.

3.1. High contrast: surface enhanced ellipsometric contrast microscopy (SEEC)

With the size of a nanostructure being well below the Rayleigh limit in a conventional optical microscope, the scattered intensity is so small that it is invisible. The scattered light intensity decreases as a function of d^6 when the object size d $<<\lambda$, the well-known phenomenon of Rayleigh scattering.Since scattering is uniform in all directions, by increasing the contrast, nanoparticles can become visible in the far field.

One way of increasing the contrast is to reduce the background signal using an antireflection-coated substrate with polarized light in the SEEC technique [34] in which a 10–100 times improvement in the contrast is achieved. Nanostructures modify the polarization of the reflected light and become visible in a crossed polarized light system. This technique is used with commercially available substrates known as Sarfus® (Nanolane, Le Mans, France) to study carbon and steroid nanotubes, chitin fibres, nanowires and gold nanostructures and dots.

Thin films and nanometrically thin structures extended laterally are often also invisible in classical microscopy due to their very low contrast, even using phase contrast microscopy. The visibility depends on the difference between the refractive index of the structure to be visualized and that of the background, and the extent of the structure laterally across the field. These can become visible with the SEEC technique, such as with lipid layers and vesicles, lectin layers and cell structures. A successful study has been performed on bacterial motility by observation of just such a thin layer consisting of the slime trails left by the bacteria [15]. This study involved observing the nanometer-thick extracellular matrix (slime) trails left by moving bacteria(delta-proteobacteriumMyxococcusxanthus) in water. Two techniques were used. The first, phase contrast microscopy, was used to show the bacteria themselves (figure 4(a)). Then SEEC was used to show up the slime trails (figure 4(b)). Combining the two images (4(c)) at video rate shows up the movement of the bacteriawithout the need for labelling. This procedure was used to determine the patterns of slime secretion during different velocity phases of the moving bacteria and to analyze the slime-propulsion hypothesis. Such a high-sensitivity technique made it possible to elucidate, for the first time, how slime is deposited during motility and to demonstrate that slime does not mediate propulsion but more likely facilitates motility-driven attachment to the substrate.

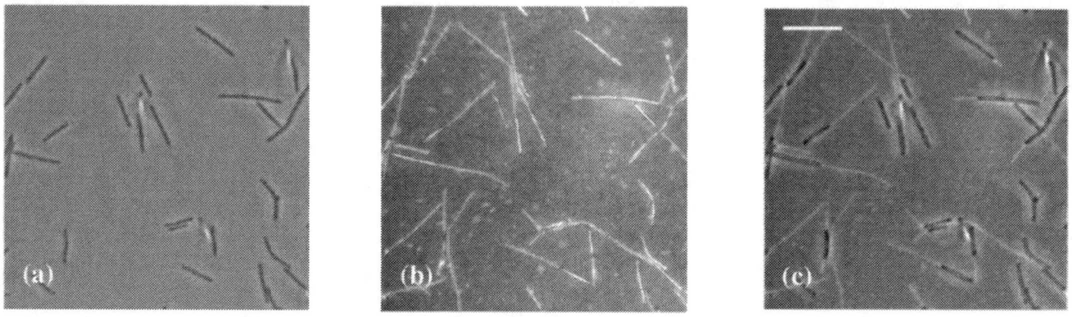

Figure 4. Observation of slime trails in the study of bacterial motility using (a) phase contrast microscopy to show the bacteria, (b) SEEC to reveal the slime trails and (c) phase contrast and SEEC combined to show the bacteria and their slime trails, scale bar: 10 μm (courtesy of Marie-Pierre Valignat, INSERM, Marseille, France).

Other applications concerning the use of SEEC for studying nanolayersare that of graphene, DNA biochips [35], self-organized monolayers of phospholipids[36], enzyme reactions [37]and in specific antibody recognition[38].The advantages of the SEEC technique are the nanometer sensitivity for studying nanolayers and nanoparticles using full field unlabelled imaging in real time and surface roughness and layer thickness measurements by calibrating the z-axis.

3.2. High contrast: strobed phase-contrast microscopy

While the previously mentioned slime trails were not visible with phase contrast microscopy, given the right conditions, even imaging atomic-sized structures becomes possible, such as dislocations in semiconductor crystals when extended in one of the imaging dimensions and associated with a sufficient variation in refractive index of the strain field[39].

The study of the properties of 5 nm thick unilamellarlipid membranes in the form of giant vesicles is important to help understand complex cell functions, the role of vesicle formation and transport, and the development of synthetic cells.While not directly visible perpendicularly to the membrane, the edges of such vesicles do becomevisible in phase-contrast microscopy, with the refractive index being about 10% higher than that of the aqueous surroundings. The dynamic thermal shape fluctuations of giant vesicles (10 μm radius, figure 4(a)) have been studied using strobed phase contrast microscopyto reduce the measurement uncertainties of the bending elasticity of lipid membranes [40]. Since thermal fluctuations are very complex, resulting from the addition of many different vibration modes, the higher frequency movements become blurred during the acquisition time of a single standard video image (40 ms). Stroboscopic illumination with a xenon flashlamp to freeze this movement reduces the blurring, as illustrated by the single sample image in figure 4(a). The positions of the membraneare measured from the minimaof the intensity profiles through the centre of the near circular vesicle. The time dependent fluctuations are measured from repeated measurements acquired at frame rate (25 images per second) to obtain a long image sequence (10^4 images).

A further improvement in the measurements of the bending modulus to reduce the uncertainty resulting from various artifacts (vesicles with defects) has been achieved using more sophisticated measurement procedures. Selection criteria are used on the series of images of moving vesicles to retain only certain images that satisfy the requirements of the applied theory [16]. These criteria are defined from certain parameters to test four quadrant arc shapes within the region of interest (ROI) that represents a ring containing the vesicular contour delimited by two concentric circumferences with internal radius R_{int} and external radius R_{ext} ($R_{int}<R_{ves}<R_{ext}$). These delimit the contour, where R_{ves} is the vesicle radius, (figure 4(b)).This procedure has been successfully used to measure the bending moments of 1-stearoyl-2-oleoyl-sn-glycero-3-phosphocholine lipid membranes, as well as to estimate the friction between the monolayers making up the bilayer.

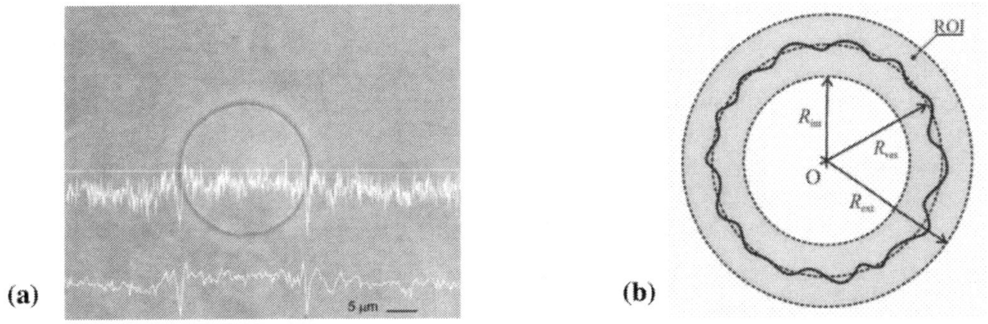

Figure 4. Measurement of dynamic thermal shape fluctuations of a giant vesicle using strobed phase contrast microscopywith (a) a single sample image of a vesicular contour with intensity profiles before (upper) and after processing (lower) and (b) a region of interest (ROI) containing the vesicular contour used to define selection criteria of the measured vesicles (courtesy of J. Genova et al. [16, 40], Laboratory of Liquid Crystals, ISSP, BAS, Sofia, Bulgaria).

3.3. Phase measurement: high resolution coherence scanning interferometry (CSI)

Information from nanostructures can also be obtained by measuring the phase of the light in interference microscopy. CSIis commonly used for measuring nanometric surface roughness (along the optical axis) of materials using quantified interferometry in an optical microscope equipped with specific interference objectives [41].

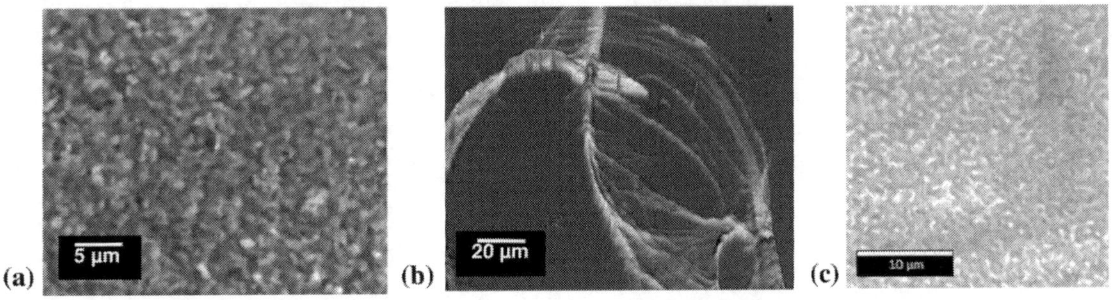

Figure 5. Nano surface roughness measurements using CSI of(a) PPHMDS layer dipped in Fibronectin withR_t = 1.57 µm (Leitz-Linnik), (b) fibres of fibronectin (R_t = 30 – 300 nm) on PPHMDS layer used for new biomaterials (ZygoNewView 7200) and (c) self-organized circular nano ripple structures on stainless steel with period = 1 µm and R_t = 600 nm (Leitz-Linnik) [19]

The results in figure 5(a) and (b) show some different measurements of a PPHMDS (polymerized hexamethyldisiloxane) layer dipped in fibronectin, both commonly used in biomaterials development. The local fine roughness of the layer is given in figure 5(a) and the fibre structures shown in figure 5(b) [18]. A second application example is given in figure 5(c) of self-organizednanostructured "ripples" made on stainless steel with a fs pulsed IR laser (λ = 1030 nm, 300 fs pulses). The results show circular nano ripples made using circularly polarized light, with nanostructures having a period of 1 µm, and a total roughness (peak-valley roughness) of R_t = 600 nm [23]. Variations in the laser parameters allow the modification of the colour aspect and wetting behavior of the surface. The technique has been further developed for real time measurement using a high speed camera and FPGA processing to achieve a 3D image acquisition rate of over 20 images per second [42].

CSI can also be used for imaging interfaces and structures buried under transparent layers but in this case, the signal to noise ratio can be very poor due to low reflectivity, noise and artefacts [43]. The depth profile images in tomographic analysis, also known as full-field OCT [44]can also be degraded so much that structures near to the Rayleigh limit in size are lost in the noise [45]. By means of a series of noise reduction techniques, in [20] we demonstrated how much finer details can be observed of small pores buried in a Mylar polymer film of 3.6 µm thickness. This material is important for use as an insulator in electronics and micro-electronics. The system used is an adapted Leitz-Linnik interference microscope with x50 objectives (NA = 0.85) and a CCD camera. By using image averaging to reduce temporal noise, subtraction of the "dark" image to reduce the fixed pattern noise of the camera and division by the "flat field" image of a white test target to reduce image defects from dust on the optical components, much higher quality tomographic profile images are obtained.

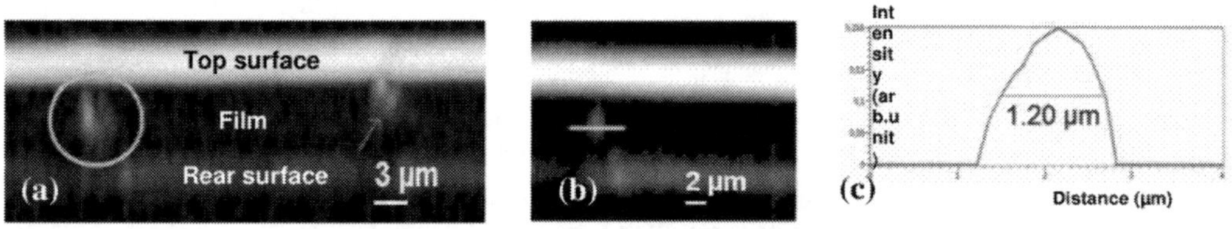

Figure 6. Improved defect analysis in a 3.6 µm thick Mylar film using high resolution tomographic CSI showing (a) depth profile (XZ) image after 10x averaging showing improved details of defects (circle and arrow) previously buried in noise,(b) depth profile (XZ) image after 100x averaging and dark and flat correctionsand (c) improved profile of small pore from line in (b)[20]

The results given in figure 6(a) show the improvement after averaging 10 images and in (b) after 100 images together with dark and flat corrections (in a different place from (a))[20]. In this way, near-diffraction limited sized structures initially lost in noise can be observed (figure 6(c)).

3.4. Deconvolution techniques:Through-focus Scanning Optical microscopy (TSOM)

The third method of nanodetection consists of using deconvolution techniques, applicable more in the case of regular small structures. For example, TSOM is a new method that uses an ordinary optical reflection microscope at visible wavelengths to give subnanometer dimensional sensitivity in the critical dimension measurement of nanostructures in micro-electronics[21].

By scanning through the focus of the microscope and sampling the XY images as a function of Z, the observation of the XZ image near to the focus reveals a large amount of information concerning the size of nanostructures, with sub-nm sensitivity. Great care is paid in optimizing the image acquisition conditions (averaging, noise reduction, calibration…) and in normalizing the images so that they can be compared between each other or with a library of images from calibrated standards.

The results in figure 7(a) show a typical result of a differential TSOM image of 40-50 nm wide lines in Tri-gate FinFETtransistors (Intel). A comparison of the intensity profiles (averaged over 8 lines) illustrates the difference (green profile) between a 44.2 nm wide line (blue profile) and a 55.3 nm wide line (red profile), that have a uniform height of 71.4 nm. The images were made at a wavelength of λ = 546 nm and an imaging objective with NA = 0.8. Although the optical lateral resolution limit is only 0.42 μm, the cross section of the diffraction pattern of the 40 nm wide line is observable because of the high contrast provided by the optimized imaging and acquisition.The results in figure 7(b) show finite difference time domain (FDTD) simulations to predict how TSOM would perform for measuring structural perturbations for 22 nm and 16 nm node transistor fins. The results show that TSOM would have sufficient sensitivity to several critical parameters such as fin width, sidewall angle (SWA), fin height, high-k (dielectric constant) material layer thickness, BOX (buried oxide) recess, and pitch walk (aperiodicity of fin walls). The TSOM technique can also be used for measuring the size of nanoparticles as small as 20 nm in diameter and to determine the number of particles in clumps of 1 to 4 nanoparticles of 120 nm in size.

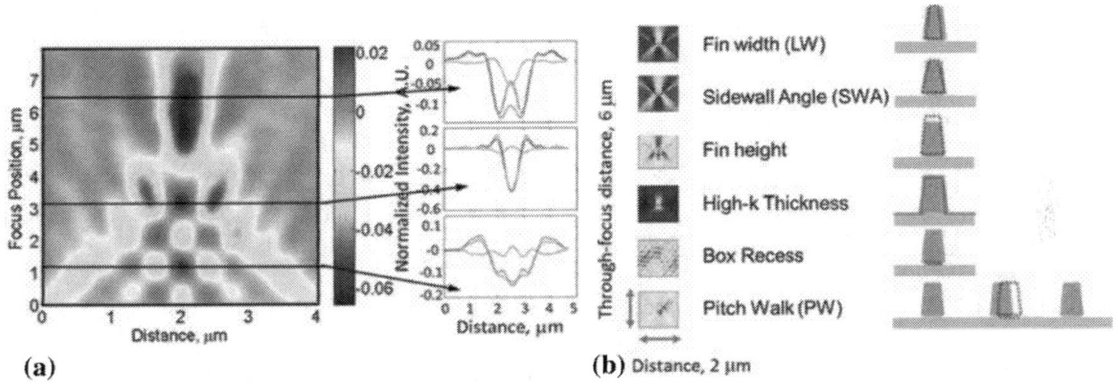

Figure 7. Results of critical dimension measurement of nanometricfins in microelectronics using TSOM;(a) difference in measurements using differential TSOM (green profile) between 44.2 nm wide line (blue profile) and 55.3 nm wide line (red profile) and (b) simulation images for differential TSOM (left) for different parameter perturbations of sub-22 nm node lines (right).

4. Conclusions

In this review we have presented an improved classification scheme of the many optical nanoscopy techniques that exist and in particular underlined the emerging far field superresolution and nanodetection techniques that do not require labelling. Superresolutionhas been illustrated with TDMshowing a resolution of 97 nm on 3D views of diatoms and SMON for imaging viruses of less than 100 nm in size. Nanodetection techniques using contrast improvement have been illustrated with the SEEC technique for real time imaging of bacteria motility and strobed phase contrast for improved measurement of the bending moments of lipid membranes as well as to estimate the friction between the monolayers making up the bilayer. The use of phase measurement using interference microscopy in high resolution tomographic imaging has been improved using image acquisition/processing of μm pores in Mylar polymers. Finally, deconvolution techniques have been shown using TSOM and the careful control and calibration of intensity profiles for critical dimension measurement in nanostructures for microelectronics.

We thus demonstrate that there are a growing number of far field optical nanoscopy techniques that do not require labelling but that are capable of resolving nanostructures below 100 nm in size or providing useful information of nanostructures down to nm and sub-nm measurement sensitivity.

Acknowledgments
Thanks are extended to P. Pfeiffer for the measurements on the fibronectin.

References

[1] Ehrenberg M 2014 The Royal Swedish Academy of Sciences, Stockholm, Sweden
[2] Klar T A, Jakobs S, Dyba M, Egner A and Hell SW 2000 *Proc. Natl. Acad. Sci. USA* **978** 206

[3] Dickson RM, Cubitt A B, Tsien RY and Moerner W E 1997 *Nature* **388** 355
[4] Betzig E, Patterson GH, Sougrat R, Wolf Lindwasser O, Olenych S, Bonifacino J S, Davidson M W, Lippincott-Schwartz J and Hess H F 2006 *Science* **313** 1642
[5] Montgomery P C 1990 *Nanotechnology* **1** 54
[6] Fillard JP 1996 *Near Field Optics and Nanoscopy* (Singapore: World Scientific)
[7] Montgomery PC, Montaner D, Anstotz F and Serio B 2012 *J. Phys. Conf. Ser.* **398** 012001
[8] Montgomery P C and Leong-Hoi A 2015 *J. Nanotech., Sci. Appl.* **8** 31
[9] Bretschneider S, Eggeling C and Hell SW 2007 *Phys. Rev. Lett.* **98** 218103
[10] Gustafsson MG. 2005 *Proc Natl Acad Sci U S A* **02** 13081
[11] Montgomery PC, Serio B, Anstotz F and Montaner D 2013 *Appl. Surf. Sci.* **281** 89
[12] Debailleul M, Georges V, Simon B, Morin R and Haeberlé O 2009 *Opt. Lett.* **34** 79
[13] Li L, Guo W, Yan Y, Lee S, and Wang T 2013 *Light: Sci. & Appl.* **2** e104
[14] Lecler S, Takakura Y and Meyrueis P 2005 *Opt. Lett.* **30** 2641
[15] Ducret A, Valignat MP, Mouhamar F, Mignot T and Theodoly O 2012 *Proc. Natl. Acad. Sci. USA* **109** 10036
[16] Genova J, Vitkova V and Bivas I 2013 *Phys.Rev. E Stat Nonlin Soft Matter Phys.* **88** 022707
[17] Pecheva E, Montgomery P, Montaner D and Pramatarova L 2007 *Langmuir* **23** 3912
[18] Pramatarova L, Radeva E, Pecheva EHikov T, Krasteva N, Dimitrova R, Spassov T, Mitev D, Montgomery P and Sammons R 2011 *On Biomimetics*, Ed. Pramatarova L (InTech) **14** pp 297-320
[19] Hairaye C, Mermet F, Engel T, Montgomery PC and Fontaine J 2014 *J. Phys.Conf. Ser.* **558** 012063
[20] Leong-Hoi A, Claveau R, Flury M, Uhring W, and Montgomery P 2015 *Proc. SPIE* **9528** 952807
[21] Attota R, Bunday B and Vartanian V2013 *Appl. Phys. Lett.* **102** 222107
[22] Born M and Wolf E 1980 *Principles of Optics: Electromagnetic Theory of Propagation, Interference and Diffraction of Light* 6th ed. (Oxford: Pergamon Press)
[23] Mashanov GI, Tacon D, Knight AE, Peckham M and Molloy J E 2003 *Nat. Methods* **29** 142
[24] Mahou P, Vermot J, Beaurepaire E, Supatto W 2014 *Nat.Methods* **11** 600
[25] Bewersdorf J, Schmidt R and Hell SW2006 *J. Microscopy* **222** 105
[26] Van Putten E, Akbulut D, Bertolotti J, Vos W, Lagendijk A and Mosk A 2011 *Phys. Rev. Lett.* **106** 193905
[27] Choi Y, Kim M, Yoon C, Yang T D, Lee K J and Choi W 2011 *Opt. Lett.* **36** 4263
[28] Liu Z, Lee H, Xiong Y, Sun C and Zhang X 2007 *Science* **315** 1686
[29] Cotte Y, Toy F, Jourdain P, Pavillon N, Boss D, Magistretti P, Marquet P and Depeursinge C 2013 *Nat. Photonics* **7** 113
[30] Liu H, Bailleul J, Simon B, Debailleul M, Colicchio B, Haeberlé O 2014 *Appl. Opt.* **53** 748
[31] Lecler S, Haacke S, Lecong N, Crégut O, Rehspringer J and Hirlimann C 2007 *Opt. Expr.* **15** 4935
[32] Guo W, Wang ZB, Li L, Liu Z and Luk'yanchuk B 2008 *Nanotechnology* **19** 455302
[33] Ritter JG, Veith R, Veenendaal A, Siebrasse JP and Kubitscheck U 2010 *PLoS One* **5** 11639
[34] Ausserré D and Valignat M P 2007 *Opt.Expr.* **15** 8329
[35] Monot J, Petit M, Lane SM, Guisle I, Léger J, Tellier C, Talham D R and Bujoli B 2008 *J. Am. Chem. Soc.* **130** 6243
[36] Gunnarsson A, Bally M, Jönsson P, Médard N and Höök F 2012 *Anal. Chem.* **84** 6538
[37] Egea AMC, Metivier M, Croguennoc P, Remaud-Simeon M and Vieu C 2013 *BioNanoScience* **4** 37
[38] Souplet V, Desmet R and Melnyk O 2007 *J. Pept. Sci.* **13** 451
[39] Montgomery PC and Fillard JP 1989 *Electron. Lett.* **25** 89
[40] Genova J, Zheliaskova A, Vitkova V and Mitov MD 2009 *J. Optoelectron. Adv. Mater.* **11** 1222
[41] Schmit J, Reed J, Novak E and Gimzewski JK 2008 *J. Opt. Pure Appl. Opt.* **10**(6) 064001

[42] Montgomery P, Anstotz, F, Johnson G and Kiefer R 2008 *J. Mater. Sci. Mater. Electron.* **19** 194
[43] Benatmane A and Montgomery PC 2004 *Eur. Phys. J. Appl. Phys.* **27** 411
[44] Dubois A, Grieve K, Moneron G, Lecaque R, Vabre L and Boccara AC 2004 *Appl. Opt.* **43** 2874
[45] Montgomery P C, Montaner D and Salzenstein F 2012 *Proc. SPIE* **8430** 843014

Journal of Physics: Conference Series **682** (2016) 012011 doi:10.1088/1742-6596/682/1/012011

Technologies for deposition of transition metal oxide thin films: application as functional layers in "Smart windows" and photocatalytic systems

K Gesheva[1], T Ivanova[1], G Bodurov[1], I M Szilágyi[2,3], N Justh[2], O Kéri[2], S Boyadjiev[3], D Nagy[4] and M Aleksandrova[5]

[1] Central Laboratory of Solar Energy and New Energy Sources at the Bulgarian Academy of Sciences, "Tzarigradsko chaussee" 72 Blvd., 1784 Sofia, Bulgaria
[2]Budapest University of Technology and Economics, Department of Inorganic and Analytical Chemistry, Szent Gellért tér 4., Budapest, H-1111, Hungary;
[3]MTA-BME Technical Analytical Chemistry Research Group of the Hungarian Academy of Sciences, Szent Gellért tér 4., Budapest, H-1111, Hungary;
[4]University of Edinburgh, The King's Buildings, Mayfield Road, Edinburgh, EH9 3JL, UK
[5]Technical University Sofia, „St. Kliment Ohridski" 8 Blvd., 1756 Sofia, Bulgaria

E-mail: kgesheva@yahoo.com

Abstract. "Smart windows" are envisaged for future low-energy, high-efficient architectural buildings, as well as for the car industry. By switching from coloured to fully bleached state, these windows regulate the energy of solar flux entering the interior. Functional layers in these devices are the transition metals oxides. The materials (transitional metal oxides) used in smart windows can be also applied as photoelectrodes in water splitting photocells for hydrogen production or as photocatalytic materials for self-cleaning surfaces, waste water treatment and pollution removal. Solar energy utilization is recently in the main scope of numerous world research laboratories and energy organizations, working on protection against conventional fuel exhaustion. The paper presents results from research on transition metal oxide thin films, fabricated by different methods - atomic layer deposition, atmospheric pressure chemical vapour deposition, physical vapour deposition, and wet chemical methods, suitable for flow-through production process. The lower price of the chemical deposition processes is especially important when the method is related to large-scale glazing applications. Conclusions are derived about which processes are recently considered as most prospective, related to electrochromic materials and devices manufacturing.

1. Transition metal oxides

Transition metal oxides (TMO) are claimed to be one of the most interesting classes of solids, exhibiting varieties of properties, structures and applications [1]. Their properties are determined by the unique nature of their outer d-electrons. Transition-metal oxides can be insulators, semiconductors, metals, or undergo semiconductor-metal transitions. Transition metal oxides possess unusual and useful electronic, optical and magnetic properties. Many of these properties strongly depend on materials defects like vacancies, dislocations, stacking faults and grain boundaries [2].

TMOs are widely investigated due to their numerous applications. They are used as electrodes in electrochemical processes, functional components in technologically important catalytic processes. For example, they are used in selective oxidation, selective reduction and dehydrogenation. The catalytic properties of TMOs are defined of two main factors: these oxides exist in various crystallographic forms with closed stoichiometries with metal ions revealing different valence states and on the other hand, TMOs possess possibility for easy surface oxidation and reduction, which is related to high densities of cations and anions vacancies.

Other interesting application includes gas sensors [3-9], where a gas sensitive oxide (such as WO_3 or MoO_3) responses to the presence of certain gas by changing e.g. their electrical properties (resistance). The layered structured TMO (MoO_3,V_2O_5, MnO_2) can play role in lithium micro batteries, based on reverse injection of Li^+ ions in the host crystal lattice. The transition metal oxides can be used in high density memories, optical devices, passive display electrodes, magnetic recording and holographic devices, fuel cells, micro electronic circuits.

In this paper, the electrochromic and photocatalytic properties of TMO will be briefly reviewed.

2. Electrochromic devices

The chromogenic phenomenon is a reverse optical change under external influences. Electrochromism is a physical phenomenon describing reversible optical absorbance change of electrochemically active materials upon an application of electrical potential [10]. Recently, the electrochromic materials gain a lot of attention due to the energy saving applications and usage as smart windows [11], anti-glared rear view mirrors [12], and displays [13].

Electrochromic material (often in thin film form) is a part of electrochromic device [14], which is typically consisted of several components (see figure 1): a glass substrate covered by transparent conducting oxide (TCO); an EC film (transition metal oxides and mostly WO_3); an ion conductor (IC) (a polymer electrolyte or solid one); a counter electrode – often this is also conductive glass substrate with anodic type electrochromic functional film.

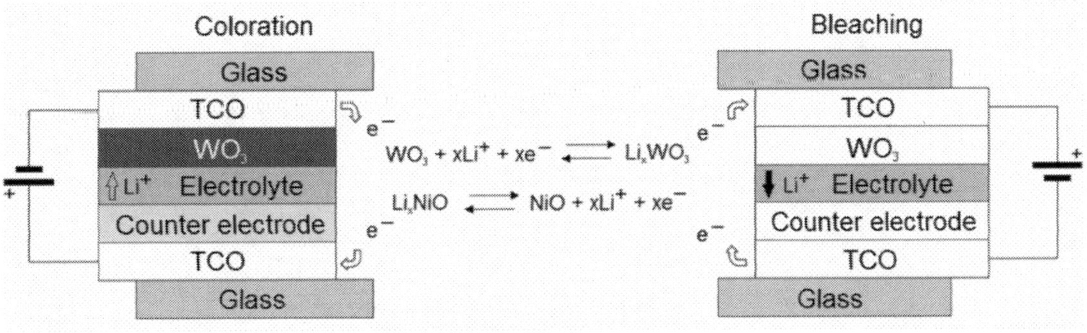

Figure 1. Schematically representation of an EC multilayer stack

The configuration of an EC device has sometimes compared to battery [14]. There are three types of electrochromic device (EC) designs: battery-like type, solution phase type and hybrid type, shown on figure 2. In the solution type design, both of the redox systems are dissolved in liquid solutions. And finally, in hybrid type design, one of the redox systems is in the form of a solid film, while the other is dissolved in a liquid solution. It is important to note here, that the solution and hybrid types are self-erasing, while the battery type EC devices have memory effect (the ability to retain colour even in open circuit, which implies high energy savings).

Cathodic type electrochromism is referred to materials that become absorptive when the alkali ions are injected or anodic electrochromism – optical absorption is decreased under ions intercalation. These cathodic and anodic oxides can be combined in devices to form smart windows resulting in a rather neutral visual appearance [15] and thus making them very suitable for general applications in architecture. The most commonly used oxides are based on Wand Ni and exhibit cathodic and anodic electrochromism. Among the other Transition Metal Oxides (TMO) used in electrochromic applications, tungsten trioxide (WO_3) [16, 17] is the most studied one due to its promising electrochromic properties such as high coloration efficiency and short switching time. Commercial electrochromic devices require improvement in the EC properties since the bright blue color of WO_3 films in the reduced state is not favorable as bronze or pale yellow colors for most building applications. It has been proved that better electrochromic performance can be achieved by the addition of dopants to WO_3 [18]. It is also reported that color neutrality can be obtained by vanadium doping in tungsten oxide. Optical absorption in these vanadium doped films is due to polaronic transition V^{4+}, V^{5+} and W^{5+}, W^{6+} in WO_3 [19]. The interaction between WO_3 and V_2O_5 is unique because of the similarities of ionic radii and the structure in their highest oxidation state [20].

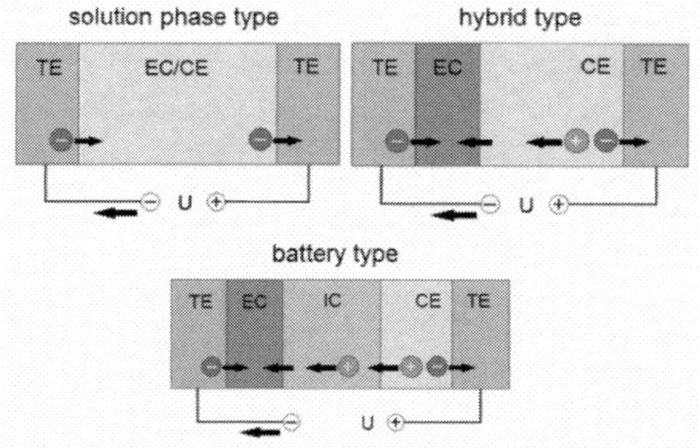

Figure 2. Different types of EC devices, where TE is transparent conducting electrode, EC - electrochromic electrode, CE – counter electrode and IC is ion conductor, respectively. The smart windows are considered from the battery type.

Another very interesting material is molybdenum oxide, which is found to be cathodic type electrochromic film with possibility for revealing great color densities. MoO_3 as a wide band gap semiconductor is a transparent in the visible spectral range and with crystal structure that allows easy intercalation/deintercalation of small ions. Molybdenum exists in different crystalline oxide modifications and can easily produced substochiometric structures. The two basic crystal phases are a unique layered structure (α - MoO_3 with orthorhombic symmetry) and metastable monoclinic β - MoO_3 (see figure 3). These phases differ in their vibrational and optical properties.

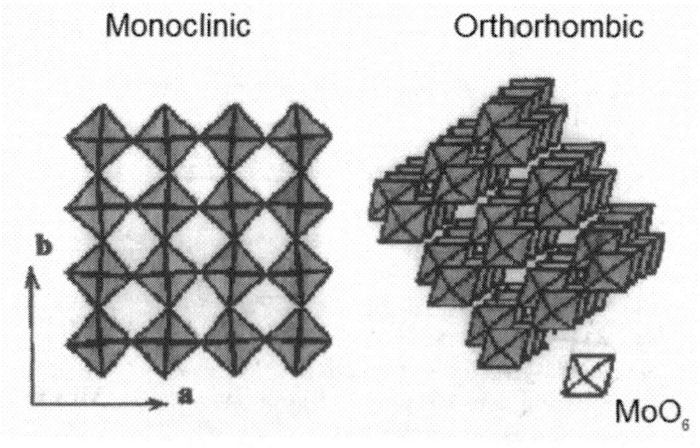

Figure 3. Schematic presentation of a) monoclinic β and b) orthorhombic α - phase of MoO_3

They exhibit different refractive indices and optical band gap energy values [21]. The orthorhombic MoO_3 possess unique two-dimensional space layered structure, consisting of distorted MoO_6 octahedra with common edges and corners. Each octahedron possess one connected oxygen atom, two O atoms are common for two octahedra and three O atoms are with common corners and are shared for three octahedra, approaching the complete stoichiometry - $MoO_1(O_{1/2})_2(O_{1/3})_3$. The double layers are characterized with sublayers from periodically arranged MoO_6, where the oxygen atoms participate inside the adjacent octahedra into and among the sublayers. The internal interaction among atoms is dominated by ionic and covalent forces, but the sublayers pair is interacted by weak van der Vaals forces [22]. This structure gives opportunity for injection of different donor ions into interlayered spaces and favors EC properties. In contrast, the metastable monoclinic β - phase has a cubic ReO_3 - like structure, consisting of three-dimensional rows from shared MoO_6 octahedra [23]. Spontaneous transformation $\beta \rightarrow \alpha$ phase occurs at temperatures of 370-400°C.

The main characteristic of MoO_3 films can be summarized as follows [24, 25]: excellent electrochemical stability; low cost; cyclic stability; a stronger and more uniform absorption of light in its colored state (molybdenum bronze); a better open-circuit memory than a majority of TMOs; molybdenum bronze is more akin to the sensitivity in human visual perception.

Binary combinations of oxides have been investigated to improve or modify the optical and electrical properties of different oxides and even to tailor their properties in predictable way. Possible beneficiary effects for mixing two electrochromic materials are to obtain increased coloration efficiency, improved durability, color neutrality, a larger switching potential range or faster reaction kinetics [26]. Mixed MoO_3-WO_3 films can show higher electrochromic efficiencies as a result of enhanced electron intervalence transfer between Mo^{5+} and W^{6+} states, and W^{5+} to W^{6+} transitions [27]. Optically, it is expected that the absorption peak of the tungsten/molybdenum mixed oxide films is shifted notably toward the eye-sensitive peak, and therefore electrochromic performance of the films is significantly increased [28].

Our group has been investigated MoO_3, WO_3, mixed Mo-W and W-V oxides films with respect to their optical and electrochromic behaviour [29]. The chosen technological approach for deposition of electrochromic functional films is APCVD (atmospheric pressure chemical vapour deposition).

3. Atmospheric pressure chemical vapour deposition - technology

CVD technology can be applied for obtaining a wide spectrum of thin film materials such as metals, semiconductor III-V compounds, dielectric oxides, perovskite heterostructures, magnetics and conductors. It is especially useful for fabrication of nanosized layers, nanoparticle and nanotubes (graphene and carbon nanotubes), free standing diamond films, etc. Recently, it has been found that although CVD is traditionally not compatible with polymer thin-film deposition, through modifications of the CVD reactants or reactor designs, polymer depositions at low temperatures and low-energy inputs can be achieved, making CVD a widely used technique for polymer deposition [30, 31]

CVD possesses some advantages over physical deposition methods and other chemical technologies such as:

- Broad group of high purity depositing thin films
- Possibility for covering flat and complex substrate morphologies
- Selective growth on chosen part of the substrates
- Different properties by varying deposition parameters
- Film deposition at low temperatures
- Preparation of single layer, multilayered structures, composites and functional coatings
- APCVD is compatible with industrial manufacturing, and for electrochromic application it can be integrated in in-line glass production [32].
- In order to achieve greater deposition rates and to lower deposition temperatures CVD processes can be enhanced by plasmas, ions, lasers, photons etc means, which can even further to broaden the materials that can be deposited by this technology.

On the other hand, this method also shows some disadvantages:

- Complexity of chemical reactions into CVD reactors
- Possible contaminations from precursors to the film structures
- Limitation of substrate choice due to possible diffusions or alloying
- Careful choice of precursors as they must be volatile at low temperatures, chemically stable, not toxic or poisonous, well designed for surface reaction to appear.
- Some by-products of CVD reactions can be hazardous (CO_2, HF or H_2)

In real technological CVD reactors, it is common to have heterogeneous and complex reactions near to heated substrates. In order to predict and obtain information for deposition under specific conditions, a thermodynamic modulation is usually performed, including data for free energy of formation of every gaseous and solid components. But as it is mentioned for a CVD process, there is a multi component and multi phase system, which makes its theoretical description very difficult [33]. Modelling can then be used to predict growth rate, etching rate, surface morphology and doping, all key aspects when doing CVD of electronic device structures.

The APCVD equipment used in the present research for deposition of MoO_3, WO_3, Cr_2O_3, mixed Mo-W and W-V oxides films consists of horizontal quartz reactor with cold walls. A graphite susceptor covered with SiC coating is used as substrate holder and the heating is realized with the help of high frequency (HF) generator. The used precursor is placed in special sublimator, immersed into silicon oil bath, heated by magnetic stirrer at certain temperatures. The precursor vapors are carried from the sublimator to the CVD reactor by an argon flow through heated Teflon tubes. From a separate gas line, O_2 enters into the reactor. In the CVD reactor around the hot substrates gas reaction condition is created at a local thermodynamic equilibrium, which leads to oxide deposits.

This configuration of CVD reveals some advantages:

- The cold wall reactor allows to obtain deposits only on the heated substrates, and the control of substrate temperature is easier
- The atmospheric pressure allows more homogenous reactions and appearence of quasi thermodynamical equilibrium in areas near the substrates
- Varying technological parameters (substrate and sublimator temperatures, gas flow rates (Ar and O_2), deposition time) can lead to films with different properties

4. APCVD MoO_3-WO_3 films – electrochromic characteristics

Metal oxide films of MoO_3, WO_3, and mixed MoO_3-WO_3 films were deposited by carbonyl CVD process at atmospheric pressure. The powder source consisted of the two carbonyl precursors in a certain ratio ($Mo(CO)_6$:$W(CO)_6$=1:4) and this mixture was heated in silicon bath at the temperature of 90°C. For the comparative study the deposition temperature was kept at 200°C because only at this temperature it was possible for the three kinds of metal oxide films to be deposited. Visually, the as-deposited films are slightly coloured: bluish (WO_3), yellowish (MoO_3) and brownish (MoO_3-WO_3). The deposition time was kept constant (40 min) and because of the different growth rates the film thickness was 300 nm for MoO_3, 400 nm for WO_3 and 120 nm for MoO_3-WO_3. As a result, mixed films were successfully deposited and they were visually transparent and uniform [34].

The mixed oxide films are expected to improve the electrochromic behavior. The doping of MoO_3 in WO_3 structure leads to different electrochromic effect. Deposition of CVD MoO_3-WO_3 films from mixed precursor $Mo(CO)_6$:$WO(CO)_6$= 1:1 were previously performed keeping the optimal gas flow ratio 1/32 [35]. For these samples, the optical modulation varies in the range of 30-50% and the estimated color efficiency values were 25-30 cm^2/C in electrolyte 1M $LiClO_4$ + PC (propylene carbonate). These values are much smaller than those of MoO_3-WO_3 films deposited from precursor mixture 1:4, in which $Mo(CO)_6$ prevails, showing colour efficiency CE = 114 cm^2/C and optical modulation around 70% at a wavelength λ = 550 nm [36]. Investigations were carried out on these Mo-W oxide films in Li^+ containing electrolyte at different scan rates. The results show that increasing scan rate (corresponding to faster intercalation/deintercalation rates) results in smaller values of the electrochromic characteristics. It was determined that the colour efficiency values vary in the range 40

– 77 cm^2/C and 18 – 30 % optical modulation at a scan rate 50 mV/s. The highest electrochromic color efficiency is obtained at scan rate (10 mV/s) at wavelength 550 nm - the maximum of the human eye sensitivity. The effect can be proposed to originate from the molybdenum oxide presence into the films.

Electrochemical studies were also performed to reveal the behavior of CVD MoO_3-WO_3 films under different ion insertion mechanisms. The measurements were carried out to observe the influence of two type proton electrolytes - 1 M H_2SO_4 and 1 M H_3PO_4, comparing with 1 mol/l $LiClO_4$+PC electrolyte. It must be pointed out that MoO_3 films are found to be stable in acidic electrolytes and WO_3 films are destroyed, but mixed oxide films show no change of its electrochromic properties after long term cycling and therefore, they are also successful studied in proton type liquid electrolytes. The estimated electrochromic characteristics are shown at figure 4. The smallest values are found for phosphorous acid electrolyte. The obtained electrochromic parameters manifest strong wavelength dependence. When the used electrolyte was 1 M H_2SO_4+glycerine the highest values are 84 cm^2/C (λ=750 nm), 100.5 cm^2/C (λ=700 nm) and 120 cm^2/C (λ=650nm) and there they exceeded even those of Li electrolyte. In the same time in the spectral range 500-600 nm the best results are observed for1 mol/l $LiClO_4$+PC [37].

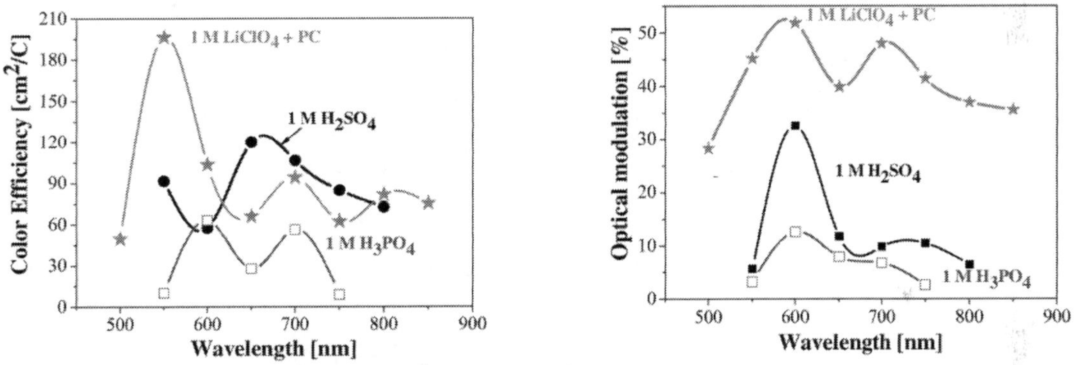

Figure 4. Electrochromic characteristics of CVD MoO_3-WO_3 films, derived from cyclic voltametric measurements in three types of electrolytes [37].

5. Flexible transparent conductive substrates

In the recent years the optoelectronic devices produced on flexible substrates have become very popular due to their light weight, thinness and portability in comparison to the glass based devices. The most studied applications (electrochromic displays, electroluminescent light-emitting diodes, solar cells) contain multilayer stack of different materials deposited on plastic substrate of polyethylene terephthalate (PET) [38-40]. One of the most spread, commercially developed transparent electrode for the glass based version of these devices is indium-tin-oxide (ITO). It has been found that ITO is brittle for the flexible analogues and cracks are revealed at low tensile strain, multiple bending or folding [41]. In contrast to ITO, the conductive polymer poly(3,4-ethylenedioxythiophene) polystyrene sulfonate (PEDOT:PSS) is an perspective alternative for transparent electrode in the flexible optoelectronic devices due to its superior mechanical properties, such as high elasticity, coefficient of thermal expansion closer to that of the plastic substrates, as well as cheaper and easier processing [42]. The low free surface energy of the flexible substrates such as PET, polyethylene sulfonate (PES) or polyethylene naphtalate (PEN), leads to poor adhesion of the PEDOT:PSS coatings and therefore additional surface modification is required, such as plasma or solvent treatment [43]. Such methods, however, change the chemical composition of the polymer surface, increase the surface roughness, or may induce thermal stress in the plastic substrate. To avoid these drawbacks Aleksandrova et. al. suggested UV treatment of PET, before deposition of PEDOT:PSS [44].

For PET surface modification, a mercury lamp with a power of 200 W and a wavelength of 365 nm was used. For comparison of the properties of UV/PET-PEDOT: PSS electrodes with those of the conventional ones, ITO films with same thickness of 35 nm were RF sputtered on PET. To demonstrate the electrode efficiency of UV treated PET/PEDOT:PSS, a simple single layer organic light emitting device (OLED) was prepared by spin coating of a 150 nm electroluminescent film of poly[2-methoxy-5-(2-ethylhexyloxy)-1,4-phenylenevinylene] (MEH-PPV) on the PEDOT:PSS (spin coated as well). Aluminum top contacts were thermally evaporated in vacuum via shadow mask.

The mechanical stability in term of bendability of the UV treated and non-treated PET with deposited PEDOT:PSS films were estimated by lab-made electromechanical setup for tension test (figure 5). The strain applied to the substrate is related to the force intensity generated by the electromechanical system. The force is alternative and its frequency can be tuned. For these experiments a force with an intensity of 25 N from peak to peak and constant frequency of 50 Hz was supplied to the sample. The total number of bends was obtained by setting different time durations of the applied cyclic force. A four point probe FPP5000 resistance measurement system was used to measure the change in the sheet resistance of the films after different numbers of mechanical loads. Optical microscopy at x500 magnification was used to observe cracks revealed in the films after bending.

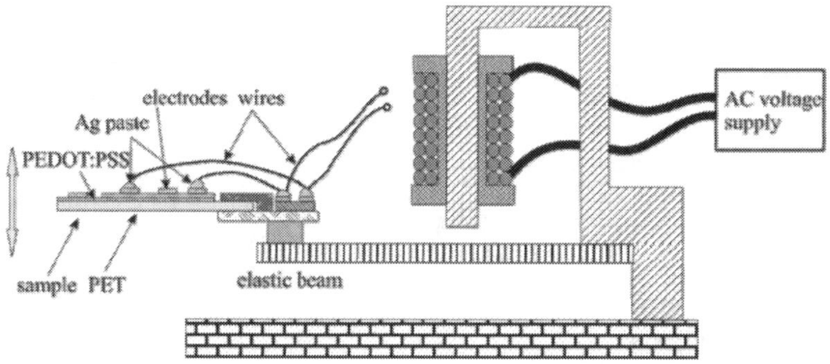

Figure 5. Experimental setup for dynamic mechanical loading applied on the flexible substrates coated by transparent conductive films [43].

XPS measurements showed that UV exposure influences the carboxyl bonds strength from the PET surface and modifies the oxygen functional groups [45]. It was proved that UV treatment causes incorporation of oxygen groups on the PET surface, which results in increasing free energy and improves the surface wettability. Higher wettability resulted in better distribution of the polymeric film. XPS analysis also confirmed that the exited non-bonded oxygen atoms on the PET surface possess energy close to the energy of the oxygen atoms in the bonds with sulfur atoms in the PSS chain, which exceeds the energies of the PSS– H^+ and C–O–C bonds in the PEDOT backbone [46]. This is an indication for their oxidation from the oxygen atoms of the PET surface and bonding with the surface functional groups. This is in agreement with a greater adhesive strength at the PEDOT: PSS/PET interface, and a more favorable behavior in multi-bendable devices is expected for the PEDOT:PSS deposited on the UV treated PET. It was found that PEDOT:PSS film adhesion to the PET substrate is stronger on the UV treated substrate surface. At 50 N/cm^2 tensile force and a peel angle of 90° the coating still remained on the PET surface. For the untreated substrates the films peeled off from the surface at typically 19 N/cm^2, probably due to lack of uniformity of the distributed solution at spin coating on hydrophobic surface. It results in formation of irregularities in the PEDOT:PSS film, that can cause strain forces and can be reason for poor adhesion.

Figure 6 shows the change of the film's sheet resistance under repeated bends and allows estimation of the films stability. According to the bending tests, the PEDOT:PSS film deposited on UV treated PET exhibits only 3.8% resistance change in the entire range of bending cycles. The initial resistance of the samples was close to that of conventional ITO (52.4 Ω/sq and 50.5 Ω/sq for ITO) and stays stable after repeating mechanical loading. This can be explained by the higher adhesion strength at the PET interface. The weaker adhesion of PEDOT:PSS deposited on untreated PET leads to a less stable electrical behavior as a greater resistance change of 8.7% is observed. The initial resistance of 57 Ω/sq was the highest measured. The ITO transparent coating shows the greatest change in its resistance of 23.5 %.

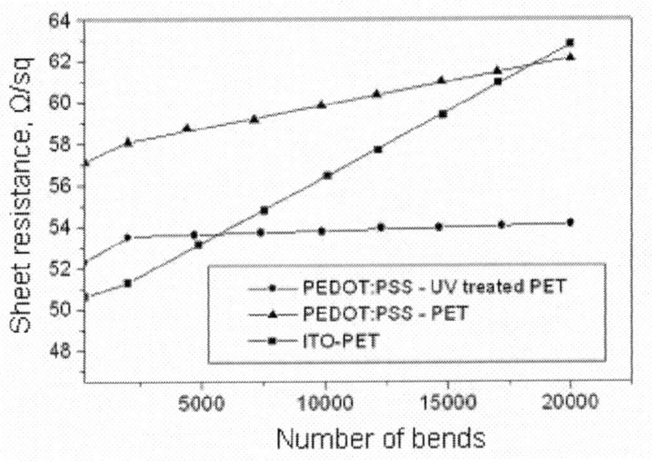

Figure 6. Sheet resistance change at different numbers of bends for PEDOT:PSS deposited on UV treated and untreated PET. ITO on PET was used as a reference [44].

Figure 7 presents microscopic images of the PEDOT:PSS films deposited on UV treated and untreated PET, as well as of ITO deposited on PET. As can be seen when 5000 cycles of mechanical loading was applied to the samples no significant effect was observed in the UV treated PEDOT:PSS film's morphology (figure 7 a). This is in agreement with the results obtained for the stable sheet resistance and the higher adhesive strength. Bending cycles induced cracks in the ITO coating that propagated laterally to the loading direction, leading to fast physical degradation and instability in the sheet resistance (figure 7 b).

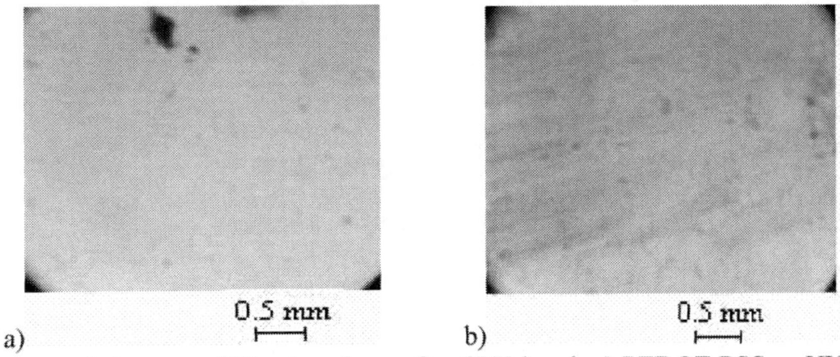

Figure 7. Microscopic images of films' surfaces after 5000 bends a) PEDOT:PSS on UV treated PET; b) ITO on PET [43].

Figure 8 represents the current-voltage characteristics of single layer flexible OLED device with PEDOT:PSS electrode on UV treated and non-treated PET, as well as with conventional ITO.

Both curves are similar to that of diode structure, but for the PEDOT:PSS based device the turn on voltage is higher and the current at given constant voltage is smaller due to the higher sheet resistance in comparison to ITO.

Optimization of the emissive recombination is out of the aim in this work, as the main indication for working sample is the current, flowing through the interfaces. Nevertheless, this characteristic proves the successful fabrication of flexible OLED with material alternative to ITO for transparent conductive electrode, which is more stable at repeating bends.

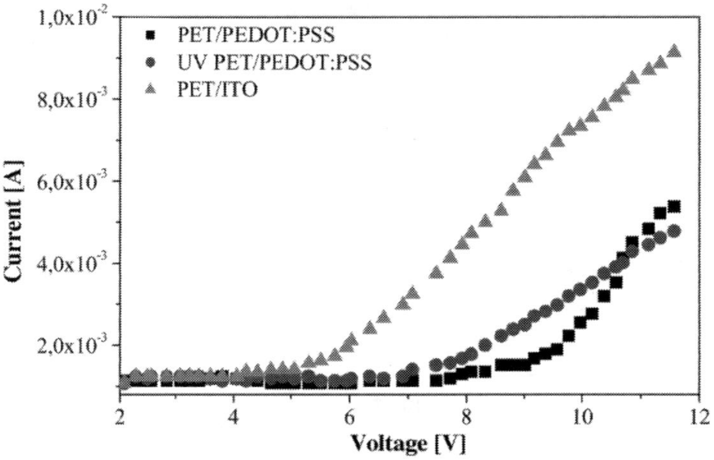

Figure 8. Current -voltage characteristics of the flexible OLED with PEDOT:PSS anodes on UV treated and non-treated PET substrate. Device with ITO anode was used as control sample [45].

6. Photocatalysis

Civil, commercial and industrial sectors are all facing environmental problems related to hazardous wastes, toxic air and contaminated waters. The elimination of toxic and hazardous chemicals has become an international priority. Due to this interest in application of photocatalysis for these environmental problems has grown exponentially, which enables waste water treatment, pollutant decomposition and self-cleaning of surfaces by the use of sunlight.

Transitional metal oxides are the most prominent photocatalytic materials. When the energy of an absorbed photon matches or exceeds the bandgap energy of the TMOs, an electron (e^-) is promoted from the valence band into the conduction band, leaving a hole behind (h^+). On the one hand, these electrons and holes can participate in redox reactions of decomposing pollutants, or they can react with water, which leads to the production of reactive $\cdot OH$ or $\cdot O_2^-$ species, which can also react with hazardous molecules. On the other hand, the electrons and holes can recombine easily, which might result in lowering the photocatalytic efficiency.

TMO materials in photocatalysis were already used for the remediation of a wide variety of contaminants (e.g. alkanes, alkenes, phenols, etc.). In many cases, complete mineralization of the contaminant has been reported. The most common photocatalysts are WO_3, ZnO, TiO_2, SnO_2. TiO_2 is the most widely-spread TMO photocatalyst because it has the best conduction band and bandgap energy levels for water splitting. The only disadvantage of TiO_2 is that its absorption range is in UV, while e.g. WO_3 absorbs also visible light. To increase the efficiency of photocatalysis (and especially of TiO_2), an option is doping with heteroatoms (C, N, S), dye molecules or other organic compounds, as well as with a visible light active TMO, which widen the absorptions range more to the visible spectrum as well. Another way to reach better photocatalytic activity is decreasing the recombination rate of photogenerated electrons and holes by making composites of the TMO with noble metals (e.g. Au, Ag), carbon nanostructures, or with other TMO materials [47-51].

7. Atomic layer depositon (ALD)

Besides CVD another way to deposit TMO thin layers, which can be used as photocatalysts or electrochromic materials is atomic layer deposition. The beginning of the theory of ALD dates back to the 1950s to St. Petersburg, where then Aleskovskii and Koltsov made the first experiments in the 1960s with the method they called molecular layering (ML). Independent of them in Helsinki, Suntola

Journal of Physics: Conference Series **682** (2016) 012011 doi:10.1088/1742-6596/682/1/012011

and coworkers also developed a method named atomic layer epitaxy (ALE), and it was them in the 1970s, who enabled ALE to be used widely in the industry. As the dimensions decreased in the semiconductor industry, it led to the demand for atomic level control of conformal, thin film deposition (especially of the high-κ oxides); thus, now the method called ALD started to gain increased international interest in the 1990s. Then in the 2000s ALD also reached nanotechnology, and was used to fabricate sensors, solar cells, catalysts, nanocoatings, nanocomposites, etc.

ALD is able to meet the needs for atomic layer control and conformal deposition using sequential, self-limiting surface reactions. A general ALD reaction has four consecutive steps: (1) the first precursor is pulsed into the reactor, which chemisorbs on the surface of the substrate; (2) the unreacted precursor and reaction by-products are removed by an inert gas purge and/or by evacuation; (3) the second precursor is pulsed, which reacts with the first precursor solely on the surface; (4) the unused second precursor and the reaction by-products are removed.

The advantages of ALD are precise thickness control at monolayer level and self-limiting growth. The film thickness can be planned easily by the number of the ALD cycles and the self-limiting aspect of ALD leads to excellent step coverage and conformal deposition. The composition of the film can also be controlled precisely. ALD can use solid, liquid or gaseous precursors (elements, molecules or radicals), and the main requirement is that they have to be volatile. ALD can be used to deposit elements (Au, Ag, Pt, Ru, Fe, etc.), binary compounds (e.g. oxides, nitrides, carbides, sulfides, fluorides, like Al_2O_3, TiO_2, ZnO, HfO_2, GaN, TaC, CaS, SrF_2, etc.), ternary compounds (e.g. $MgAl_2O_4$, TiAlN), polymers, organic and hybrid materials. The usual reaction temperature is between 100-400°C, and the ideal reaction temperature zone is often described by the so-called ALD window. Usually ALD reactions take place in vacuum, but it is also possible to grow thin films by atmospheric pressure ALD or at lower temperatures.

ALD and CVD have many similarities, yet there are several distinctive features. In ALD the precursors enter the reactor separately, thereby eliminating gas-phase reactions. ALD film growth frequently occurs at lower temperatures compared to CVD processes because of the elimination of the gas-phase reactions. In ALD there is no shadow effect, which might reduce the growth rate if there is an object between the source and the target [52-56].

There are only few ALD reactions yet, which can be used to deposit WO_3 and MoO_3. There are two molybdenum precursors reported for MoO_3: $C_{12}H_{30}N_4Mo$ [57] and $Mo(CO)_6$ [58]. Likewise there were two kind of tungsten precursors used for WO_3: $W(CO)_6$ [59-61] and $W_2(NMe_2)_6$ [62]. The oxygen precursors were H_2O for thermal ALD and ozone (O_3) for plasma ALD processes, both in the case of WO_3 and MoO_3. The MoO_3 has a tight ALD window with the range of 152-172 °C, while there were successful depositions of WO_3 films from 150-300 °C.

8. Photocatalytic thin films and nanostructures prepared by ALD

For photocatalytic water-splitting MoO_3–WO_3 composite films were successfully prepared by APCVD [63]. In contrast, in the few studies which described the deposition of MoO_3 and WO_3 by ALD, the photocatalytic properties were not studied, and there are also no studies yet of using ALD materials as elelctrochromic layers. Nevertheless, there are a large number of papers, which report about the photocatalytic use of ALD deposited materials. Regarding the photocatalytic thin films and nanostructures prepared by ALD, it was TiO_2 as TMO material, which was mostly grown by ALD.

On the one hand, photocatalytic TiO_2 thin films were grown by ALD on various planar substrates. The composition of such thin films can be finely tuned by ALD, which was demonstrated by e.g. doping with S by adding a third (H_2S) precursor besides $TiCl_4$ and H_2O, or with F atoms by exchanging the $TiCl_4$ precursor with TiF_4. Due to the doping, the absorption of TiO_2 was shifted to the visible range, and this resulted in photocatalytic activity also under visible light illumination [64-65].

On the other hand, and this is the more frequent case, various nanostructured substrates (e.g. nanoparticles, nanotubes, nanowires, nanopatterned surfaces, etc.) were also covered with photocatalytic TMO layers by ALD. Although photocatalytic WO_3 and MoO_3 layers were not deposited yet by ALD, WO_3 nanowires prepared by electrospinning [66] were used as substrates for

ALD TiO_2 thin films. The as-obtained WO_3/TiO_2 core/shell nanofibers (figure 9) were both UV and visible light active photocatalysts [67]. By combining electrospinning and ALD it is also possible to prepare photocatalytic magnetic particle loaded nanotubes [68], while by using electrospinning, ALD and hydrothermal synthesis photocatalytic hybrid ZnO nanoneedle decorated polymer fibres can be produced [69]. Photocathodes to produce hydrogen through water photoelectrolysis can be prepared by covering silicon nanobelts arrays by thin TiO_2 protection layer using ALD [70]. A very interesting example of self-cleaning photocatalytic surfaces was obtained, when TiO_2 layers were grown on lotus leaves to yield a superhydrophobic and photocatalytic bionanocomposite with tuneable properties [71].

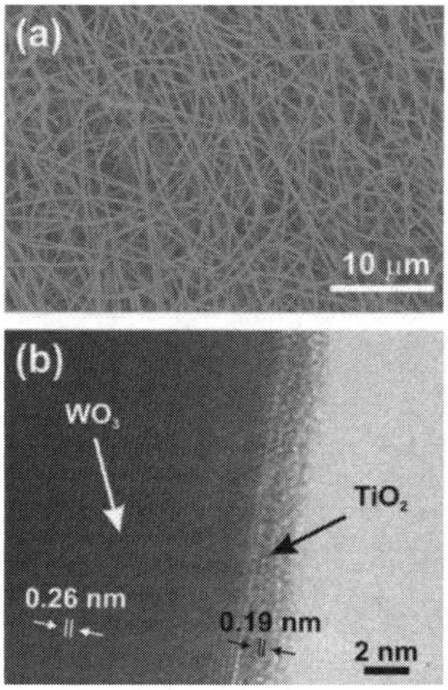

Figure 9. (a) SEM image of electrospun WO_3 nanofibers; (b) HRTEM image of the photocatalytic WO_3/1.5 nm TiO_2 core/shell nanofibers composite [67]. (Copyright Wiley-VCH Verlag GmbH & Co. KGaA. Reproduced with permission.)

9. Conclusions

In this paper the application transitional metal oxides were presented both in electrochromic devices and photocatalytic reactions. The thin films of TMOs were grown by CVD, APCVD or ALD. For comparison results of Physical Vapor Deposition are also given. The theory of these thin films deposition methods was discussed, and numerous examples were presented in which MoO_3, WO_3 and TiO_2 layers were grown for electrochromic and photocatalytic applications.

Acknowledgements

This paper was presented at INERA Conference Workshop "Light in Nanoscience and Nanotechnology 2015", October 20-22th, 2015, Hissar, Bulgaria. The Conference is part of the Program of INERA REGPOT Project of Institute of Solid State Physics, Bulgarian Academy of Sciences.

I. M. S. thanks for a János Bolyai Research Fellowship of the Hungarian Academy of Sciences and an OTKA-PD-109129 grant. S. Boyadjiev thanks the Postdoctoral Fellowship programme of the Hungarian Academy of Sciences (2013-2015).

References

[1] Rao C N R 1989 *Ann Rev Phys Chem* **40** 291
[2] Kung H H 1989 *Transition Metal Oxides: Surface Chemistry and Catalysis*, Elsevier Science, NY
[3] Li X L, Lou T J, Sun X M and Li Y D 2004 *Inorg Chem* **43** 5442

[4] Boyadjiev S, Georgieva V, Vergov L, Baji Z, Gáber F and Szilágyi I M 2014 *J Phys Conf Ser* **559** 012013
[5] Szilágyi I M, Saukko S, Mizsei J, Tóth A L, Madarász J and Pokol G 2010 *Solid State Sci* **12** 1857
[6] Szilágyi I M, Wang L, Gouma P I, Balázsi C, Madarász J and Pokol G 2009 *Mater Res Bull* **44** 505
[7] Szilágyi I M, Madarász J, Pokol G, Király P, Tárkányi G, Saukko S, Mizsei J, Tóth A L, Szabó A and Varga-Josepovits K 2008 *Chem Mater* **20** 4116
[8] Szilágyi I M, Saukko S, Mizsei J, Király P, Tárkányi G, Tóth A L, Szabó A, Varga-Josepovits K, Madarász J and Pokol G 2008 *Mater Sci Forum* **589** 161
[9] Balázsi C, Wang L, Zayim E O, Szilágyi I M, Sedlackova K, Pfeifer J, Tóth A L and Gouma P I 2008 *J Eur Ceram Soc* **28** 913
[10] Kao S Y, Lin Y, Hu C W, Leung M and Ho K 2015 *Solar Energ Mater Solar Cell* **143** 174
[11] Sabry M, Eames P C, Singh H and Wu Y 2014 *Solar Energ* **103** 200
[12] Granqvist C G 1995 *Handbook of Inorganic Electrochromic Materials*, Elsevier Science, Amsterdam
[13] Cheng C P, Kuo Y, Cheng C H and Zheng Zh W 2014 *Solid State Electr* **99** 16
[14] Heckner K H and Kraft A 2002 *Solid State Ionics* **152-153** 899
[15] Granqvist C G 2014 *Thin Solid Films* **564** 1
[16] Tuna O, Sezgin A, Budakoglu R, Turkuz S and Parlar H 2015 *Vacuum* **120** 28
[17] Gogova D, Stoyanov G and Gesheva K 1996 *Renewable Energy* 8 546
[18] Meenakshi M, Gowthami V, Perumal P, Sivakumar R and Sanjeeviraja C 2015 *Electrochim Acta* **174** 302
[19] Li Z, Liu H, Liu Y, Yang S and Yan Liu Y 2010 *Proc SPIE* **7654** 765411
[20] Najbar M, Camra J, Białas A, Wesełucha-Birczyńska A, Borzecka-Prokop B, Delevoye L and Klinowski J 1999 *Phys Chem Chem Phys* **1** 4645
[21] Livage J and Ganguli D 2001 *Solar Energ Mater Solar Cell* **68** 365
[22] Stjerna B, Olsson E and Granqvist C G 1994 *J Appl Phys* **76** 3797
[23] Kuwabara K, Sugiyama K and Ohno M 1991 *Solid State Ionics* **44** 313
[24] Shen Y, Yang Y, Hu F, Xiao Y, Yan P and Li Zh 2015 *Mater Sci Semicond Proc* **29** 250
[25] Lin T N, Lin Y H, Lee C T, Han S and Weng K W 2015 *Thin Solid Films* **584** 341
[26] Patil C E, Tarwal N, Jadha P, Shinde P, Deshmukh H, Karanjkar M, Moholkar A, Gang M, Kim J and Patil P 2014 *Current Appl Phys* **14** 389
[27] Madhavi V, Kumar P J, Kondaiah P, Hussain O M and Uthanna S 2014 *Ionics* **20** 1737
[28] Lin Y S, Tsai T H, Lu W H and Shie B S 2014 *Ionics* **20** 116
[29] Gesheva K A 2007 *Thin Film Optical Coatings for Effective Solar Energy Utilization* Nova Science, NY, USA
[30] Gozde O I, Coclite A M and Gleason K K 2012 *Rep Prog Phys* **75** 016501
[31] Seidel S, Riche C and Gupta M 2011 *Chemical Vapor Deposition of Polymer Films. Encyclopedia Of Polymer Science and Technology*, John Wiley & Sons, Weinheim, Germany
[32] Drosos C and Vernardou D 2015 *Solar Energ Mater Solar Cell* **140** 1
[33] Pedersen H and Elliott S D 2014 *Theor Chem Acc* **133** 1476
[34] Gesheva K, Szekeres A and Ivanova T 2003 *Solar Energ Mater Solar Cell* **76** 563
[35] Ivanova T. Gesheva K, Kalitzova M, Marsen B, Cole B and Miller E L 2007 *Mater Sci Eng B* **142** 126
[36] Gesheva K A , Cziraki A,. Ivanova T and Szekeres A 2007 *Thin Solid Films* **515** 4609
[37] Gesheva K A, Ivanova T, Marsen B, Cole B, Miller E L and Hamelmann F 2007 *Surf Coat Technol* **201** 9378
[38] Liu Q, Dong G, Xiao Y, Gao F, Wang M, Wang Q, Wang S, Zuo H and Diao X 2015 *Mater Lett* **142** 232

[39] Wang Y J, Ouyang S H, Lu J G and Shieh H P D 2015 *SID DIGEST* 557
[40] Lee U J, Lee S-H, Yoon J, Oh S J, Lee S H and Lee J K 2013 *Solar Energ Mater Solar Cell* **108** 50
[41] Alzoubi K, Hamasha M M, Lu S, Sammakia B 2011 *J Display Technol* **7** 593
[42] Chou C S, Chou Ch Sh, Kuo Y T and Wang C P 2013 *Adv Powder Technol* **24** 336
[43] Cuong N K, Saeki N, Kataoka S and Yoshikawa S 2002 *Hyomen Kagaku* **23** 202
[44] Aleksandrova M, Kurtev N, Videkov V, Tzanova S and Schintke S 2015 *Microel Eng* **145** 112
[45] Wang G F, Tao X M, Xin J H and Fei B 2009 *Nanoscale Res Lett* **4** 613
[46] Aleksandrova M 2016 *Microel Intern* **33** in press
[47] Hoffmann M R, Martin S T, Choi W and Bahnemann D W 1995 *Chem Rev* **95** 69
[48] Hashimoto K, Irie H and Fujishima A 2006 *Jap J Appl Phys* **44** 8269
[49] Liu G, Wang L, Yang H G, Cheng H M and Lu G Q 2010 *J Mater Chem* **20** 831
[50] Czakkel O, Geissler E, Szilágyi I M and László K 2013 *Nanomater Env* **1** 23
[51] Szilágyi I M, Fórizs B, Rosseler O, Szegedi Á, Németh P, Király P, Tárkányi G, Vajna B, Varga–Josepovits K, László K, Tóth A L, Baranyai P and Leskelä M 2012 *J Catal* **294** 119
[52] George S M 2010 *Chem Rev* **110** 111
[53] Kim H, Lee H B R and Maeng W J 2009 *Thin Solid Films* **517** 2563
[54] Knez M, Nielsch K and Niinistö L 2007 *Adv Mater* **19** 3425
[55] Szilágyi I M, Nagy D 2014 *J Phys Conf Ser* **559** 012010
[56] Devi A 2013 *Coord Chem Rev* **257** 3332
[57] Bertuch A, Sundaram G, Saly M, Moser D and Kanjolia R 2014 *J Vac Sci Technol A* **32** 01A119
[58] Diskus M, Nilsen O and Fjellvåg H 2011 *J Mater Chem* **21** 705
[59] Mamun M, Zhang K, Baumgart H and Elmustafa A A 2015 J Solid State Sci Technol **4** P398
[60] Nandi D K and Sarkar S K 2014 *Energy Proc* **54** 782
[61] Malm J, Sajavaara T and Karppinen M 2012 *Chem Vapor Dep* **18** 245
[62] Dezelah IV C L, El-Kadri O M, Szilágy, I M, Campbel, J M, Arstila K, Niinistö L and Winter C H 2006 *J Am Chem Soc* **128** 9638
[63] Gesheva K A, Ivanova T, Marsen B, Zollo G, Kalitzova M 2008 *J Cryst Growth* **310** 2103
[64] Pore V, Ritala M, Leskelä M, Areva S, Järn M and Järnström J 2007 *J Mater Chem* **17** 1361
[65] Pore V, Kivelä T, Ritala M and Leskelä M 2008 *Dalton Trans* **45** 6467
[66] Szilágyi I M, Santala E, Heikkilä M, Kemell M, Nikitin T, Khriachtchev L, Räsänen M, Ritala M and Leskelä M 2011 *J Therm Anal Calorim* **105** 73
[67] Szilágyi I M, Santala E, Heikkilä M, Pore V, Kemel, M, Nikitin T, Khriachtchev L, Räsänen M, Ritala M and Leskelä M 2013 *Chem Vapor Dep* **19** 149
[68] Santala E, Kemell M, Leskela M and Ritala M 2009 *Nanotechnology* **20** 035602
[69] Kayaci F, Vempati S, Ozgit-Akgun C, Biyikli N and Uyar T 2014 *Appl Catal B* **156** 173
[70] Bao X Q and Liu L 2014 *J Power Source* **268** 677
[71] Szilágyi I M, Teucher G, Härkönen E, Färm E, Hatanpää T, Nikitin T, Khriachtchev L, Räsänen M, Ritala M and Leskelä M 2013 *Nanotechnology* **24** 245701

Journal of Physics: Conference Series **682** (2016) 012012 doi:10.1088/1742-6596/682/1/012012

High energy electron-beam irradiation effects in Si-SiO_x structures

D Nesheva[1*], V Dzhurkov[1], M Šćepanović[2], I Bineva[1], E Manolov[1], S Kaschieva[1], N Nedev[3], S N Dmitriev[4] and Z V Popović[2]

[1] Institute of Solid State Physics, Bulgarian Academy of Sciences,72 Tzarigradsko Chaussee Blvd,1784 Sofia, Bulgaria

[2] Center for Solid State Physics and New Materials, Institute of Physics, University of Belgrade, Pregrevica 118, Belgrade 11080, Serbia

[3] Institute of Engineering, Autonomous University of Baja California, Benito Juarez Blvd. s/n, C.P.21280, Mexicali, B.C., Mexico

[4] Joint Institute for Nuclear Research, Flerov Laboratory of Nuclear Reactions Dubna, Moskow region 141980, Russia

E-mail: nesheva@issp.bas.bg

Abstract. Homogeneous SiO_x films (x=1.3, 200 nm and 1000 nm thick) and composite a-Si-SiO_y films (y ~ 1.80) containing amorphous Si nanoparticles have been prepared on crystalline (c-Si) substrate. A part of the films was irradiated at temperature below 50 °C by 20 MeV electrons with two different fluences (7.2×10^{14} and 1.44×10^{15} el.cm^{-2}). Atomic force microscopy (AFM), Raman spectroscopy and capacitance (conductance) - voltage ($C(G)$-V) measurements on Al/c-Si/SiO_x/Al or Al/c-Si/(a-Si-SiO_y)/Al structures were used to get information about the irradiation induced changes in the surface morphology, the phase composition in the film bulk and at the Si-SiO_x interface. The AFM results show that the electron irradiation decreases the film surface roughness of the films annealed at 250 °C. The Raman scattering data imply appearance of amorphous silicon phase and some structural changes in the oxide matrix of the homogeneous SiO_x films. In the composite films electron beam stimulated decrease of the defects at the a-Si/SiO_y interface has been assumed. The initial $C(G)$-V results speak about electron induced formation of electrically active defects in the SiO_y matrix of the composite films.

1. Introduction

Electron or ion irradiation of solids normally causes formation of atomic defects in a solid state target and spoils its properties. It may result in irradiation-induced amorphization, recrystallization and annealing. The radiation induced changes of electrical properties of solid-state materials and devices may cause an electronic device or system to fail. Therefore during the last few decades there was significant activity to explore such alterations in various types of semiconductor devices, in particular in those based on metal-oxide-silicon (MOS) structures [1, 2]. The basic radiation problem in a MOS transistor is appearance of radiation-induced trapped charge, built up in the gate oxide and traps right at the Si/SiO_2 interface, which causes a shift in the threshold voltage. If this shift is large enough, the device cannot be turned off, even at zero volts applied. The interface traps are localized states with energy levels in the Si band-gap and their occupancy is determined by the applied voltage, giving rise to a voltage-dependent threshold shift [1, 2].

Irradiation may also cause formation of amorphous nanosized domains and nanocrystals with either the original composition and crystal structure or new nanophases formed by decomposition of the target material. [3, 4]. Evolution of atom displacement cascades, radiation-induced defect generation and accumulation, as well as radiation-induced segregation are regarded as mechanisms involved in the process of nanophase formation. An area of research activity in this field is formation of nanosized Si particles in silicon oxide films which are prospective for development of new functional materials and devices suitable for usage in intense radiation fields in spacecraft, medicine, nuclear power plants, etc.

Electron irradiation has been carried out in high voltage transmission electron microscopes. Formation of amorphous or crystalline Si nanoparticles in a silicon oxide thin film matrix has been reported [5-7]. Irradiation with a high intensity convergent 100 or 200 keV electron beam at very high fluences (10^7-10^9 C×m^{-2} i.e. 10^{22}-10^{24} el×cm^{-2}) has been performed under ultrahigh vacuum conditions and ambient [6, 7] or high temperature [5]. It has been assumed that valence electron ionization is the key factor for the transformation from SiO_2 to amorphous Si while beam heating and knock-on atom displacement are responsible for the transformation from amorphous to crystalline Si [7].

To the best of our knowledge, the information on the effect of high energy electrons (> 10 MeV) on the processes of the nanoparticle growth in silicon oxide matrices and the crystallization/amorphization of Si nanoparticles is very limited [3]. Our previous investigations on c-Si/SiO_2(20 nm thick) structures irradiated at ambient temperature in air by 20 MeV electrons with a fluence of 1.2×10^{15} el×.cm^{-2} have shown silicon nanoparticle formation [3, 8, 9]. On the other hand the 20 MeV electron irradiation of 200 and 1000 nm thick homogeneous SiO_x ($x = 1.3$) layers on c-Si substrate with a fluence of 2.4×10^{14} el.cm^{-2} have not given rise to Si nanoparticle formation in the SiO_x layers [10].

In this work, the effect of 20 MeV electron irradiation with higher fluences (7.2×10^{14} el×cm^{-2} and 1.44×10^{15} el×cm^{-2}) on the surface morphology, the phase separation process and the charge trapping in two types of structures, c-Si/SiO_x and c-Si/(a-Si-SiO_y), is investigated.

2. Experimental details

2.1 Sample preparation

Silicon suboxide SiO_x layers with an initial composition of $x = 1.3$ [11] and thickness of 200 and 1000 nm were prepared by thermal evaporation of SiO at a vacuum of 1×10^{-3} Pa on p or n-type (100) c-Si substrates ((7.2-10) and (4-6) Ω×cm for p or n-type Si, respectively) maintained at room temperature. Prior to film deposition the silicon wafers were cleaned chemically using a standard procedure for the microelectronics industry. The deposition rate and film thickness were monitored by a calibrated quartz microbalance system. All as-deposited layers were annealed at 250 °C for 30 min in an Ar atmosphere to keep them stable at ambient conditions. This annealing procedure does not give rise to formation of pure silicon phase [11]. In a part of the samples, amorphous Si nanoparticles were grown in the SiO_x layers by an additional furnace annealing at 700 °C in argon for 60 min, thus producing a-Si-SiO_y composite layers [12]. The annealing process was used not only for growing of silicon nanoparticles but also to form simultaneously a dielectric layer with tunneling thickness at the interface with the silicon wafer.

Both kinds of structures, c-Si/SiO_x ($x = 1.3$) and c-Si/(a-Si-SiO_y) ($y \sim 1.8$ [13]), were irradiated with 20 MeV electrons with two different fluences, $F_1 = 7.2\times10^{14}$ el×cm^{-2} and $F_2 = 1.44\times10^{15}$ el×cm^{-2}. The irradiation was carried out in Microtron MT-25 in FLNR, Joint Institute for Nuclear Research, Dubna. The beam current was about $I_e = 5$ μA. The sample temperature was controlled during the entire irradiation process and kept close to room temperature.

2.2 Sample characterization

Atomic force microscopy (AFM) technique was used to study the surface morphology of the films.

The measurements were carried out in tapping mode using a Multimode V microscope (Bruker, Santa Barbara, CA). The image resolution was 512 lines per scan direction for all measurements and the scan rate was 1 Hz. Aluminum coated silicon cantilevers TAP150-Al-G (Budget Sensors, Bulgaria) with a nominal resonance frequency of 150 kHz and a typical force constant of 5 N/m were used; the tip radius is less than 10 nm. All images were just flattened and no further processing was carried out. Roughness measurements were performed using Nanoscope 7.30 programme.

Stokes micro-Raman scattering measurements were performed in the backscattering geometry by a Jobin-Yvon T64000 triple spectrometer system, equipped with a confocal microscope and a nitrogen-cooled CCD detector. Spectra were collected by using the 532 nm (2.33 eV) line of a mixed Ar^+/Kr^+ ion laser at room temperature in the air. A low output laser power (≤ 5 mW) was used to avoid local sample heating and subsequent phase separation.

High frequency capacitance-voltage (*C-V*) and parallel conductance-voltage (*G-V*) measurements at 1 MHz were carried out to study the electron irradiation induced changes of the interface and oxide defects in MOS capacitors with a-Si-SiO_y composite layers. The MOS capacitors were prepared by depositing thin Al layers as top (gate) and back contacts. The top Al metallization was carried out through a mask and MOS capacitors with area of ~ 2×10^{-3} cm^2 were formed. The structures were characterized electrically by using Agilent B1500A Semiconductor Device Analyzer and GwInstec LCR-8110G Meter provided with a PSWI6 0-7.2 programmable power supply.

3. Results and Discussion

3.1 AFM results

Three-dimensional 5×5 μm^2 AFM surface images of 1000 nm thick non-irradiated homogeneous and composite a-Si-SiO_y films, as well as a 20 MeV electron irradiated with a fluence of 7.2×10^{14} el×cm^{-2} homogeneous SiO_x film are depicted in figure 1. It is seen that the non-irradiated homogeneous layer (figure 1 (a)) has a relative rough, uniform surface with a root mean square (RMS) roughness of 2.38 nm, probably due to both low mobility of atoms because of the low substrate temperature (300 K) and the low kinetic energy of the evaporation process. The surface of the film annealed at 700°C for 30 min is significantly smoother (figure 1(b)) with RMS value of 1.3 nm, whereas the surface of the irradiated layer displays areas having very low roughness and areas with a slightly higher roughness; the RMS determined is 1.4 nm. Hence, similarly to the furnace annealing effect, the electron irradiation causes a surface roughness decrease. The surface of the composite sample irradiated with a flux of 7.2×10^{14} el×cm^{-2} (not shown) also shows a fairly good smoothness but a few features are observed on the 5×5 μm^2 image with a height of ~ 40 nm and radius of the basis of 150 - 200 nm. The phase image of the surface with these features shows some phase lag, which may originate from different material properties. Excluding the features we obtained RMS value of 1.3 nm which equals to that of the non-irradiated composite films (annealed at 700 °C), which implies that the

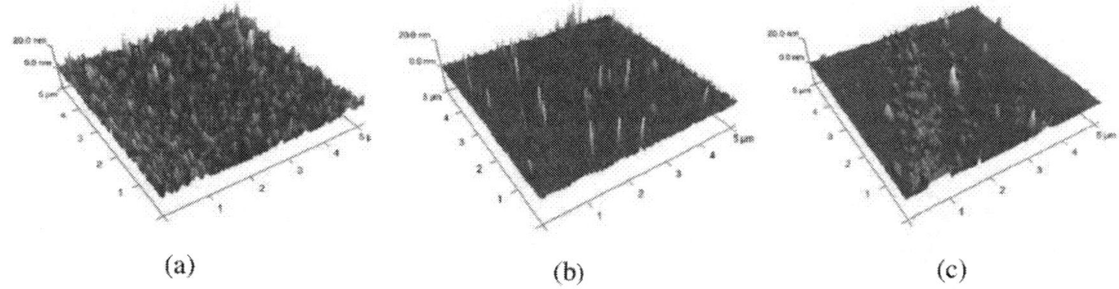

(a) (b) (c)

Figure 1. Three dimensional 5×5 μm^2 AFM surface images with 20 nm z-scale of structures with 1000 nm thick non-irradiated homogeneous SiO_x film (a), non-irradiated a-Si-SiO_y film (annealed at 700°C) (b) and 20 MeV electron irradiated homogeneous SiO_x film with a fluence of 7.2×10^{14} el×cm^{-2} (c).

irradiation of the composite layers does not cause further smoothening of their surface. The phase images of the irradiated films have revealed quite large (hundreds of nanometers) regions of different phase. The changes in phase lag often indicate changes in the properties of the sample surface. The phase signal is sensitive to variations in composition, adhesion, friction, viscoelasticity etc., the understanding of the contribution of the individual factors to the phase shift observed is not simple.

A comparison of the described AFM results with our previous results on 20 MeV irradiated c-Si/SiO_2(20 nm thick) structures [9] shows some differences. The AFM images of the c-Si/SiO_2 (20 nm) structures irradiated with a fluence of 1.2×10^{15} el$\times$cm^{-2} and a beam current $I_e \approx 9$ μA, have revealed a roughing of the surface accompanied by an increase in the crack density whereas an overall electron induced surface roughness decrease for c-Si/SiO_x structures has been observed in this study. The observed differences could be related to the 10 or 50 times greater thickness of the SiO_x layers and the approximately twice lower current ($I_e \approx 5$ μA) used in this study. It seems that at electron beam current as low as 5 μA an annealing occurs, due to energy transfer from the electron beam to the film lattice, rather than defects formation causing a roughness increase.

3.2 Raman scattering results

Raman scattering spectra were collected on non-irradiated and irradiated c-Si/SiO_x and c-Si/(a-Si-SiO_y) structures with 200 and 1000 nm thick homogeneous or composite oxide films, as well as on the top and bottom surfaces of a bare c-Si substrate. The spectra of the non-irradiated and irradiated substrates were identical and indicate lack of Raman-detectable irradiation induced changes on the substrate surface. Moreover no appreciable irradiation induced changes in the c-Si-SiO_x structures with homogeneous SiO_x films have been detected.

Figure 2 shows Raman spectra of a non-irradiated c-Si/SiO_x and two 20 MeV electron irradiated c-Si/SiO_x(1000 nm) samples with fluences of E_1=7.2×10^{14} el$\times$cm^{-2} and E_2 =1.44×10^{15} el$\times$cm^{-2}. In all spectra the strong band situated at ~ 521 cm^{-1} is due to Raman scattering from 1TO phonons of the c-Si substrate. The low intensity feature at around 300 cm^{-1} is also due to light scattering from the c-Si substrate. The weak band at ~ 430 cm^{-1} is normally related [14] to scattering from amorphous SiO_2. The wide band located between 930 and 1030 cm^{-1} is assigned to multi-phonon scattering in the c-Si silicon substrate [15, 16]. Besides, no appreciable changes have been detected in 1000 nm films irradiated with a fluence of 2.4×10^{14} el$\times$cm^{-2} [10]. It is seen from figure 2 that the irradiation with three and six times higher fluences cases some changes in the Raman spectra. In order to make these changes clearer a linear baseline was subtracted from each spectrum in figure 2 and then the spectrum of non-irradiated sample was subtracted from the spectra of the irradiated ones. The result of this subtraction, i.e. the light scattering related to the irradiation induced changes in the films, is shown in the inset of figure 2. Two weak wide bands are seen at ~ 150 cm^{-1} and 480 cm^{-1} which are typical for a-Si [17]. Hence the 20 MeV electron irradiation with 7.2×10^{14} and 1.44×10^{15} el$\times$cm^{-2} causes appearance of amorphous silicon phase in the homogeneous SiO_x films, which resembles the results

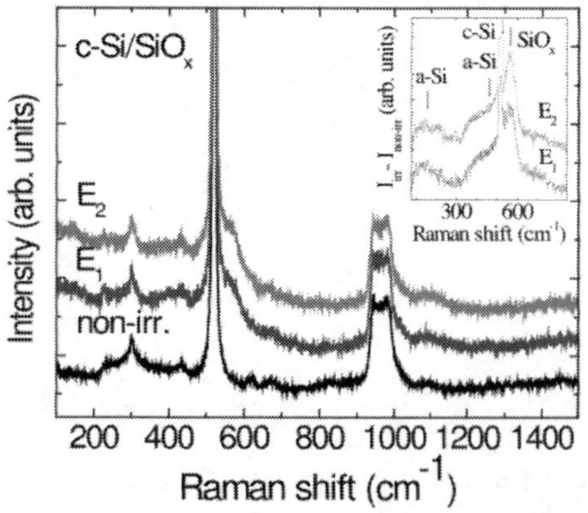

Figure 2. Raman spectra of a non-irradiated and two 20 MeV electron irradiated c-Si/SiO_x(1000 nm) samples with homogeneous SiO_x layers; irradiation fluences $E_1 = 7.2\times10^{14}$ el$\times$cm^{-2}, $E_2 = 1.44\times10^{15}$ el$\times$cm^{-2}. All spectra correspond to the same scale. The spectra of the irradiated samples practically coincide and the E_1 – spectrum is down shifted for clarity. Inset: curves obtained by subtracting the spectrum of the non-irradiated sample from the spectrum of the corresponding irradiated one.

reported in Ref.[9] though we detect amorphous silicon, not crystalline one. The formation of Si nanocrystals in the SiO_2(20 nm) film upon 20 MeV electron irradiation observed in c-Si/SiO_2 structures has been explained [9] assuming that the electron irradiation breaks the Si-O bonds, the free oxygen moves through the radiation defects and creates conditions whereby Si nanostructures are generated in the SiO_2 layer.

It can be also seen from figure 2 that a new band centered at ~ 560 cm^{-1} appears in the Raman spectra of the irradiated samples; its intensity is slightly higher in the spectrum of the sample irradiated with the higher fluence. No such band has been detected in the spectra of non-irradiated films. Such Raman scattering band has been observed in the spectra of SiO_x films prepared by thermal vacuum evaporation at similar deposition rates [18]. It has been related to specific light scattering from oxygen deficient SiO_x and considered as indication that the SiO_x films are build of combination of eight and six atom SiO rings with Raman active modes at 505 cm^{-1} and 606 cm^{-1}, respectively [19]. Due to low absorbance of the SiO_x, film thickness of 3 µm was necessary to observe the 560 cm^{-1} band [18]. As known SiO_x films prepared by thermal vacuum evaporation have a rather low density and the appearance of the 560 cm^{-1} band in the spectra of 1 µm thick films investigated here can be ascribed to an electron irradiation increase of the film density and/or increase of the amount of the eight and six atom SiO rings. It is possible that the phase changes observed in the AFM investigations are due to regions with an increased density and/or structural rearrangement.

Raman spectra of a non-irradiated and two 20 MeV electron irradiated c-Si-(a-Si-SiO_y(1000 nm)) samples with fluences of $E_1 = 7.2\times10^{14}$ and $E_2 = 1.44\times10^{15}$ el×cm^{-2} are shown in figure 3. The bands of amorphous silicon at ~ 150 cm^{-1} and 480 cm^{-1} are well seen in the spectra of all composite films indicating existence of a quite large amount of amorphous silicon phase and their intensities are comparable. Moreover, the scattered light from all composite films is superimposed on a photoluminescence (PL) background, which is related to radiative recombination in the a-Si nanoparticles [12]. As seen from figure 3 the PL background in the spectra of the irradiated composite films is higher than that in the spectrum of non-irradiated film and it is higher for the greater irradiation fluence. Since the intensities of the a-Si bands are rather similar for all composite samples, these observations can be related to irradiation induced annealing of defects at the a-Si/SiO_y interfaces and in the nanoparticle bulk rather than to an increase of the amorphous silicon phase. No photoluminescence background has been observed for the irradiated homogeneous films, which implies that either the amount of the a-Si phase or the nanoparticle size should be quite small.

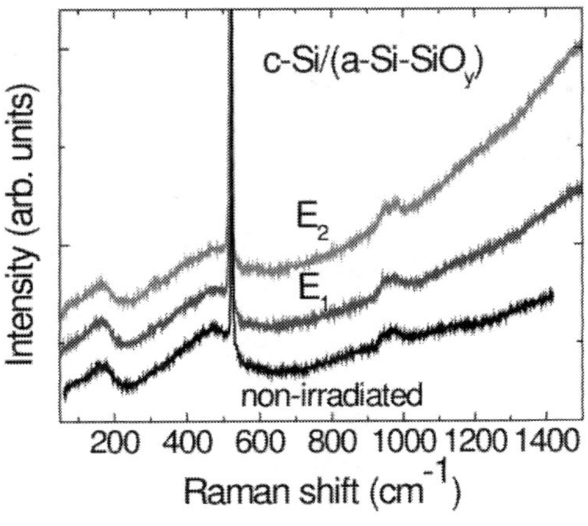

Figure 3. Raman spectra of a non-irradiated and two 20 MeV electron irradiated c-Si/(a-Si-SiO_y(1000 nm)) samples with fluences of $E_1 = 7.2\times10^{14}$ el×cm^{-2} and $E_2 = 1.44\times10^{15}$ el×cm^{-2}. All spectra correspond to the same scale. The spectra of the irradiated samples practically coincide and the E_2 – spectrum is up-shifted for clarity.

3.3 Electrical characterization

The furnace annealing at 250°C for 30 min is not effective in reduction of the electrically active structural defects, the oxide layer thickness in the MOS structures is large and the *C-V* curves measured on both non-irradiated and irradiated samples with homogeneous SiO_x films have shown rather complex behavior that requires further investigation. Therefore, here we will concentrate on the behavior of the structures with a-Si-SiO_y composite films, which experienced an annealing in argon for 60 min at 700°C. Figure 4(a) shows high frequency *C-V* hysteresis curves of non-irradiated and irradiated with a fluence of 1.44 el×cm^{-2} n-MOS capacitors with 1000 nm thick composite oxide layer. The curves on the non-irradiated sample were measured in two voltage ranges: (+4 V to -3 V) and back (curve 1) and ±40 V (curve 2). In order to avoid charging of the structures, the first measurement on each MOS capacitor investigated was carried out in a narrow enough range. In all measurements, scanning means that the gate bias was first swept from positive (accumulation regime) to negative (inversion regime) voltage and then in the reverse direction. Practically the curve 1 does not show hysteresis when compared to that of curve 2. The hysteresis of the ± 40 V scan is predominantly due to a shift of the *C-V* characteristic towards more negative voltages but a shift to more positive voltages is also seen (see also the positions of the maximums of the parallel conductance shown in figure 4(b), curves 2, measured simultaneously with the *C-V* curves in a single voltage scan. The shift to more negative voltages can be related to positive charging of both a-Si nanoparticles and traps in the SiO_y matrix, whereas the positive shift may be due to traps charged with electrons.

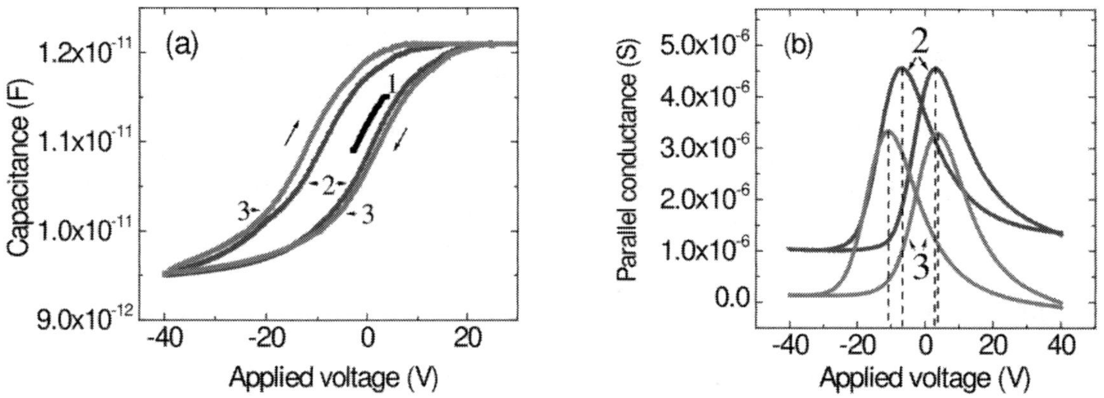

Figure 4. (a) *C-V* characteristics measured at 1 MHz of non-irradiated (curves 1, 2) and 20 MeV electron irradiated (curves 3) samples with 1000 nm thick composite a-Si-SiO_y layer, measured by sweeping the gate voltage in the (+3, -4 V) range, curve 1, and ± 40 V range, curves 2, 3; (b) Equivalent parallel conductance vs. gate bias measured simultaneously with the corresponding *C-V* characteristics.

A comparison of the *C-V* hysteresis curves of the irradiated n-MOS capacitor with 1000 nm thick composite oxide layer (figure 4, curves 3) with the curves of the non-irradiated one shows that there is no change of the capacitance value in accumulation. The result indicates that the electron irradiation did not cause appreciable changes of film thickness and dielectric constant. In addition, the curve slopes are quite similar which indicates that no significant increase of the interface traps occurred upon electron irradiation. However the electron irradiation causes an increase of the hysteresis and this increase is mainly due to a shift of the negative (left) branch of the *C-V* characteristic towards more negative voltages. This observation is supported by the parallel conductance results shown in figure 4(b) and can be related to an increase of either positive charge traps in the SiO_y matrix or the density of the a-Si nanoparticles. Since the Raman scattering results have not indicated an appreciable increase of the a-Si phase in the composite layers (Figure 3), one can infer that the electron irradiation of a-Si-SiO_y composite films causes an increase of the concentration of electrically active defects in the oxide matrix.

It is seen from figure 4(b) that the shape of the equivalent parallel conductance G vs. gate bias of both irradiated and non-irradiated samples is similar. The parallel conductance peak observed in the transition region between flatband conditions and weak inversion corresponds to energy loss due to carrier recombination through interface states [20]. If defects in the oxide layer contribute to the loss, they should be situated very close to the c-Si wafer/dielectric interface in order to be able to respond to the measuring frequency of 1 MHz. The similarity of the parallel conductance curves implies that the interface quality i.e. the defect density at the c-Si wafer/SiO_y interface of both samples is similar which is in agreement with the conclusion based on the similar slopes of the C-V curves (figure 4(a)).

As reported in Ref.[21] the 23 MeV electron irradiation of c-Si-SiO_2 (20 nm) MOS structures generates positive charges in the oxide and at the Si-SiO_2 interface. The former result is in good agreement with the assumed increase of positive charge traps in the SiO_y matrix of a-Si-SiO_y composite films whereas the latter is not confirmed by the initial results on c-Si/a-Si-SiO_y structures. This could be due to the fact that the suboxide layer deposited on the native oxide of the c-Si substrate somehow impedes interface defect creation but this point will be clarified in a future study.

4. Conclusions

Homogeneous, 200 nm and 1000 nm thick, SiO_x films (x=1.3) on crystalline (c-Si) substrate and composite a-Si-SiO_y films containing amorphous Si nanoparticles have been irradiated with 20 MeV electrons with two different fluences (7.2×10^{14} and 1.44×10^{15} el.cm^{-2}). The amorphous Si nanoparticles were grown by an annealing of a part of the homogeneous SiO_x films in an argon atmosphere at 700 °C for 60 min. The AFM study of the effect of the electron irradiation on the surface morphology of both kinds of films have shown that the irradiation causes a significant decrease of the surface roughness of the homogeneous films, annealed at 250 °C, but no change in the surface roughness of the composite films, annealed at 700 °C, has been observed. These changes have been assigned to electron beam induced annealing. The Raman scattering data gave us basis to conclude that the electron irradiation causes formation of amorphous silicon phase in the homogeneous layers whereas in the composite layers it most likely improves the interface between the Si nanoparticles and the oxide matrix. The results from C(G)-V measurements on irradiated MOS structures with composite films speak about formation of electrically active defects in the oxide matrix of the films.

Acknowledgements

This work is supported in the frames of Grant № 41/30.01.2013/p.17 from JINR, Dubna, Russia and Project INERA under the contract REGPOT 316309.

5. References

[1] Fourches N T and Nenoi M 2012 *Current Topics in Ionizing Radiation Research* **32** 741 (Rijeka: InTech)
[2] Oldham T R 2003 *IEEE Trans. Nucl. Sci.* **50** 483
[3] Kaschieva S and Dmitriev S N 2010 *Radiation Defects in Ion Implanted and/of High-Energy Irradiated MOS Structures* (New York: Nova Science Publishers)
[4] Krasheninnikov A V and Nordlund K 2010 *J. Appl. Phys.* **107** 071301 and references therein
[5] Takeguchi M, Tanaka M and K Furuya 1999 *Appl. Surf. Sci.* **146** 257
[6] Chen G S, Boothroyd C B and Humphreys C J 1998 *Phil. Mag.* A **78** 491
[7] Du X-w, Takeguchi M, TanakaM and Furuya K 2003 *Appl. Phys. Lett.* **82** 1108
[8] Kaschieva S, Gushterov A, Angelov Ch and Dmitriev S N 2012 *J. Phys.:Conf. Series* **356** 012005
[9] Kaschieva S, Gushterov A, Angelov Ch and Dmitriev S N 2014 *J. Phys.: Conf. Series* **514** 012039
[10] Nesheva D, Dzhurkov V, Scepanovic M, Bineva I, Manolov E, Kaschieva S, Nedev N, Dmitriev S N and Popovic Z V 2014 *J. Phys.: Conf. Series* **558** 012045

[11] Nesheva D, Bineva I, Levi Z, Aneva Z, Merdzhanova T and Pivin J C 2003 *Vacuum* **68** 1
[12] Nesheva D, Raptis C, Perakis A, Bineva I, Aneva Z, Levi Z, Alexandrova S and Hofmeister H 2002 *J. Appl. Phys.* **92** 4678
[13] Donchev V, Nesheva D, Todorova D, Germanova K and Valcheva E 2012 *Thin Solid Films* **520** 2085
[14] Rahmani A, Benoit M and Benoit C 2003 *Phys. Rev.* B **68** 184202
[15] Temple P A and Hathaway C E 1973 *Phys. Rev.* B **7** 3685
[16] Borowicz P, Latek M, Rzodkiewicz W, Łaszcz A, Czerwinski A and Ratajczak J 2012 *Adv. Nat. Sci.: Nanosci. Nanotechnol.* **3** 045003
[17] Iqbal Z and Veprek S 1982 *J. Phys.* C **15** 377
[18] Monticone E, Rossi A M, Rajteri M, Gonnelli R S, Lacquaniti V and Amato G 2000 *Phil. Mag.* B **80** 523
[19] Pai P G, Chao S S, Takagi Y, Lucovski G 1986 *J. Vac. Sci. Technol.* A **4** 689
[20] Nicollian E H, Brews J R 1982 *MOS Physics and Technology* (New York: Wiley–Interscience)
[21] Kaschieva S, Halova E, Vlaikova E, Alexandrova S, Valcheva E and Dmitriev S 2006 *Plasma Processes and Polymers, Special Issue: Highlights from the 14th International Summer School on Vacuum, Electron and Ion Technologies* **3** 237

2D Raman band of single-layer and bilayer graphene

V N Popov
Faculty of Physics, University of Sofia, BG-1164 Sofia, Bulgaria

E-mail: vpopov@phys.uni-sofia.bg

Abstract. We present a computational study of the 2D Raman band of single-layer and bilayer graphene within a density-functional-based non-orthogonal tight-binding model. The phonon dispersion is derived perturbatively and the 2D band intensity is calculated in fourth-order quantum perturbation theory within this model. The 2D band intensity is enhanced through resonant processes in which the laser excitation matches an electronic transition and the energy and momentum of the scattered phonons match the difference of those of pairs of electronic states. As a result, the 2D band is dispersive, i.e., its position depends on the laser excitation. Here, we calculate the shift and shape, as well as the dispersion rate, of the 2D band for both single-layer graphene and bilayer graphene. The results are compared to available experimental data.

1. Introduction

The successful synthesis of few-layer graphene [1] and the theoretical predictions of its amazing electronic properties[2,3] have triggered an avalanche of theoretical and experimental work. Few-layer graphene has been identified as a promising material with prospective application in nanoelectronics [4].Various experimental techniques are currently applied to the characterization of the few-layer samples but Raman spectroscopy has obvious advantages as an inexpensive and nondestructive technique of choice [5].

Few-layer graphene is a two-dimensional structure of stacked graphene layers. In the simplest case of a single-layer graphene (SLG), there is only one Raman-active phonon, which gives rise to the G band in the Raman spectra. In perfect graphene, the latter are dominated by the two-phonon 2D band, which is several times more intense than the G band. The 2D band is observed because of resonant processes and is, therefore, *dispersive*, meaning dependence of the Raman band position on the laser excitation. The theoretical description of these resonant processes has been proposed more than a decade ago [6], but realistic calculations of the position, line shape, and dispersion rate of SLG have been reported only recently [7,8].

The bilayer graphene (BLG) consists of two graphene layers with either AA, or AB stacking. Most efforts have been applied to the study of the more stable AB-stacked structure. The two layers are bound together by weak Van der Waals interactions, resulting in splitting of the doubled electron and phonon dispersions. Although the splitting is comparatively small, the split components of an electronic band or phonon branch have different symmetries, yielding modified resonant processes for the 2D band. Recently, the calculated 2D band of BLG has been reported [9] and a systematic study of the two-phonon Raman bands of BLG and their dispersion rates has been performed [10]. While a significant amount of work has already been done on the clarification of the underlying scattering processes, a comparative investigation of these processes and the calculated 2D bands is still lacking.

Journal of Physics: Conference Series **682** (2016) 012013 doi:10.1088/1742-6596/682/1/012013

Here, we report a comparative study of the 2D band of SLG and BLG, revealing the similarities and differences in the resonant processes, responsible for these bands.

2. Theoretical

The electronic band structure is calculated here within a density-functional-based non-orthogonal tight-binding model (NTB) [11]. The dependence of the Hamiltonian and overlap matrix elements on the radial coordinate is taken over from ab-initio studies on carbon dimers [12,13].Only matrix elements between atomic orbitals of type s, p_x, p_y, p_z are considered. The Slater – Koster scheme is used for the angular dependence of the matrix elements. The model is appended with repulsive potentials, also derived within the ab-initio approach. The total repulsive energy describes part of the electronic correlations, which is not included in the band energy. The possibility to calculate the total energy allows for performing relaxation of the atomic structure. This step is indispensable for phonon dispersion calculations.

The phonon dispersion is calculated using a perturbative approach within the NTB model [14]. This approach makes utilizes electron-phonon matrix elements, which are derived within the NTB model [15]. The Raman intensity I of any two-phonon Raman band can be calculated in fourth-order perturbation theory within the NTB model by use of the expression [8,10]:

$$I \propto \sum_f \left| \sum_{c,b,a} \frac{M_{fc} M_{cb} M_{ba} M_{ai}}{\Delta E_{ic} \Delta E_{ib} \Delta E_{ia}} \right|^2 \delta\left(E_i - E_f\right) \quad (1)$$

Here, $E_{iu} = E_u - E_i$, where $E_u (u = a,b,c)$ are the energies of the intermediate states of the system of photons, electrons, holes, and phonons. E_i is the initial state energy, equal to the laser excitation energy E_L. M_{ai} and M_{fc} are the momentum matrix elements for creation and recombination of an electron-hole pair, respectively. M_{ba} and M_{cb} are the electron-phonon matrix elements for scattering between intermediate states. The electron-photon and electron-phonon matrix elements are calculated explicitly within the NTB model [15]. The summation over the intermediate states runs over all valence and conduction bands, and over all electron wavevectors **k**. The summation over the final states runs over the considered pair of phonon branches and over all phonon wavevectors **q**. For both summations, convergence is reached with a 400 × 400 mesh of **k** and **q** points in the Brillouin zone. The calculations are performed for parallel scattering configuration, backscattering geometry, and Stokes processes.

3. Results and discussion

The atomic structure of AB-stacked BLG is shown in figure 1. It has a unit cell of four atoms, while the unit cell of SLG has only two atoms. Both structures have hexagonal symmetry and share a common Brillouin zone (figure 1).

The electron dispersion is readily derived by solving the non-orthogonal tight-binding equation of rank equal to the product of the number of atoms per unit cell (2 or 4) and the number of atomic orbitals per atom (4). The calculated electron dispersion of SLG and BLG close to the Fermi energy, taken as zero, is shown in figure 2 and figure 3, respectively. In the energy range of ±1.5 eV only π bands are present. In the case of SLG, these orbitals cross at the Fermi energy, forming a valence band (π band) and a conduction band (π* band). It is most amazing that the π bands have linear dispersion in the vicinity of the Fermi energy at the two non-equivalent special points K and K', and therefore the effective mass of the electrons is zero. This results in an effective wave equation of the Dirac – Weil type, which is often used in the simplified description of transport and optical phenomena. Away from the Fermi energy, there is a deviation from the linear behavior of the bands, known as band warping effect. The conduction and valence bands form cones above and below the plane of the two-dimensional wavevectors, called *Dirac cones*, and the common points of these cones are called *Dirac points*.

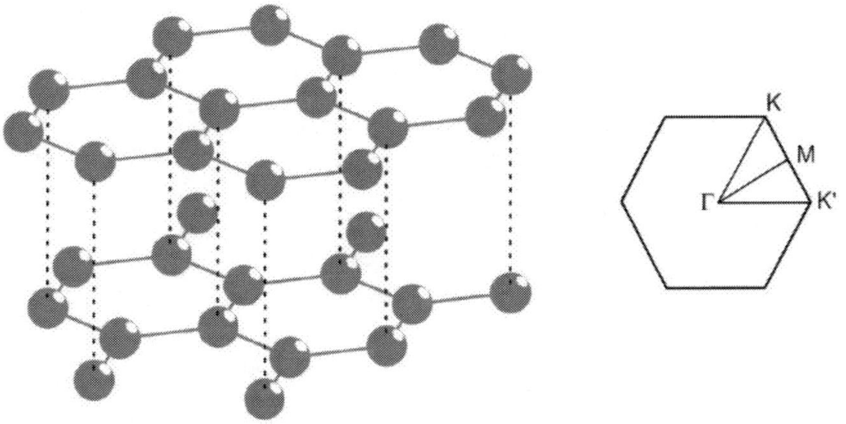

Figure 1. Left: The atomic structure of BLG consisting of two AB-stacked layers of SLG. Right: The Brillouin zone of SLG and BLG with special points.

BLG consists of two SLG layers, bound together by weak Van der Waals interactions. In the approximation of negligible interlayer interactions, the electronic dispersion of BLG is identical to that of SLG with the only difference that the number of electronic states is doubled. The account for the interlayer interactions leads to splitting of this double degeneracy of the energy levels (figure 3). However, there is a qualitative difference of the behavior of the bands of the two structures close to the Dirac points. Namely, the linear dispersion disappears and the split bands acquire parabolic wavevector dependence. This qualitatively different behavior of the bands is important at small excitations, e.g., for transport phenomena, but plays almost no role at higher energy excitations, e.g., optical phenomena. In the latter case, the most important feature is the band splitting.

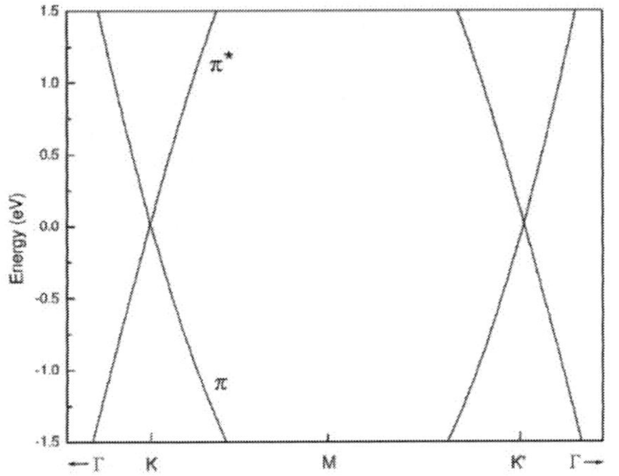

Figure 2. Electronic band structure of SLG along the KMK' direction close to the Fermi energy.

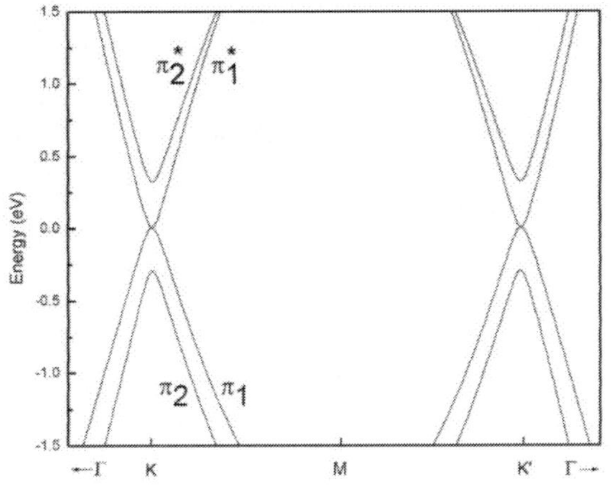

Figure 3. Electronic band structure of BLG along the KMK' direction close to the Fermi energy.

The phonon dispersion is obtained from the dynamical equation for the two structures. This equation is a matrix equation of rank equal to the product of three and the number of atoms in the unit cell (2 or 4). The calculated phonon dispersion of SLG and BLG is shown in figure 4 and figure 5, respectively. The dispersion has a number of phonon branches, denoted according to the atomic vibrational pattern. The usual notation includes the letters "L", "T", and "Z" for longitudinal, transverse, and out-of-plane, respectively. The second letter "A" and "O" denotes acoustic and optical, respectively. In BLG, the number of branches doubles in comparison with SLG and the superscript "+" and "–" is added for the branch withsymmetric and antisymmetric atomic displacement with respect to the center of symmetry, respectively. The splitting of the branches, due to interlayer interactions, is normally of about few cm^{-1}. Larger splitting is present for the ZA branch, but it is not important for the current study of the 2D band, which originates from the TO branch.

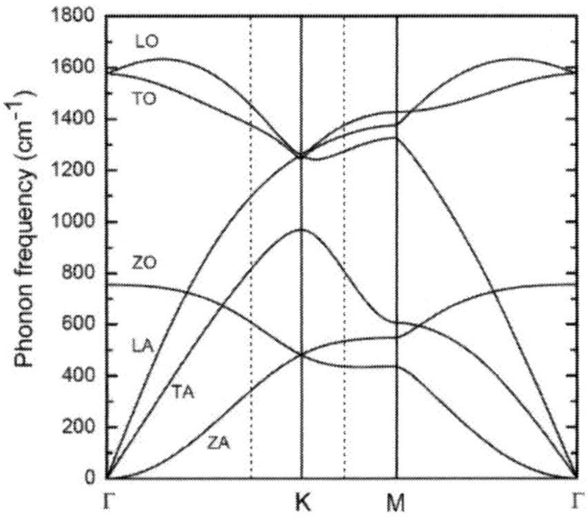

Figure 4. Phonon dispersion of SLG with the displacement pattern of the branches.

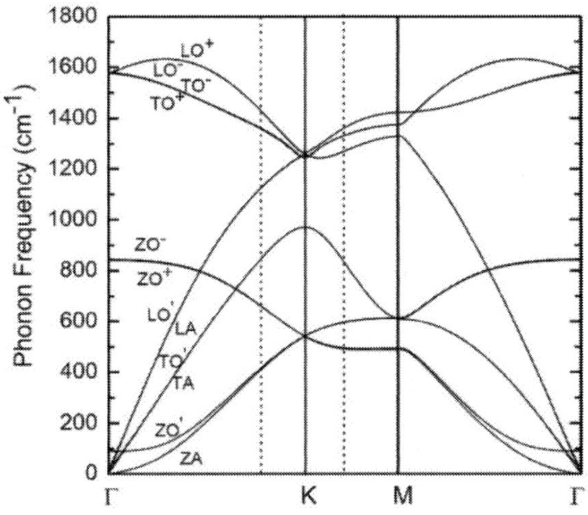

Figure 5. Phonon dispersion of BLG with the displacement pattern of the branches.

The electron and phonon dispersions are both important for the calculation of the 2D band intensity, as it is clearly seen from equation 1. In order to analyze the results, obtained by using this equation, we remind that expression inside the absolute value symbol is the scattering amplitude. The latter can be represented by a corresponding diagram, consisting of virtual processes. In the present case, the scattering process can be interpreted as a sequence of the following virtual processes: electron-hole creation, scattering of an electron (or a hole) by a phonon, a consecutive scattering of an electron (or a hole) by a phonon with opposite momentum with regard to the first phonon, and final electron-hole annihilation. There are altogether eight such scattering processes. For optical processes, the only available electronic states for scattering are those of one and the same Dirac cone at the K or K' point (intravalley processes) and those of Dirac cones at the K and K' points (intervalley processes). In this paper, all types of scattering processes are considered. A simplified schematic of one of the eight scattering processes in SLG and BLG is displayed in figure 6 and figure 7, respectively.

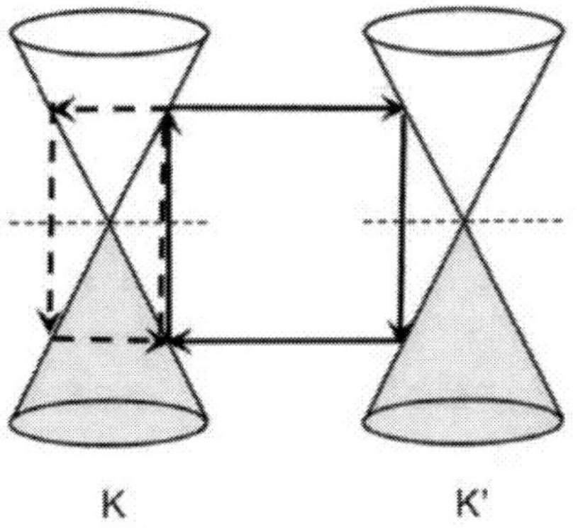

Figure 6. Dirac cones of SLG with examples of two scattering processes, drawn by dashed and solid lines in the cases of intravalley and intervalley virtual processes, respectively.

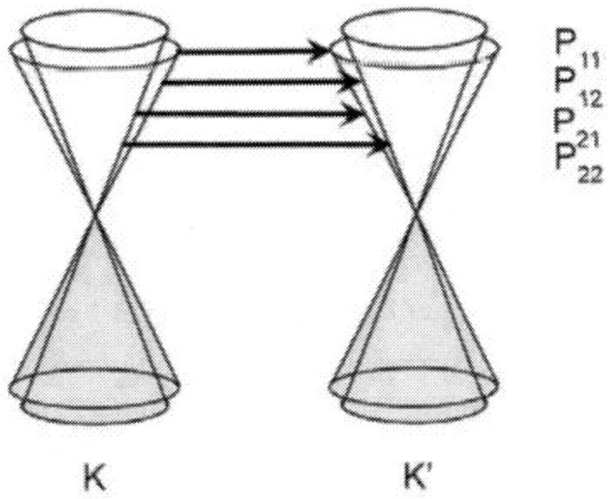

Figure 7. Dirac cones of BLG with illustration of the four types of virtual processes of electron scattering by a phonon.

In the case of BLG, the number of Dirac cones doubles and there are already four types of electron scattering processes instead of only one in SLG, as seen in figure 7. All these processes are symmetry-allowed. The following notation for these processes is adopted: P_{11}, P_{12}, P_{21}, and P_{22} (see figure 7). The doubling of the number of the Dirac cones increases the number of the electron excitation processes four times ($\pi_1 \rightarrow \pi_1^*, \pi_1 \rightarrow \pi_2^*, \pi_2 \rightarrow \pi_1^*$, and $\pi_2 \rightarrow \pi_2^*$). However, only the first and the fourth of them are symmetry-allowed. All allowed scattering processes are taken into account in the present calculation.

Looking back at equation 1, it is straightforward to perceive that some of the scattering processes will have resonant contribution to the Raman intensity for vanishing denominator of this equation. This is possible if one, two, or three of the differences $E_{iu} = E_u - E_i$ vanish. Such resonances are usually termed single, double, triple resonance. The explicit calculations show that the dominant contribution to the intensity comes from triple resonances [7,8,10].

Equation 1 can be used to obtain all two-phonon Raman bands arising from phonons with opposite momenta of all pairs of phonon branches. Bands, due to pairs from the same branch, are termed *overtone bands*, while bands, due to pairs of phonons from different branches, are *combination bands*. The intensity of a given two-phonon band depends crucially on the matrix elements in the numerator of equation 1 with the major factor determining the intensity being the electron-phonon matrix element. It has been argued that the phonons of the TO branch in the vicinity of the K and K' point have the largest electron-phonon coupling [7]. Therefore, it can be expected that the 2TO overtone and combination bands will dominate the second-order Raman spectra. This is indeed the case and the most intense band is the 2D band, which consists of these bands. The remaining, less intense two-phonon bands are discussed elsewhere [7,8,10].

The calculated 2D bands of SLG at $E_L = 1.96$ eV and BLG at $E_L = 2.33$ eV are shown in figure 8 and figure 9, respectively. The 2D band of SLG is almost symmetric in shape. The reason for the slight asymmetry of this band is due to the two types of contributions to this band. The calculations reveal that these two contributions come from TO phonons along the ΓK and KM directions. The former contribution yields about 90% of the integrated 2D band intensity. The approximate symmetry of the 2D band is due to the approximate mirror symmetry of the TO branches with respect to the K point. The increasing deviation from this symmetry with increasing the laser excitation brings about visually asymmetric 2D band shape [7,8]. The 2D band of BLG usually has an asymmetric shape because of the various contribution to this band [9,10]. The quantitative prediction of this band has been a long-standing problem. Initially, it has been assumed that the four types of scattering processes P_{11}, P_{12}, P_{21}, and P_{22} give rise to contributions with well-resolved Raman shift and the dispersion rates of the four components of the 2D band have been derived [16]. Recently, it has been argued that the contributions of processes P_{12} and P_{21} cannot be resolved and that the experimental 2D bands should be fitted only by three Lorentzians [9,10]. The results, presented in figure 9 confirm this conclusion. It can also be seen in figure 9 that the phonon branches $TO^{\pm}$ give rise to two overtone bands $2TO^{+}$ and $2TO^{-}$ and a combination band $TO^{-}TO^{+}$, all of them being of comparable intensity but of different shape. The combination band has been overlooked in Ref.[9] resulting is a slightly different band shape.

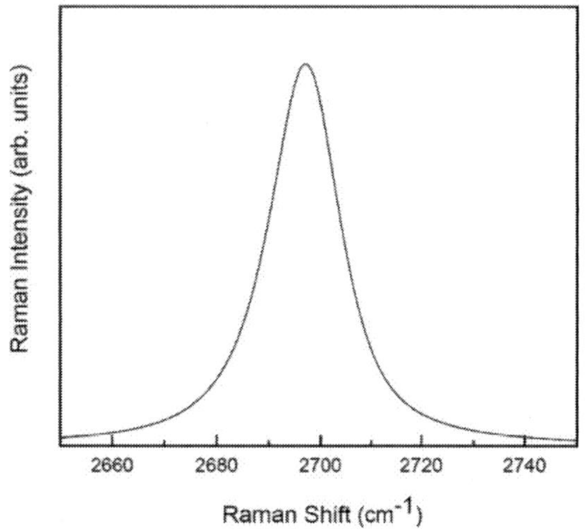

Figure 8. 2D band of SLG at EL = 1.96 eV.

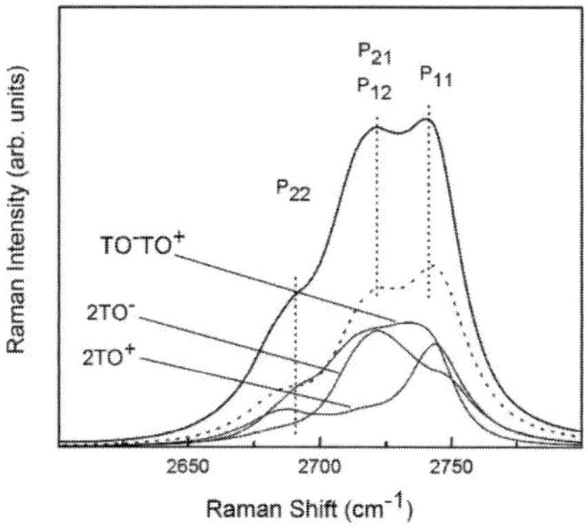

Figure 9. 2D band of BLG at EL = 2.33 eV.

Finally, the 2D band position of SLG and BLG has a clear dependence on the laser excitation (not shown). In the case of SLG, the calculated dispersion rate of89 eV/cm^{-1} agrees well with the experimental values [7]. The three components of the 2D band of BLG were found here to have close values of the dispersion rate with an average value of98 eV/cm^{-1}, which is higher than the experimental value by a few percent [5].

The reported results on the 2D Raman band of SLG and BLG can be used in the characterization of few-layer graphene. The theoretical prediction of the two-phonon bands will further facilitate the characterization of graphene samples.

Acknowledgments

The author would like to acknowledge the financial support from EU Seventh Framework Programme project INERA under grant agreement number 316309.

References

[1] GeimA K and NovoselovK S 2007 *Nature Materials* **6** 183
[2] Castro Neto AH, Guinea F, Peres NMR, Novoselov KS and Geim A K 2009 *Rev. Mod. Phys.* **81** 109
[3] McCann E and Koshino M 2013 *Rep. Prog. Phys.* **76** 056503
[4] BonaccorsoF, SunZ, HasanT and FerrariA C 2010 *Nature Photonics* **4** 611
[5] MalardL M, PimentaM A, DresselhausG and DresselhausM S 2009 *Phys. Rep.* **473** 51
[6] ThomsenC and ReichS 2000 *Phys. Rev. Lett.* **85** 5214

[7] VenezuelaP, LazzeriM and MauriF 2011*Phys. Rev.* B **84** 035433
[8] Popov V N and LambinPh 2012 *Eur. Phys. J.* B **85** 418
[9] Herziger F, Calandra M, Gava P, May P, Lazzeri M, Mauri F and Maultzsch J.2014 *Phys. Rev. Lett.***113** 187401
[10] PopovV N 2015 *Carbon***91** 436
[11] PopovV N and HenrardL 2004 *Phys. Rev.* B **70** 115407
[12] PorezagD, FrauenheimT, KöhlerT, SeifertG and Kaschner R1995 *Phys. Rev.* B **51** 12947
[13] Popov V N and Van Alsenoy Ch 2014*Phys. Rev.* B **90** 245429
[14] PopovV N and LambinPh 2006 *Phys. Rev.* B **73** 085407
[15] Popov VN, Henrard L and Lambin Ph 2005*Phys. Rev.***72**035436
[16] Malard LM, Nilsson J, Elias DC, Brant JC, Plentz F, Alves ES, Castro NetoA H and Pimenta M A2007 *Phys. Rev.* B**76** 201401(R)

PECVD polymerised coatings on thermo-sensitive plastic support

D Mitev[1], E Radeva[1], D Peshev[2], M Cook[3], L Peeva[3]

[1] REGPOT-2012-2013-1 NMP INERA Project, Institute of Solid State Physics-BAS, 72 Tzarigradsko Chaussee Blvd., 1784 Sofia, BULGARIA
[2] University of Chemical Technology and Metallurgy, Department of Chemical Engineering, 8 Kl. Ohridsky Blvd., 1756 Sofia, Bulgaria
[3] Chemical Engineering and Chemical Technology, Imperial College London, Exhibition Road, London SW7 2AZ, U.K.

Email: dpmitev@issp.bas.bg

Abstract. With this paper we describe several alternative approaches aimed at polymer-on-polymer coatings, produced by Plasma Enhanced Chemical Vapour Deposition (PECVD) technique. In our case three types of input compounds were applied for varied "cold plasma" PECVD deposition of nano-thick functional layers: namely hexamethyldisiloxane (HMDSO), pentane and toluene onto plastic substrates. The output characterisations through SEM imaging, ATR-FTIR, EDX, AFM and contact angle measurements, proved the methods' feasibility and properties of the plasma-polymerised coatings.

1. Introduction

During the last three decades applications of Chemical Vapour Deposition (CVD) and Plasma Enhanced Chemical Vapour Deposition (PECVD) technologies show constant growth in various fields of scientific research, followed by technology and medicine [1-3]. In particular PECVD is an outstanding alternative for depositing a range of thin films at lower temperatures than those applied in CVD reactors; this often without settling for a lesser coating quality. The application of "cold plasma" PECVD equipment provides an excellent opportunity to etch, modify and coat thermally sensitive substrates, including porous and nanostructured plastics. In the same time plasma-assisted polymerisation renders possible an utilisation of "non-conventional" organic monomers for highly branched and cross-linked thermoset films having pronounced resistance towards solvents and strong adhesion to the substrate. The creation of composite materials, or layers, provides the researchers with an ability to combine the desirable properties of each of the materials involved in the layered composite. The versatility of plasma polymer depositions in aspects of production of insoluble and infusible films, electrically conductive polymers, controllably crosslinked networks and their potential to reduce environmental, health, and safety issues associated with solvents are well described [4].

Plasma-enhanced CVD depositions onto porous and soft plastic materials are of particular interest not only due to wide area of polymer applications in medicine, separation science and industry. They also attract attention because of the "coating challenge" related to the thermally sensitive and susceptible to stress damage flexible substrates. Deposition of SiOx thin films on polyethyleneterephtalate (PET) as a barrier anti-oxygen coating for packaging industry was researched by [5]; precursor combination of oxygen-diluted hexamethyldisiloxane (HMDSO) was used with

Journal of Physics: Conference Series **682** (2016) 012014 doi:10.1088/1742-6596/682/1/012014

varied concentration in MW and RF plasmas. Along with plasma-deposition of HMDSO onto Si wafers, substrates of polylactic acid films were used by [6]. In [7] a common mechanism of plasma polymerization of aromatic compounds is proposed, with conclusions based on PECVD of benzaldehyde, benzyl alcohol, styrene, and benzyl chloride. [8-9] studied the plasma polymerization of toluene, on inorganic copper and silicon substrates though. Hydrogenated carbon films deposited by PECVD of methane are described in [10]. PECVD of hexane aliphatic hydrocarbon onto PET was researched by [11], in parallel with HMDSO-based coatings.

With present work we demonstrate a practical application of PECVD method, aimed at "cold plasma" depositions on sheets of porous thermo-sensitive polymers and their characterisation. The scopes of this paper are especially the surface coatings and their properties. Substrates here will be denoted as S1÷S3 due to related ongoing work; their properties will be topic of following distinct article.

2. Experimental part

2.1. Materials

All reagents used were of analytical grade quality: hexamethyldisiloxane (Merck-Schuchardt, Germany), pentane (Fluka AG, Buchs, Switzerland), toluene (Valerus Ltd, Bulgaria). Distilled water (GFL Typ 2004, Burgwedel, Germany) was used throughout this work where necessary.

2.2. PECVD coatings

Coatings were deposited using RF plasma chamber of type B 30.2,VEB Hochvakuum Dresden, Germany).

Polymer samples were placed horizontally onto an intermediate fluoropolymer plate, between the two working electrodes layered with aluminium foil (figure 1). The vapour pressure of the used precursors was high enough to ensure introduction without carrier gas. The depositions occurred within 85-100 Pa working pressure, with RF powers of several watts depending of the plasma stability. The thicknesses of the coatings were simultaneously compared by those derived with quartz resonator frequency changes [12], and visually, trough SEM images of corresponding cross sections.

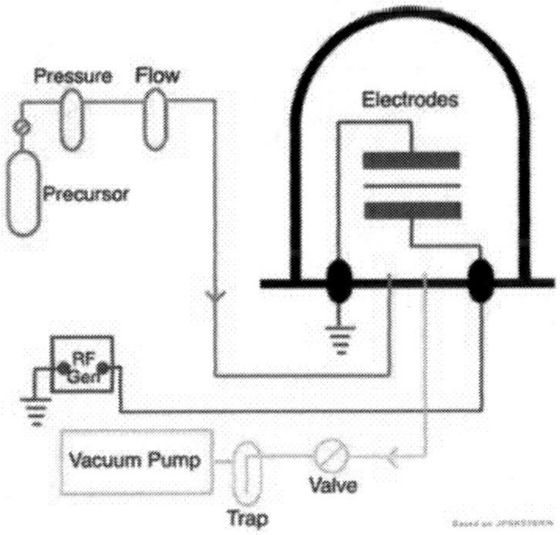

Figure 1. PECVD setup.

2.3. Instruments

Following instrumentation was used throughout the work:

SEM. For higher resolution images, LEO Gemini 1525 FEGSEM (Karl Zeiss), with an accelerating voltage of 5 kV, was used. Prior to analysis for both SEM analysis (here and EDX), samples were

Journal of Physics: Conference Series **682** (2016) 012014 doi:10.1088/1742-6596/682/1/012014

sputter coated (Q150T S turbo-pumped sputter coater, Quorum Technologies Ltd., Ashford, UK) with 10 nm chromium under an Ar atmosphere (0.02 mbar).

Infrared spectra. The FTIR analysis was performed using a Spectrum 100 FT-IR Spectrometer fitted with Universal ATR Accessory. The scanning range was 600-4000 cm^{-1} and the resolution was 1 cm^{-1}. "Net" absorption spectra of the differences between the plasma-covered and initial plain polymers were collected and analysed, because of the picture of pure related changes.

EDX. Especially for the Si-containing films, corresponding SEM-EDX elemental analysis was performed using a JEOL JSM 6010 LA. For EDX mapping an accelerating voltage of 20 kV was used.

Contact angle measurements were performed with an Easy-Drop Instrument (Kruess) at room temperature. 2 microlitres of DI water was deposited on the surface, and analysed via the sessile drop method. A minimum of 6 samples from different spots were obtained for each film.

AFM. The analyses were performed on a Multimode 8 (Bruker, USA) AFM equipped with E-type pizzo scanner. Samples were attached to a magnetic sample disc using carbon adhesive tabs. Scans were conducted using a silicon tip on nitride lever (MSNL-10, Bruker). The cantilever resonance frequency is in the range 90-160 kHz, with a nominal spring constant of 0.6 N/m. Nanoscope analysis 1.5 was used to process the obtained images.

3. Results and Discussion

Films having different surface properties and degree of hydrophobicity were developed with the course of our work. Depending on the different durations of each procedure not only the thicknesses vary, but also varied surface morphology was observed.

3.1. Polysiloxane plasma-deposition

An application of pure HMDSO vapour, unlike the SiOx films from HMDSO/oxygen mixtures [13], outputs siloxane polymer having distinctive organic nature. As visible on figure 2, siloxane coatings together with the pentane-based ones display intrinsic hydrophobic nature.

Figure 2. Contact angles for plain substrate S1 and the plasma-deposited films type H (20 min), P (5 min) and T (20 min deposition). Corresponding standard deviations are (from left): 7.8°; 3.7°; 3.4°; 5.0°.

PECVD deposited siloxanes show dense well defined surface layers onto the initial polymers. After 20 minutes long deposition in average 110-120 nm thick film develops (figure 3); with parallel quartz resonator proof measurement of 110 nm. Some surface peculiarities show up throughout the procedure though. Depending on the coating time, the surface morphology of the primary layers is characteristic with granular surface nanostructures of varying sizes and quantities, randomly distributed throughout the surface of the siloxane deposit. Figure 3b clearly presents such topography after 20 minutes deposition time: with particles being on the surface, fallen off their nests during the SEM preparations, and some partially immersed into the layer. After total of 40 minutes PECVD treatment however the final layer displays smoother look, with less but rounded granular structures. Besides of these heterostructures, AFM reveals rather smooth siloxane surface: with unevenness of the plain S1 substrate within 0-15.1 nm (area 1x1 μ), coated for 20 min specimens vary within 0-12 nm (1x1 μ), 0-18.2 nm (2x2 μ; 0.6 N/m cantilever) and 0-17.6 nm (2x2 μ; 4.0 N/m cantilever).

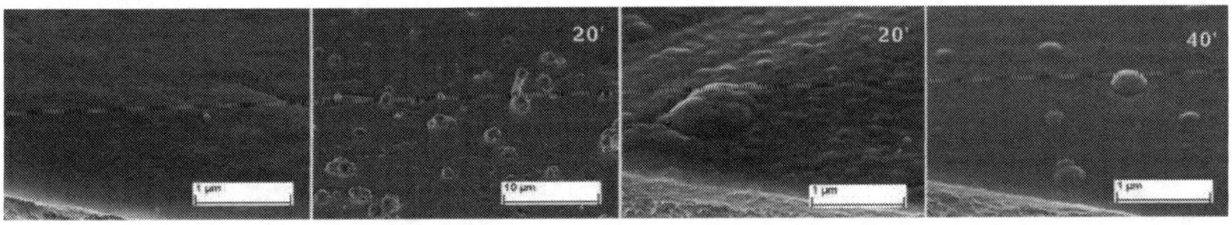

Figure 3. Plain polymer substrate S1 (left), and plasma coated (HMDSO 20 and 40 min; right). Magnification 5.00 and 50.00k.

Previous researches connect the topographic nature of the siloxane plasma films with the type of the precursor: HMDSO/O_2 mixture (rough) and HMDSO (even) [13-14]. Our results partially verify this conclusion; averaged EDX of 20 and 40 minutes-long deposited coatings reveals an increase of Si and O content with the time (Table 1). However, SEM-EDX mapping testifies about enhanced further presence of these two elements at the places of granular structures; according to this a mixed-mode mechanism of the depositions could be proposed here. Most probably, the nanogranules of interest (rather inorganic SiOx/SiOCH-type) firstly develop into the working volume of PECVD instrument, and subsequently fall onto the organic siloxane film, becoming further partially dissolved and immersed. It is still to be researched where the oxygen arises in the deposition process; with possible sources: adsorbed onto the porous substrate, or through specific plasma decomposition.

Table 1. EDX content of the siloxane coated vs. plain surface

	Material								
	Plain S1			HMDSO 20'			HMDSO 40'		
Element	Mass %	At %	SD	Mass %	At %	SD	Mass %	At %	SD
C	89.2	91.67	0.01	79.64	84.24	0.01	59.61	67.54	0.01
O	10.8	8.33	0.03	19.15	15.21	0.04	35.23	29.96	0.04
Si	-	-	-	1.21	0.55	0.01	5.16	2.5	0.01

Figure 4. EDX-mapping of Si and O content for the surface of 40' siloxane coating.

In order to investigate the composition of the surface plasma polymers, Fourier transform infrared spectroscopy was applied. Several typical for the siloxanes adsorptions are distinctively pronounced here (Figure 5): Si-O-Si asymmetric stretching vibration mode around 1035 cm^{-1} and bending vibration around 800 cm^{-1}; strong peaks of methyl groups (Si-$(CH_3)_3$ rocking vibration) at about 845 cm^{-1} and symmetric bending at 1265 cm^{-1}; surface Si-OH groups situated around 3300 cm^{-1} for some of the coatings [13, 15-16].

The characteristic absorption bands between 2847 and 2960 correspond to the asymmetric and symmetric stretching CH_x vibration (all sp^3-hybridized CH_3 and CH_2). All in all, we observe the pique at around 1687-1735 cm^{-1} to appear and rise simultaneously with the –OH vibration at around 3300 cm^{-1}, so we associate this quite strong for some coatings feature with bending mode $\delta_{(OH)}$ of surface hydroxyl groups. The intensity of Si-related absorptions generally increases with the enhancement of Si content of the coatings, which is not necessarily connected with the higher thickness of the layer, but also with the degree of "inorganication" and the presence of SiOx/SiOCH-inclusions.

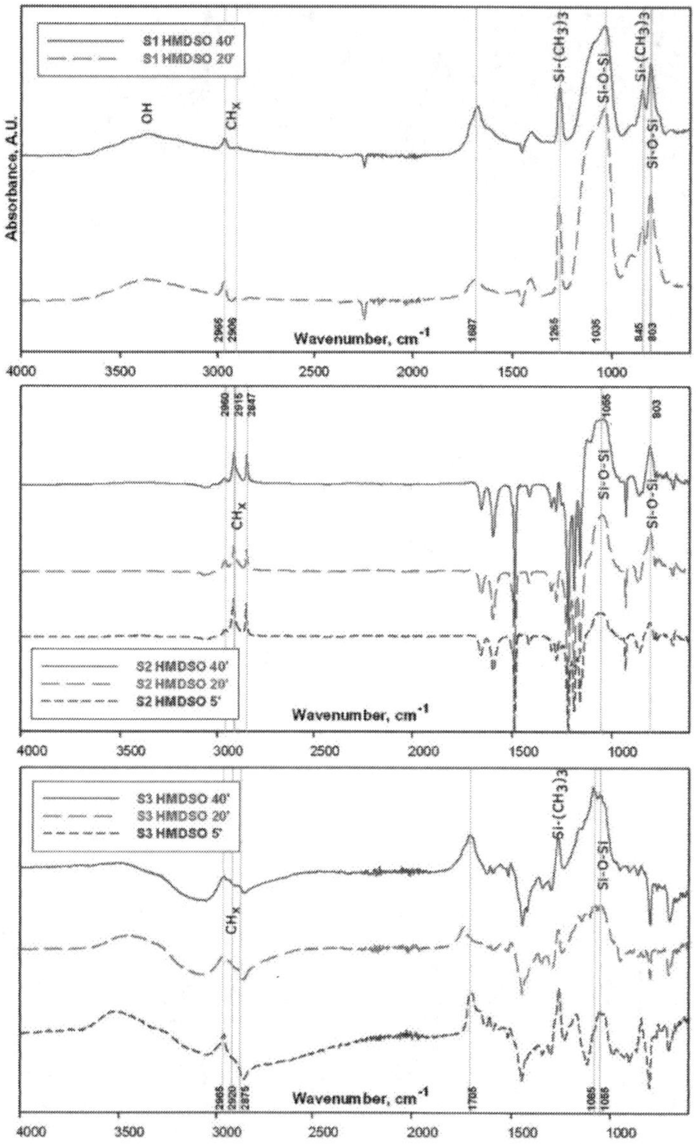

Figure 5. ATR-FTIR "net" spectra of siloxane coatings on substrates S1÷S3.

3.2. Pentane and Toluene plasma depositions

Aliphatic and aromatic hydrocarbons were also applied as plasma polymer-forming precursors, under conditions that we used for siloxane coatings. Figure 6 clearly displays the most prominent change following the plasma deposition of these types: namely enhancement of the peaks related to hydrocarbon groups, and especially the CH_x stretches around 2800-3000 cm^{-1}. These alterations are pronounced after short 5 min long PECVD treatment; longer coating times make them far stronger (Figure 6-upper). Looking at the pentane coating's FTIR spectra for substrate S1 (Figure 6-lower, top), we see it as a sole indicator of the surface change.

A comparison between the spectra of pentane and toluene coatings on S2 substrate reveals strong similarity (Figure 6-lower). Most likely this can be evidence for decomposition of aromatic hydrocarbons in PECVD plasma conditions, where the deposition of plain hydrocarbon chains and fragments occurred by the "pentane mechanism". Similar conclusions have been also drawn by other authors [7-8].

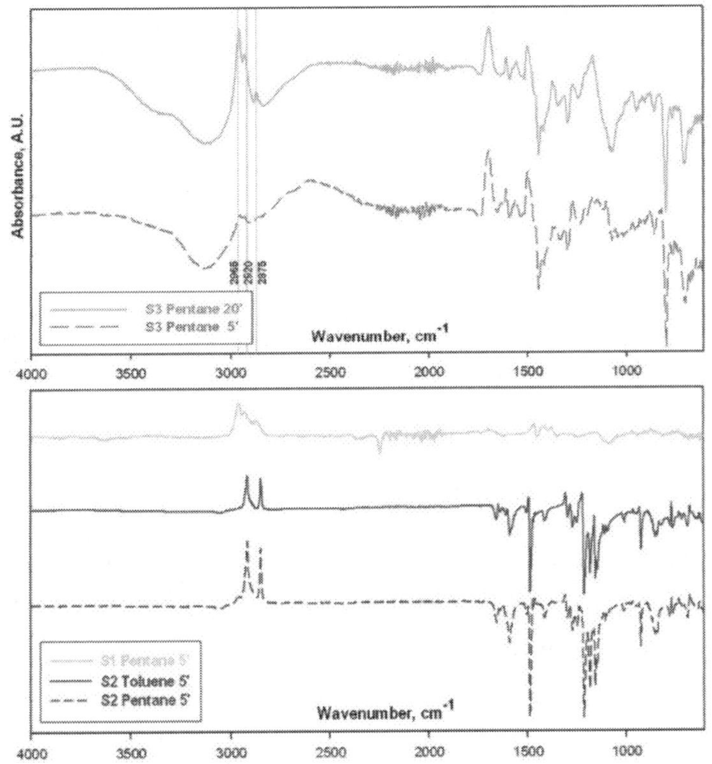

Figure 6. ATR-FTIR "net" spectra of pentane and toluene coatings on substrates S1÷S3.

Corresponding SEM images for pentane coatings show homogenous surface with no inclusions, although having more "rugged" appearance (Figure 7). Toluene-derived layers, similar to the pentane ones, demonstrate looser structure unlike the siloxanes, and sharp well defined interface can not be observed at the cross sections.

Figure 7. Plain polymer substrate S2 (left), and plasma coated (Pentane 5 and 20 min; right). The thicknesses of the coatings are 12 and 85 nm (quartz resonator). Magnification 100.00k.

4. Conclusions

With this work we studied the deposition and properties of three types of PECVD polymerised films, under different duration conditions. Low-temperature plasma processes rendered possible the application of three porous thermo-sensitive substrates. Surface coatings of siloxane type are dense and well differentiable on SEM microscopy images; they exhibit dual-mode mechanism of the deposition combined with possible hetero-inclusions within the layers. In addition, siloxane FTIR characterisations are substrate-dependent, which must be taken into account when analyzing. On the contrary, films derived by aliphatic hydrocarbons are rather loose, but uniform polymer deposits, as found also by [10] for methane. Although having some specifics, toluene-based films show close proximity with the pentane-derived ones.

Acknowledgements
REGPOT-2012-2013-1 NMP INERA Project is kindly acknowledged.

References

[1] Pierson H 1992 *Handbook of chemical vapor deposition (CVD), principles, technology, and applications* (Park Ridge, NJ: Noyes) p 436

[2] Kumar M and Ando Y 2010 Chemical vapor deposition of carbon nanotubes: A review on growth mechanism and mass production *Journal of Nanoscience and Nanotechnology* **10** 3739-3758

[3] Martin PM, 2009 *Handbook of Deposition Technologies for Films and Coatings Science, Applications and Technology* (Norwich, N.Y.: Elsevier) 936 pp

[4] Alf M, Asatekin A, Barr M et al. 2010 Chemical vapor deposition of conformal, functional, and responsive polymer films *Advanced Materials* **22** 1993-2027

[5] Grüniger A, Bieder A, Sonnenfeld A et al. 2006 Influence of film structure and composition on diffusion barrier performance of SiOx thin films deposited by PECVD *Surface and Coatings Technology* **200** (14-15) 4564-4571

[6] Chaiwong C, Rachtanapun P, Sarapirom S and Boonyawan D 2013 Plasma polymerization of hexamethyldisiloxane: Investigation of the effect of carrier gas related to the film properties, *Surface & Coatings Technology* **229** 12-17

[7] Asandulesa M, Topala I, Pohoata V et al. 2013 Chemically polymerization mechanism of aromatic compounds under atmospheric pressure plasma conditions *Plasma Processes and Polymers* **10** (5) 469-480

[8] Yu Y, Kim J, Cho S and Boo J 2003 Plasma-polymerized toluene films for corrosion inhibition in microelectronic devices *Surface and Coatings Technology* **162** 161-166

[9] Cho S and Boo J 2012 Characteristics of multilayered plasma-polymer thin films using toluene and TEOS by PECVD *Microelectronic Engineering* **89** 19-22

[10] Capote G, Freire F, Jacobsohn L and Mariotto G 2004 Amorphous hydrogenated carbon films deposited by PECVD in methane atmospheres highly diluted in argon: effect of the substrate temperature *Diamond and Related Materials* **13** 1454-1458

[11] Kuzminova A, Shelemin A, Petr M et al. 2013 Barrier Coatings on Polymeric Foils for Food Packaging, *WDS'13 Proc. of Contributed Papers, Part III* ed Jana Šafránková and Jiří Pavlů, (Prague: MATFYZPRESS) pp 128-133

[12] Bishop C 2015 *Vacuum Deposition onto Webs, Films and Foils* 3rd edition (Kidlington, Oxford: Elsevier) pp 104-105

[13] Boutamine M, Bellel A, Sahli S et al. 2014 Hexamethyldisiloxane thin films as sensitive coating for quartz crystal microbalance based volatile organic compounds sensors *Thin Solid Films* **552** 196-203

[14] Purohit V, Mielczarski E, Mielczarski J and Akesso L 2013 Evidence of coexistence of micro and nanoporosity of organo-silica polymeric films deposited on silicon by plasma deposition *Materials Chemistry and Physics* **141** (2-3) 602-612

[15] Lee K, Cho S, Jo S et al. 2008 Hydrophobic coating of Y_2O_3:Eu phosphors by using HMDSO/toluene plasma at low pressure and their wettability *Journal of Korean Physical Society* **53** (2) 631-635

[16] Bousquet A, Granier A, Goullet A and Landesman J 2006 Influence of plasma pulsing on the deposition kinetics and film structure in low pressure oxygen/hexamethyldisiloxane radiofrequency plasmas *Thin Solid Films* **514** 45-51

Dielectric and electrical characterization of 5CB nematic liquid crystal doped with silver nanoparticles

G B Hadjichristov[1,*], Y G Marinov[1], A G Petrov[1], L Marino[2] and N Scaramuzza[2]
[1] Georgi Nadjakov Institute of Solid State Physics, Bulgarian Academy of Sciences, 72 Tzarigradsko Chaussee, Sofia 1784, Bulgaria
[2] Dipartimento di Fisica, Università della Calabria, Via Pietro Bucci, Cubo 31C, 87036 Rende (CS), Italy

E-mail: georgibh@issp.bas.bg

Abstract. Thin films (25 μm) of nanocomposites formed from nematic liquid crystal (LC) pentylcyano-biphenyl (5CB) doped with 0.5% wt. silver (Ag) nanospheres with a mean diameter ~ 10 nm were characterized by electrical measurements and dielectric spectroscopy in the frequency range from 1 mHz to 100 kHz. By using LC cells in which the electrodes are stripes of 1.2 mm width separated by a distance of 50 μm, it was possible to study the properties of Ag-5CB nematic nanocomposites by applying the external electric field either in the plane or perpendicular to the plane of the films. The results were compared with those obtained for undoped nematic 5CB in identical configurations of the experiments.

1. Introduction

The interplay between liquid crystals (LCs) and metal nanoparticles (NPs) mixed with LCs, i.e. metal-NPs/LC nanocomposites, is challenging as well as fascinating for creating novel functional materials [1-9]. Owing to their unique properties, the resulting hybrid materials hold great promise in many applications such as displays, optics, optoelectronics & electrooptics, sensors, nonvolatile memory and tunable metamaterials [10-20]. In recent years, the research on nanocomposites formed by the mixing of nematic LCs with NPs of noble metals have attracted considerable attention due to their interesting electrical, optical and electro-optical properties, useful for practical devices [21-26].

The focus of the study presented here is on the dielectrical and electrical characterization of nanocomposites formed from the room-temperature nematic LC pentylcyano-biphenyl (5CB) doped with 0.5% wt. silver (Ag) NPs (nanospheres with a mean diameter ca. 10 nm). Nematic AgNPs/5CB nanocomposite films with a thickness of 25 μm were studied within LC cells with striped electrodes, and data were obtained by applying the external electric field either in the plane or perpendicular to the plane of the AgNPs/5CB nanocomposite layer. The dielectric spectroscopy technique was used, previously applied for the investigation of continuous LC layers [27-29] and soft-solid nanocomposite systems [18,30].

* To whom any correspondence should be addressed.

Journal of Physics: Conference Series **682** (2016) 012015 doi:10.1088/1742-6596/682/1/012015

2. Experimental

The nanocomposites studied here were formed from the room-temperature nematic LC pentylcyano-biphenyl (5CB) doped with a small amount (0.5 wt.%) of AgNPs (nanospheres with a mean diameter of about 10 nm from PlasmaChem GmbH). The nematogen 5CB (obtained from Merck) was used without any special purification. At room temperature, 5CB has a stable nematic phase. This compound exhibits, on heating, the nominal phase sequence: Crystal–N (22.5° C), N–I (35.3° C), as reported by the manufacturer. 5CB has a relatively large positive dielectric anisotropy ($\Delta\varepsilon \sim 10$ at 20° C and frequency 1 kHz [31]).

The nanocomposite AgNPs/5CB mixture in the isotropic phase was injected by capillary forces into planar LC cells. They were assembled with commercial parallel-oriented patterned glass plates for LC cells manufactured by E.H.C Co., Ltd.,Tokyo, Japan, for LC cell type with a trade name FF(88) (figure 1 a). The parallel-rubbed glasses (1.1 mm-thick) were coated by the manufacturer with ultrathin (20~40 nm) conductive layer of indium-tin-oxide (ITO) with a resistance of 100 Ω $^{-1}$, and polyimide LX-1400 (Hitachi-KASEI) overcoating with a thickness to within ~20 nm, rubbed with a polyester fiber. The active area of the cells was 1 cm × 1 cm. For cell construction, Mylar spacers with a thickness d = 25 μm were used. The specifics of the prepared LC cells are their stripe-like parallel ITO electrodes, as schematically shown in figure 1(b). In this arrangement, the electrode stripes had a length L = 1 cm and width W = 1.2 mm, and were separated from each other by a gap G = 50 μm. For the sake of reference measurements, identical cells were filled with pure 5CB. We have examined two supply schemas, hereafter called "configuration I" (figure 1 c) and "configuration II" where two neighbouring electrodes were used (figure 1 d). With the former configuration, the external electric field and the electric current flow are perpendicular to the film plane, whereas with the latter they are in-plane.

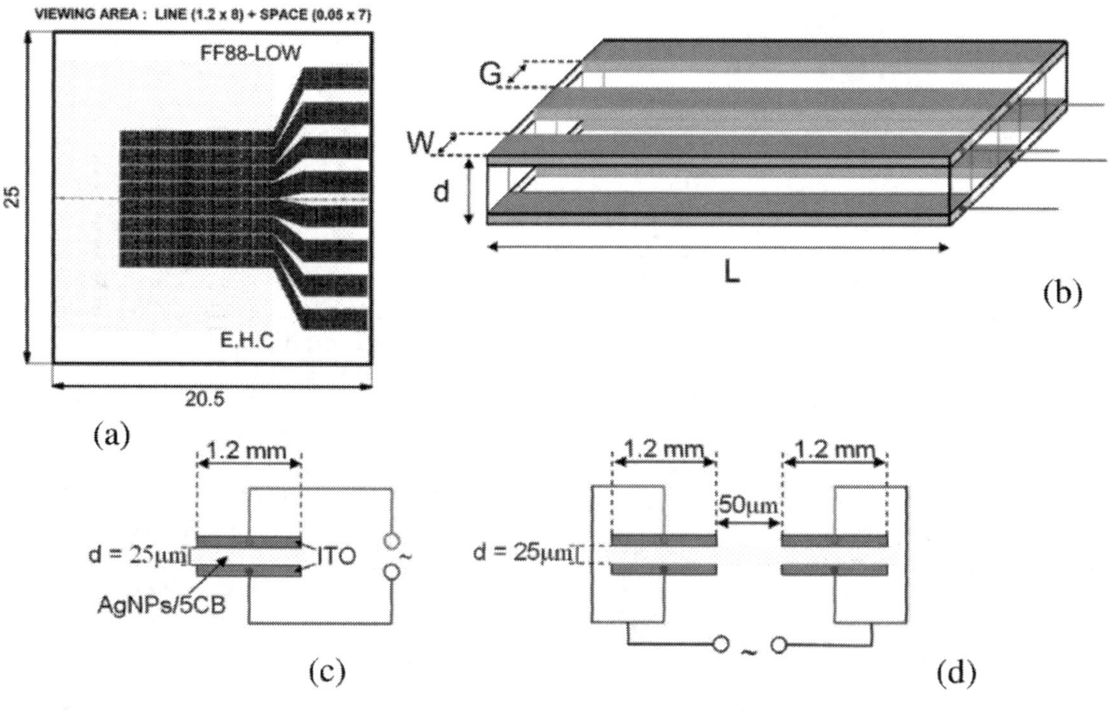

Figure 1. (a) ITO coated glass plate FF88; (b) Schematic view of a LC cell with layered stripe electrodes – ITO nanolayers – the area colored in gray. The Ohmic contacts and the wires of two segments are sketched (on the left). The supply schemas examined here: "configuration I" (c) and "configuration II" (d).

The electrical response of AgNPs/5CB nanocomposite films was studied by direct current (DC) and alternating current (AC) measurements. The DC experiments were carried out by using a computer-controlled Keithley 617 Programable Electrometer with the capability of measuring resistance up to 10^{15} Ω. The frequency response of the films was investigated by a phase-sensitive lock-in technique. For this purpose, a sinusoidal AC voltage from the built-in generator of a Stanford Research Systems SR830 DSP lock-in amplifier was applied to the ITO electrodes of the cells. The sampling signal was acquired by a computer that controls the frequency sweep, in our case in the range 1 Hz – 10 kHz. For the recording of frequency spectra of the amplitude of the electric current via AgNPs/5CB nanocomposite films, the interval between the data acquirement was equal to 4 s (the same applies to the data obtained in the course of the DC measurements).

The characterization of the AgNPs/5CB nanocomposite film by dielectric spectroscopy was performed by the apparatus described in [28,29]. Raw data for the complex impedance were obtained by use of a computer-controlled system: EG&G Princeton Applied Research Potentiostat/Galvanostat Model 273A coupled with EG&G Princeton Applied Research Model 5210 Lock-In Amplifier controlled by the impedance software M398 in order to acquire the real and imaginary parts of impedance in the frequency range of 1 mHz – 10 kHz. In this case, the probe voltage was 0.5 V_{rms} in the sinusoidal waveform. The corresponding electric field strength was smaller than the Freedericksz transition threshold of the 5CB LC. Each measurement was repeated at least 3 times in order to verify the values of the measured dielectric permittivity. The measurements were carried out under ambient environment at room temperature.

3. Results and Discussion

By optical microscopy, homeotropic orientation was observed for the AgNPs/5CB nanocomposite films in the LC cells considered here, in contrast to the planar alignment for pure 5CB in identical cells. Clearly, this is due to the NPs dopants that change the anchoring conditions.

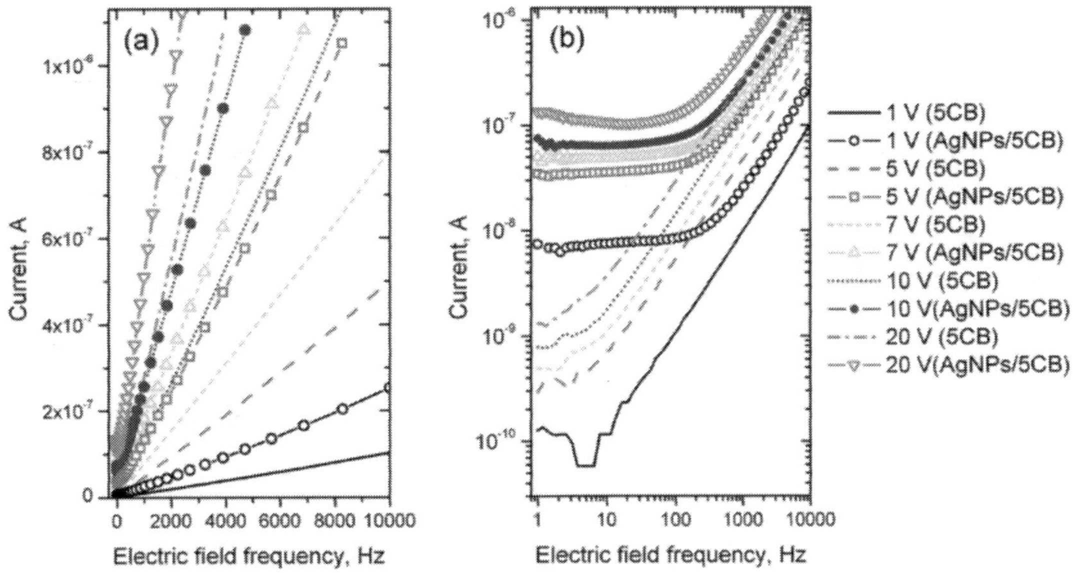

Figure 2. (a) The amplitude of the complex electric current versus the frequency of the applied AC field, as measured for the AgNPs/5CB composite (symbols) and pure 5CB (lines) in identical cells with stripe electrodes with the configuration I under identical experimental conditions. Data obtained at various voltages denoted on the left; (b) The same with a log-log scale.

Figure 2 presents results from measurements consisting in determining the frequency dependent electric current flowing through the AgNPs/5CB nanocomposite films when the electric field was orthogonally applied to the films (configuration I, see figure 1 c). Due to the presence of metal NPs, the electrical conductivity of the AgNPs/5CB nanocomposite is enhanced as compared to that of the undoped 5CB. This is reasonable and is in accordance with other similar results, e.g. the significant increase of the ionic conductivity in the case of composites of AuNPs with 5CB, compared to that of the pure 5CB sample [13,14]. At the applied voltages in the range from 1 V to 20 V, we observed at this measurement scheme an enhancement by a factor of ~ 2.5 (figure 2 a) and about two orders of magnitude in the region of lower frequency (best visible in figure 2 b). This effect was confirmed by supplementary DC measurements (figure 3).

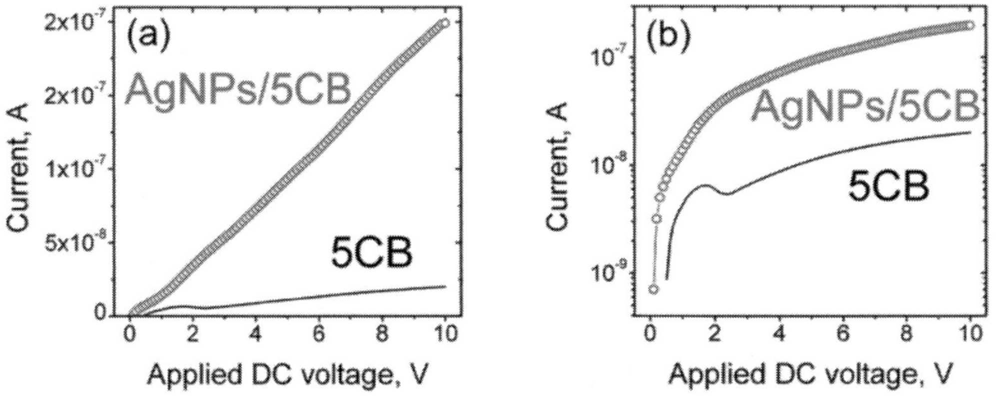

Figure 3. (a) DC current-voltage characteristics of AgNPs/5CB composite (circles) and pure 5CB (line) in identical cells with stripe electrodes for the configuration I under identical experimental conditions. (b) The same with a log-log scale.

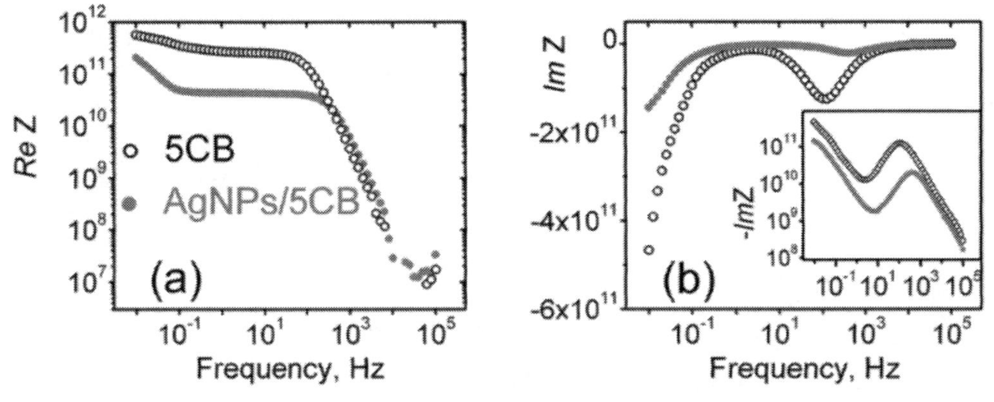

Figure 4. Real (a) and imaginary (b) parts of complex impedance measured for AgNPs/5CB composite (solid circles) and pure 5CB (open circles) under identical experimental conditions (at the configuration I) as a function of the frequency of the applied electric field. In both cases, the temperature was 25° C.

Figure 4 reports the impedance spectra recorded for AgNPs/5CB composite and pure 5CB under identical experimental conditions (with the configuration I). In principle, from the acquired real (*ReZ*) and imaginary (*ImZ*) parts of the complex impedance, the frequency-dependent complex dielectric function $\varepsilon^* = \varepsilon' - i\,\varepsilon''$ can be obtained by approaches described in the literature, e.g., by the procedure used in Refs. [27-29]. However, the exact determination of the real (ε') and imaginary (ε'') parts of ε^* in our case is a rather complicated task, especially for the geometry of configuration II. The problem is not easy to solve because the expressions that are normally used to derive ε' and ε'' are valid only under the assumption of a capacitor-like scheme with plane and parallel faces/electrodes and a homogenous electric field. In our system this is not absolutely true, in particular because the electric field can not be considered to be constant and uniform. A problem arises also from the neighboring LC volumes which have dielectric characteristics equal to those of the volume just affected by the applied electric field, and are also electrically anisotropic.

It is interesting to note that the presence of AgNPs in the LC 5CB involves a decrease of *ReZ* at low frequencies, as is expected from the introduction of conductive NPs, while at higher frequencies (above 100 Hz in our case) the behaviors tend to overlap due to the fact that the mobility of AgNPs in 5CB LC which is a viscous medium, tends to cancel. This phenomenon, however, does not affect the dielectric permittivity since this physical characteristic of the material depends on both the spontaneous and induced polarization in the system. As for the frequency behavior of *ImZ*, with reference to the data obtained for the 5CB LC, one may notice a usual behavior in LCs (see, for example [32]) showing that the low-frequency data are affected due to ionic conductance, whereas the high-frequency data are affected by the ITO resistance. As already implied by the *ReZ* data, the doping of 5CB with AgNPs does affect the *ImZ* even in the central part of the frequency interval measured, with a shift of the peak of the response towards the higher frequencies.

Although the above deviations result in complications and restrictions, configuration I can still be used for a rough estimation (within some uncertainty limits) of the component ε'' relevant to the dielectric loss, that is less affected by the problem described above. That is why we will concentrate our attention on this quantity. The data for dielectric permittivity was fitted by the Havriliak-Negami (H-N) function, with shape parameters (a and b) accounting for the asymmetry and broadness of the dielectric dispersion curve, according to the well established model [27,30,33] with the additions of a conductivity contribution, present at lower frequencies. The results can be compared with those obtained for undoped 5CB in identical cells keeping the same geometrical and other conditions of work. As an example, figure 5 presents the frequency-dependent ε''calculated for AgNPs/5CB composite and pure 5CB by use of expression [33,34]:

$$\varepsilon''(f) = \frac{\sigma}{\varepsilon_0 (2\pi f)^n} + \mathrm{Im}\left\{ \frac{\Delta\varepsilon}{\left[1 + i\, f/(f_0)^a\right]^{\frac{b}{a}}} \right\}, \qquad (1)$$

where f is the frequency of the applied electric field, σ is the DC ionic conductivity of the system and takes into account the free charges diffusion, f_0 is the relaxation frequency, ε_0 is the vacuum permittivity, and $\Delta\varepsilon = \varepsilon_s - \varepsilon_\infty$ is the strength of the dielectric mode. The fit parameters a and b are power indexes that define the low- and high-frequency limits of ε'', respectively). Table 1 summarizes the values of parameters corresponding to the best fit of ε'' by this model. As seen from figure 5, the calculated value of $\varepsilon''$of AgNPs/5CB nanocomposite was slightly higher than that for undoped 5CB (by a factor < 2) and is shifted towards the higher frequency.

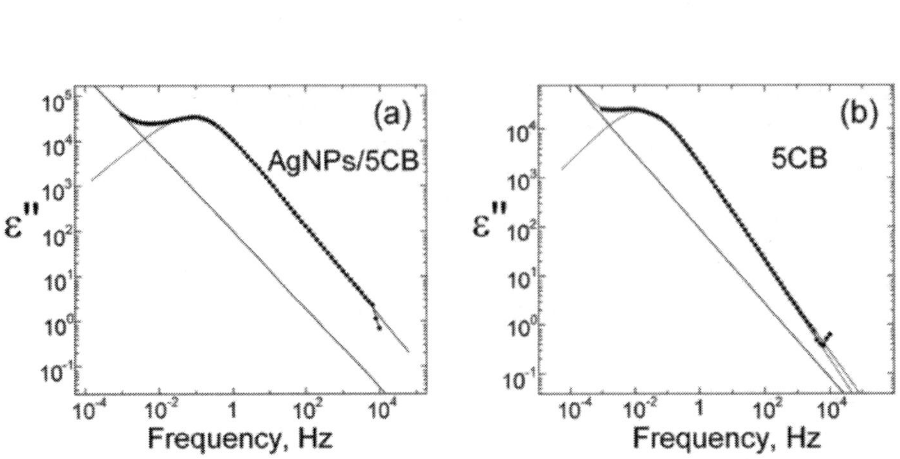

Figure 5. Comparison of the frequency-dependent imaginary part of dielectric permittivity for AgNPs/5CB composite (a) and pure 5CB (a) derived from data in figure 4 by the procedure used in Refs. [27-29]. The solid lines represent respectively the fit obtained by Eq. (1) (red line), the conductive part of Eq. (1) (black line) and the H-N function of Eq. (1) (green line).

Table 1. Best-fit parameters for ε'' (data from figure 5) as modelled by Eq. (1).

Parameters	$\log(f_0)$ Hz	$\Delta\varepsilon$	a	b	$\log(\sigma)$ [Si/cm]	n	$\log(f_{max})$ Hz
5CB	-0.91	101274	0.68	1	-10.37	0.86	-1.13
AgNPs/5CB	-1.56	70400.8	0.65	1	-10.49	0.76	-1.82

Using data from impedance measurements, one can experimentally assess and compare the electrical characteristics of AgNPs/5CB nanocomposite films by applying the external electric field either perpendicularly to the plane of the film (configuration I, see figure 1 c) or in the plane (configuration II, see figure 1 d). As an example, figure 6 shows the Cole-Cole plot (-*Im*Z vs *Re*Z) and the graphs of the loss angles (expressed in radians) at both geometries relevant to the corresponding components of the complex electric current in both orthogonal directions. From these graphs it is possibly to clearly see the difference in the electric behavior, between the AgNPs/5CB nanocomposite and the undoped LC. These characteristics indirectly indicate that the doping with AgNPs does strongly change both the conductivity and the characteristic frequencies of the system. Furthermore, the results obtained for the two orthogonal directions are completely different. This needs an additional thorough analysis, considering the actual geometry (work in progress)

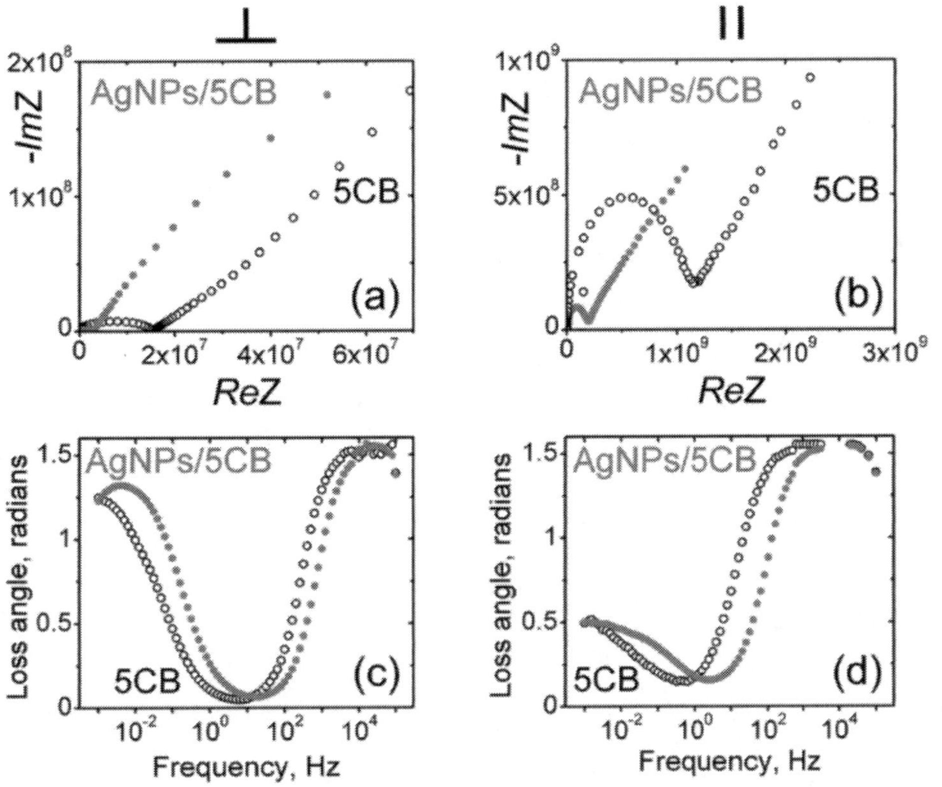

Figure 6. Cole-Cole plots (up) and frequency dependence of the loss angle (down) calculated for AgNPs/5CB nanocomposite film (solid circles) and undoped 5CB (open circles). Data obtained for the component of the electric current (a, c) perpendicularly to the film, or for the in-plane (b, d) current (i.e. at configuration I or configuration II, respectively).

4. Conclusion

We have studied the change of the electrical and dielectrical properties of room-temperature nematic pentylcyanobiphenyl (5CB) when this LC is mixed with silver nanoparticles (AgNPs) at relatively low concentration (0.5 wt.%). The effect from the nano-dopants was significant. With the electrode configuration employed here, the dielectric properties of the AgNPs/5CB nanocomposite films were characterized for two configurations of the electric current: in-plane or perpendicular. Because the dielectric permittivity and dielectric loss are the key parameters that determine the performance of electric and electro-optic devices, the results reported are useful for evaluating the feasibility of the considered nanocomposites for practical applications. Further, they can also be useful for the development of multi-segmented electronic devices for nematic-based soft electronics. Future work will be performed to calculate the dielectric permittivity in regards to configuration II.

Acknowledgments

The authors acknowledge the INERA EU project Research Potential "Research and Innovation Capacity Strengthening of ISSP-BAS in Multifunctional Nanostructures" (FP7 REGPOT-2012-2013-1) support for their participation at the LNN2015 conference. Work supported by the research project DFNI-TO2/18 "Nanostructured liquid crystals for tunable photonic devices" by the Ministry of Education and Science, National Science Fund of Bulgaria.

References

[1] Qi H, Kinkead B and Hegmann T 2008 *Adv. Funct. Mater.* **18** 212
[2] Qi H and Hegmann T 2011 *Liq. Cryst. Today* **20** 102
[3] Jamil M, Ahmad F, Rhee J T and Jeon Y J 2011 *Curr. Sci.* **101** 1544
[4] Lapanik A, Rudzki A, Kinkead B, Qi H, Hegmann T and Haase W 2012 *Soft Matter* **8** 8722
[5] Milette J, Cowling S J, Toader V, Lavigne C, Saez I M, Lennox R B, Goodby J W and Reven L 2012 *Soft Matter* **8** 173
[6] Lysenko D, Ouskova E, Ksondzyk S, Reshetnyak V, Cseh L, Mehl G H, Reznikov Y 2012 *Eur. Phys. J. E* **35** 33
[7] Infusino M, De Luca A, Ciuchi F, Ionescu A, Scaramuzza N and Strangi G 2013 *Mol. Cryst. Liq. Cryst.* **572** 59
[8] Choudhary A, Singh G and Biradar A M 2014 *Nanoscale* **6** 7743
[9] Xue C and Li Q 2014 Liquid Crystal-Gold Nanoparticle Hybrid Materials *Nanoscience with Liquid Crystals: From Self-Organized Nanostructures to Applications*, ed Q Li (Heidelberg, Springer-Verlag) chapter 4 pp 101–134
[10] Pratibha R, Park K, Smalyukh I I and Park W 2009 *Opt. Express* **17** 19459
[11] Li J, Ma Y, Gu Y, Khoo I C and Gong Q 2011 *Appl. Phys. Lett.* **98** 213101
[12] Vardanyan K K, Palazzo E D and Walton R D 2011 *Liq. Cryst.* **38** 709
[13] Prasad S K, Sandhya K L, Nair G G, Hiremath U S, Yelamaggad C V and Sampath S 2006 *Liq. Cryst.* **33** 1121
[14] Pandey A S, Dhar R, Kumar S and Dabrowski R 2011 *Liq. Cryst.* **38** 115
[15] Ha Y S, Kim H J, Park H G and Seo D S 2012 *Opt. Express* **20** 6448
[16] Fang P P, Chen S, Deng H, Scanlon M D, Gumy F, Lee H J, Momotenko D, Amstutz V, Cortés-Salazar F, Pereira C M, Yang Z and Girault H H 2013 *ACS Nano* 7 9241
[17] Acreman A, Kaczmarek M and D'Alessandro G 2014 *Phys. Rev. E* **90** 012504
[18] Marino L, Marino S, Wang D, Bruno E and Scaramuzza N 2014 *Soft Matter* **10** 3842
[19] Scharf T 2015 Self-organized plasmonic nanomaterials based on liquid crystals and metal nanoparticles *Active Plasmonic Nanomaterials* ed L De Sio (Boca Raton, Pan Stanford Publishing, CRC Press, Taylor & Francis Group) chapter 8 pp 245-279
[20] Zhang Y, Liu Q, Mundoor H, Yuan Y and Smalyukh I I 2015 *ACS Nano* **9** 3097
[21] Qi H and Hegmann T 2006 *J. Mater. Chem.* **16** 4197
[22] Khatua S, Manna P, Chang W S, Tcherniak A, Friedlander E, Zubarev E R and Link S 2010 *J. Phys. Chem.* ***114*** 7251
[23] Qi H and Hegmann T 2009 *ACS Appl. Mater. Interf.* **1** 1731
[24] Vardanyan K K, Sita D M, Walton R D, Saidel W M and Jones K M 2013 *RSC Adv.* **3** 259
[25] Hadjichristov G B, Marinov Y G, Petrov A G, Bruno E, Marino L and Scaramuzza N 2014 *J. Appl. Phys.* **115** 083107
[26] Hadjichristov G B, Marinov Y G, Petrov A G, Bruno E, Marino L and Scaramuzza N 2015 *Mol. Cryst. Liq. Cryst.* **610** 135
[27] Marino L, Ionescu A T, Marino S and Scaramuzza N 2012 *J. Appl. Phys.* **112** 114113
[28] Marino L, Ionescu A T, Pucci D, Marino S, Bruno E and Scaramuzza N 2014 *J. Appl. Phys.* **116** 114101
[29] Carbone F, Mazzulla A, Ciuchi F and Scaramuzza N 2015 *Europ. Phys. J. Plus* **130** 151
[30] Infusino M, De Luca A, Ciuchi F, Ionescu A, Scaramuzza N and Strangi G 2014 *J. Mater. Sci.* **49** 1805
[31] Ratna B R and Shashidhar R 1976 *Pramana* **6** 278
[32] Abderrahmen A, Romdhane F F, Ouada H B and Gharbi A 2008 *Sci. Technol. Adv. Mater.* **9** 025001
[33] Havriliak S and Negami S 1967 *Polymer* **8** 161
[34] Wübbenhorst M and Van Turnhout J 2002 *J. Non-Cryst Solids* **305** 40

Electrical properties of foamed polypropylene/carbon black composites

M Iliev[1], G Kotzev[2] and V Vulchev[1]

[1]Sofia University, 5, James Bourchier Blvd., 1164 Sofia
[2]Institute of Mechanics, Bulgarian Academy of Sciences, Acad. G. Bonchev St., Bl. 1, 1113 Sofia

E-mail: ozo@phys.uni-sofia.bg

Abstract. Polypropylene composites containing carbon black fillers were produced by vibration assisted extrusion process. Solid (unfoamed) composite samples were molded by conventional injection molding method, while structural foams were molded by a low pressure process. The foamed samples were evidenced to have a solid skin-foamed core structure which main parameters were found to depend on the quantity of material injected in the mold. The average bubbles' sizes and their distribution were investigated by scanning electron microscopy. It is established that the conductivity of the foamed samples gradually decreases when reducing the sample density. Nevertheless, the conductivity is found to be lower than the conductivity of the unfoamed samples both being of the same order. The flexural properties of the composites were studied and the results were discussed in the context of the structure parameters of the foamed samples.

1. Introduction

Polymers are known to be principally insulating materials. Their electrical conductivity can be enhanced by the addition of electrically conductive carbon fillers. The relationship of conductivity versus filler concentration clearly demonstrates a relatively narrow filler concentration range where a small increase in its content results in a drastic increase of the conductivity. This point called electrical percolation threshold denotes formation of a filler network that enables successful and faster conveyance of electrons throughout the samples. Further increase of the carbon filler content causes only slight increase of the electrical conductivity till a fixed value, i.e. reaching a plateau [1]. According to the conducting mechanism, to enable conductive pathways through composites the filler particles do not necessarily need any physical contact, because a small gap of several nanometers in width between them is sufficient to cause electron jumping [2]. The types of polymer matrix, conductive filler and processing parameters are main factors that influence the percolation behavior. Changing conditions for the preparation of particular polymer composites one could achieve better or worse filler dispersion or lower or higher filler concentration at the percolation point, respectively [2-5]. Lots of different methods for achieving better distribution of filler particles have been tried so far; however their successful dispersion still remains a challenge. Melt compounding is an efficient, rapid and environmentally friendly method for a good-quality dispersion of fillers in the polymer matrix.

Most mixers and plastic processing machines operate mainly under shear conditions. This way, any increase of the shear intensity in extruders could be a rational way for improving their mixing ability. Efficient melt compounding can be realized using an extrusion mixing die where the molten polymer composites will be subjected to continuously changing shear and extensional deformations created by vibrations of a moveable mandrel [6]. The polypropylene/carbon black (PP/CB) composites fabricated by this vibration assisted processing method exhibit better filler dispersion and percolation point at lower filler concentration than the composites compounded by conventional extrusion [7].

From a practical point of view it is important to produce above mentioned polypropylene composites with a possibly lower density. Structural foams are important achievements of such polymer processing. In contrast to conventional polymer foams their densities are relatively high (usually in the range of 0.6-0.9 the nominal polymer density). The structural foams are known to consist of three basic elements, unfoamed solid smooth skin, foamed core and transitional layer. This type of structure ensures a relatively greater bending strength in comparison with such strength of the unfoamed plastics.

Since foamed PP/CB composites would take place in advanced technological goods, for example in antistatic products or as a part in electromagnetic shields, it is important to know their electrical properties in comparison with the properties of unfoamed PP composites.

2. Materials and testing methods

2.1. Materials

Foamed PP/carbon filler composites were used as objects in current investigation. The polymer used is isotactic PP „Buplen“ 6531 produced by Lukoil Neftochim Co., Bulgaria, having melt flow index MFI (230, 2.16) = 4,1 *g/10 min*. PP composites were filled with a carbon black filler, Ketjen black EC-300J produced by Akzo Nobel Polymer Chemicals with specific surface area of 803 m^2/g measured by BET techniques, pore volume of 325 *ml/100 g* measured by dibuthylphthalate absorption and size of basic/primary carbonic particles of 30 - 100 *nm*. Azo-dicarbonamide type Genitron AC4 produced by Fisons Ltd. was used as a blowing agent, which decomposition temperature region falls between 200 and 230 °*C*.

2.2. Melt compounding and sample preparation,

Composite of 90 wt. % PP and 10 wt. % CB were fabricated using an extrusion system, containing a new mixing die, where the polymer melt was subjected to axial vibrations that resulted into improved dispersion of the filler in the polymer matrix [6, 7]. The melt compounding was carried out under the following process conditions: the temperature profile of the barrel was 170, 180, 190, 200, 205 ºC; screw speed of 20 rpm, vibration amplitude of 0.3 mm and vibration frequency of 10 Hz. After exiting the mixing die the extrudates were cooled and then pelletized using a Vespa 22/40 cutting machine. Plates with dimensions of 100×80×4 mm used for electrical and mechanical measurements were produced on KuASY 25×32/1 injection molding machine. Foamed samples were prepared by a low pressure injection molding process of the PP/CB composites containing 1 wt. % azodicarbonamide under the manufacturing conditions as follows: melt temperature 230 °C, mold temperature 20 °C, injection pressure 15 MPa, packing time 5 s and cooling time 20 s. Mold filling during low pressure injection molding was two-stage process. The first stage was partial filling of molds with “short-shot” of polymer melt containing gaseous phase. The second stage was full filling of the molds realized by foaming of the polymer melt after injection shot.

Table 1. Relationship of the overall density and effective cross-section *vs.* the short-shot.

Short-shot, *g*	21,9	24,5	26,4	28,3	28,9
Overall density, *kg/m*^{3}	670	750	810	870	890
Effective cross-section, %	72,8	82,6	83,8	84,9	85,4

Varying the short-shot, different sample densities were obtained (table 1). Unfoamed samples obtained by conventional injection molding process were used as reference.

2.3. Characterization

Scanning electron microscopy (SEM) observations were carried out in a secondary electron flow regime at a 10 *kV* accelerating voltage on a JEOL 5510 microscope. The examined cross sections were obtained by cutting off the samples using a metal knife of a microtome device type Reinchert, Austria and coating them with 2 *nm* layer of Au in argon atmosphere. Bubble size and bubble area occupied the cross-section of the sample were determined from the SEM images.

The DC electrical resistivity was measured at ambient temperature of 25 *°C* and humidity 50 % by the four-probe Van der Pauw technique [10]. This technique makes use of four isolated contacts on the boundary of an arbitrarily-shaped lamilar sample. In our case, the four probes were located in the four edges of the test samples. A total of eight measurements are made around the sample as shown in figure 1. Silver paste was used to ensure good contact of the sample with the copper electrodes.

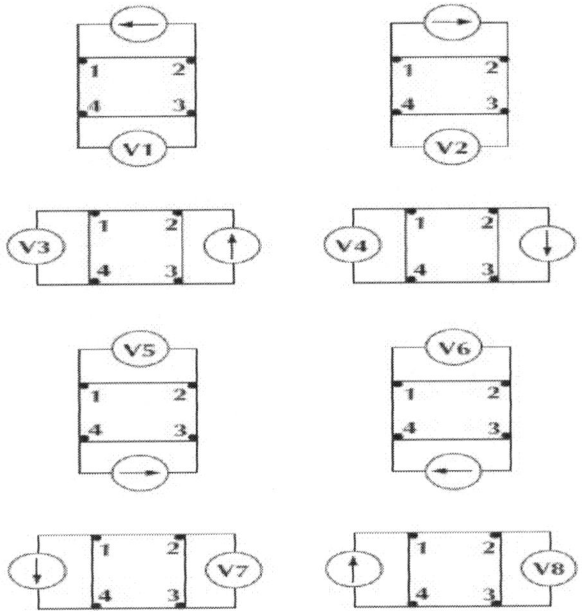

Figure 1. Van der Pauw resistivity measurements.

Flexural properties were determined according to EN ISO 178 at a test speed of 2 *mm/min.* and a temperature of 20 *°C*. Specimens with dimensions of 80×10×4 *mm* were cut from the plates and tested at a length of span between supports of 64 *mm*. Ten specimens were used for the measurements. The flexural strength, σ_{fM} and flexural strain at the flexural strength, ε_{fM} were calculated using following equations:

$$\sigma_{fM} = \frac{3FL}{2bh^2} \quad \text{and} \quad \varepsilon_{fM} = \frac{600lh}{L^2}, \tag{1}$$

where F is the maximum force applied during the test;
L is the span;
b is the width of the specimen;
h is the thickness of the specimen;
l is the deflection.

3. Results and discussion

3.1. Composites structure

From a macroscopic point of view the specimens produced by low pressure injection molding have a sandwich-like structure consisting of a solid skin and cellular core. SEM images in figure 2 show cross sections of the plates produced at varying short-shot. Parts of the cross-sections with dimension 4x4 *mm* located at the center of the plates demonstrate the different overall density of the specimens and different bubbles morphology of the core, respectively. When the overall density is relatively high (figures 2a, 2b) the bubble shape is close to spherical. In contrast, bubbles with highly extended shapes oriented along the flow direction are observed in the composites with overall density of 750 kg/m^3 (figure 2c) and 670 kg/m^3 (figure 2d). Another feature has to be noted for the sample of the lowest density: there can be clearly evidenced that in practice there is not any solid skin in its structure.
It is well-known that the resistivity of the sample is strongly dependent on its cross sectional area. That is why the effective cross sectional area, i. e. the full cross section reduced by the bubble area, was additionally calculated and presented in Table 1. From the table it can be seen that the bigger the short-shot, the higher the overall density and the effective cross-section.

3.2. Electrical properties

Our earlier experiments proved that PP/CB composites demonstrated typical relationship between resistivity and filler concentration [7, 8]. There was established the percolation threshold to depend on the melt flow index of the polymer that for the PP6531 should lie in the region between 8 and 9 wt. %. To ensure that experiments would proceed above percolation threshold foamed samples at CB concentration of 10 *wt. %* are used in current work. The volume resistivity of the samples, r measured by the four-probe Van der Pauw technique and conductivity of the samples, s calculated using equation (2) are presented in Table 2.

$$s = \frac{1}{r}. \tag{2}$$

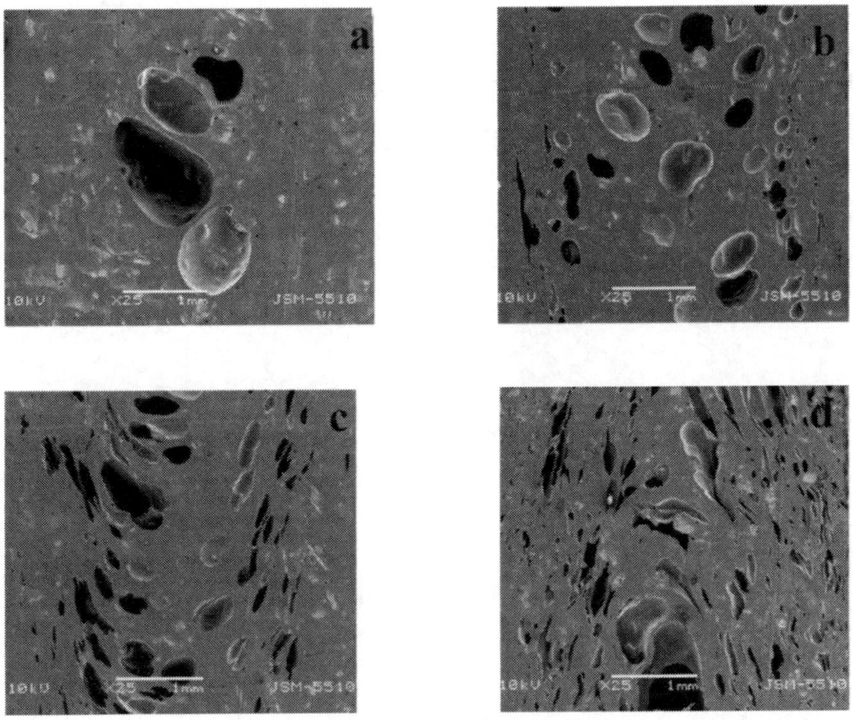

a – 890 *kg/m³*; **b** – 810 *kg/m³*; **c** – 750 *kg/m³*; **d** – 670 *kg/m³*.

Table 2. Resistivity and conductivity of the samples.

Overall density, kg/m³	Resistivity, *Ωm*	Conductivity, *S/m*
670	0.80	1.25
750	0.48	2.08
810	0.53	1.89
870	0.49	2.04
890	0.39	2.56
980	0.39	2.56

As can be seen from the table, the lower the density, the lower the conductivity. A possible reason for such behavior can be the reduced effective cross section of the foamed samples. In contrast to the foamed core, the density of the sample solid skin is close to the value of the unfoamed composite this way including a key influence on the conductivity. Since the sample conductivity depends on its cross section we have referred to a relative conductivity as conductivity divided by the effective cross sectional area. Such way calculated values are plotted *vs.* overall density and are presented on the figure 3. As is seen from the figure they are close to the unfoamed samples conductivity. The sample with the lowest density is a sharp exception. Its relative conductivity is significantly lower than the relative conductivity of the other samples.

Journal of Physics: Conference Series **682** (2016) 012016 doi:10.1088/1742-6596/682/1/012016

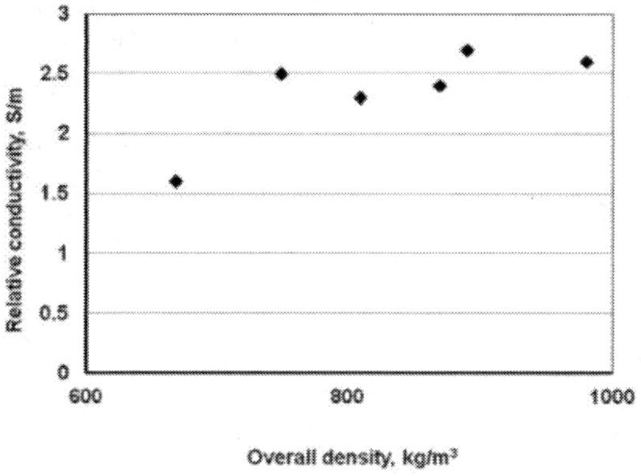

Figure 3. Relative conductivity *vs.* overall density of the samples.

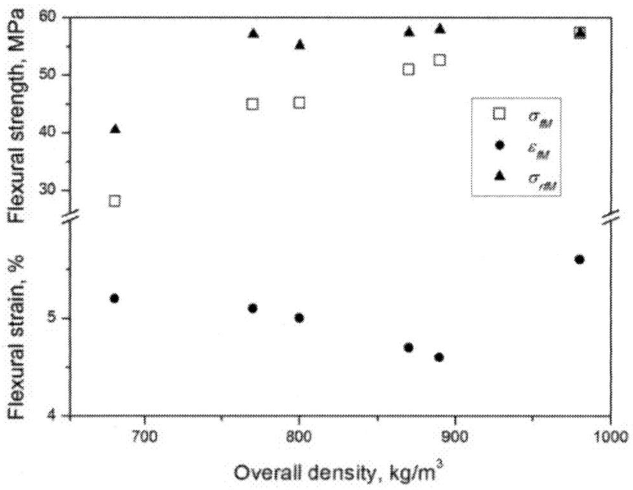

Figure 4. Flexural properties of the foamed samples.

3.3. *Flexural properties*

Some important mechanical properties enrich above presented electrical properties. In practice, during the use of moldings as constructive materials their flexural properties are the most valuable features one should examine. Such properties of foamed PP/CB composites are presented in the figure 4. It is seen that the flexural strength of foamed samples (curve 2 □) grows up with a growing overall density whilst their deformation (curve 1 ●) remains lower than that of the solid samples. Relative flexural strength calculation is a simple method used for comparing strengths of different density samples [9] following equation:

$$\sigma_{rfM} = \frac{\sigma_{fM}}{\rho / \rho_{comp}}, \quad (3)$$

where ρ is the foamed composite density, ρ_{comp} is the unfoamed composite density (in our case it is 980 kg/m^3). The values of calculated relative flexural strength are shown in figure 4, (curve 3 ▲). The structure of foamed samples and mainly existence of a skin affirm their strengths to be close or even higher than those of the unfoamed ones. Here, another exception appears again. The value of the relative flexural strength of sample with the lowest density not having any skin is too low.

4. Conclusions

The presented experimental results proved that PP/CB structural foams produced by the low pressure injection molding can be used for production of technical articles with antistatic and electro-conducting properties. The presence of a skin in studied structural foams leads to obtaining of better mechanical and electric parameters that are comparable to the properties of unfoamed products.

Acknowledgements.
The authors of this article would like to thank to INSTITUTE OF SOLID STATE PHYSICS - BAS, Project INERA/FP7-REGPOT-2012-2013-1 NMP (Research and Innovation Capacity Strengthening of ISSP-BAS in Multifunctional Nanostructures) and National Science Fund of MYES of Bulgaria Contract No. DO 02-202.

References

[1] Stauffer D and Aharnoy A 1991 *Introduction to Percolation Theory*. Taylor & Francis, London.
[2] Zhang W, Dehghani-Sanij A A and Blackburn R S 2007 Carbon based conductive polymer composites, *J Mater Sci*, **42**, 3408–3418.
[3] Du F, Scogna R C, Zhou W, Brand S, Fischer J E and Winey K I 2004 Nanotube Networks in Polymer Nanocomposites: Rheology and Electrical Conductivity. *Macromolecules*, **37**, 9048-9055.
[4] Chen G X, Li Y and Shimizu H 2007 Ultrahigh-shear processing for the preparation of polymer/carbon nanotube composites, *Carbon*, **45**, 2334–2340.
[5] Wakabayashi K, Brunner P J, Masuda J, Sheldon A, Hewlett S A, John M and Torkelson J M 2010 Polypropylene-graphite nanocomposites made by solid-state shear pulverization: Effects of significantly exfoliated, unmodified graphite content on physical, mechanical and electrical properties, *Polymer*, **51**, 5525-5531.
[6] Bulgarian Patent 2010 Reg.№ 110658. *Device for continuous mixing of polymer melts with fillers.*
[7] Kotzev G, Vulchev V, Djoumaliisky S, Natova M and Krastev R 2011 Electroconductivity and mechanical properties of melt compounded PP/CB composites, *Nanoscience & Nanotechnology*, **11**, 2017-220.
[8] Kotzev G K, Djoumaliisky S L, Natova M M, Vulchev V D and Krusteva E D May 2011 Comparative investigation of the properties of two types melt compounded PP/carbon black composites, 2nd *Int. Symp. Frontiers in polymer science*, 28 – 30, Lyon, France, P.2.050.
[9] Semerdjiev S 1982 *Introduction to structural foams*. SPS, Connecticut.
[10] Keithley Instruments Application Note Series Number 312 High Resistance Measurements.

Journal of Physics: Conference Series **682** (2016) 012017 doi:10.1088/1742-6596/682/1/012017

Simulation of light-induced degradation of µc-Si in a-Si/µc-Si tandem solar cells by the diode equivalent circuit

J A Weicht, F U Hamelmann, G Behrens

Bielefeld University of Applied Sciences, Artilleriestraße 9a, Bielefeld, Germany

JWeicht@fh-bielefeld.de

Abstract. Silicon-based thin film tandem solar cells consist of one amorphous (a-Si) and one microcrystalline (µc-Si) silicon solar cell. The Staebler - Wronski effect describes the light-induced degradation and temperature-dependent healing of defects of silicon-based solar thin film cells. The solar cell degradation depends strongly on operation temperature. Until now, only the light-induced degradation (LID) of the amorphous layer was examined in a-Si/µc-Si solar cells. The LID is also observed in µc-Si single function solar cells. In our work we show the influence of the light-induced degradation of the µc-Si layer on the diode equivalent circuit. The current-voltage-curves (I-V-curves) for the initial state of a-Si/µc-Si modules are measured. Afterwards the cells are degraded under controlled conditions at constant temperature and constant irradiation. At fixed times the modules are measured at standard test conditions (STC) (AM1.5, 25°C cell temperature, 1000 W/m²) for controlling the status of LID. After the degradation the modules are annealed at dark conditions for several hours at 120°C. After the annealing the dangling bonds in the amorphous layer are healed, while the degradation of the µc-Si is still present, because the healing of defects in µc-Si solar cells needs longer time or higher temperatures. The solar cells are measured again at STC. With this laboratory measured I-V-curves we are able to separate the values of the diode model: series R_s and parallel resistance R_p, saturation current I_s and diode factor n.

1. Introduction

The Staebler - Wronski effect describes the degradation of silicon-based thin film solar cells. The a-Si solar cell degradation depends strongly on operating temperature [1]. Light-induced degradation (LID), based on the work of Staebler and Wronski, is also observed in a-Si/µc-Si solar cells because of their amorphous layer [2]. The LID is also observed in µc-Si single solar cells [3]. In our work we show the influence of the light-induced degradation of the µc-Si layer on the diode equivalent circuit.
The behaviour of the one- or two-diode model parameters is important to characterize the behaviour of silicon-based solar cells [4]. Until now the characterization of the a-Si/µc-Si solar cells is made by a one-/two-diode model. By this characterization, the tandem solar cells are defined as single-junction solar cell. This is difficult for characterization and improvement of tandem solar cells [4], because decreasing effects cannot be assigned to a sub cell.

The performance of a-Si/µc-Si thin film solar cells in different climatic regions depends on the operating temperature condition and actual module temperature [2]. During the light-induced degradation, the operating temperature determines the amount of created dangling bonds. The light-induced degradation in a-Si/µc-Si stabilizes after the first 400h at a degradation temperature of 50°C [2].

Journal of Physics: Conference Series **682** (2016) 012017 doi:10.1088/1742-6596/682/1/012017

This degradation is visible at standard test conditions (STC), with an irradiation of 1000W/m², a cell temperature of 25°C and spectral distribution of AM1.5 [4].

2. Method

The initial current-voltage-curves (I-V-curves) of a-Si/μc-Si mini-modules (size 37.95cm²) are measured with a sun simulator. Afterwards they are degraded under controlled conditions at constant temperatures and constant irradiation (1000W/m²). At fixed times the modules are measured at standard test conditions for controlling the status of LID.

During the light-soaking of the solar cells, the I-V-curves of the solar cells were measured: The sun simulator has an irradiation of 1000W/m², AM1.5 spectrum and the cells are measured at 25°C.

In μc-Si layer in tandem solar cells, the LID is also visible [5]. This effect is weaker and higher temperatures for annealing are needed to compare to the LID in a-Si [6]. To examine the LID of the μc-Si cell, the solar cells were annealed at 150°C for several hours after the degradation. The annealing sets the parameters of a-Si back to their initial values. The μc-Si solar cells stay degraded because the healing of defects in μc-Si solar cells needs longer time or higher temperatures [3]. After annealing the a-Si/μc-Si tandem cells, they reach about 99.5% of their initial value.

The parameter for each diode equivalent circuit was extracted by the I-V-curve and a Matlab/ Simulink tool based on reference [7]. For approximation the method of Byrd et al. [8] is used. The accuracy of the solutions of the parameters by the simulation is depending on the starting parameters and how much data is inserted in the approximation: The most effective way here is to include the different status for each solar cell in one simulation: initial (a-Si and μc-Si cell are initial), annealed (a-Si cell is initial and μc-Si cell is degraded) and degraded (a-Si cell and μc-Si cell are degraded) for each different temperature.

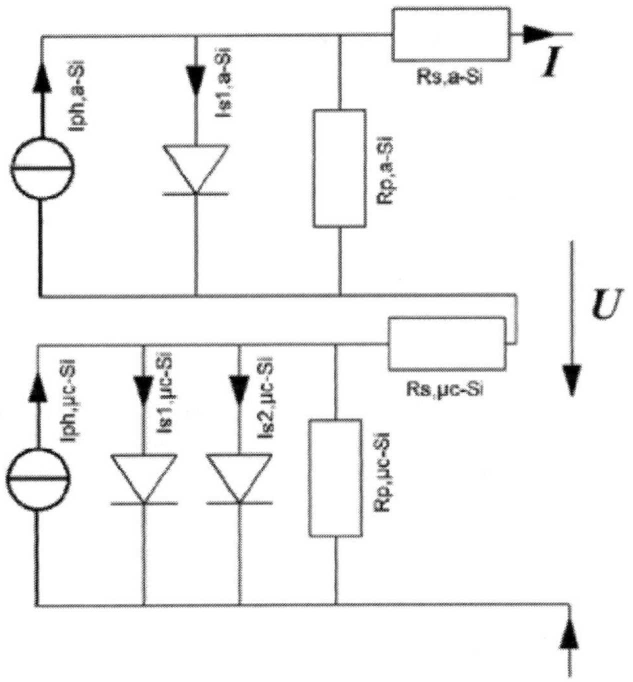

Figure 1. 3-Diode-Model: Equivalent circuit of a-Si/μc-Si tandem thin film solar cell.[11]

Journal of Physics: Conference Series **682** (2016) 012017 doi:10.1088/1742-6596/682/1/012017

The accuracy "acc" of the simulated data is defined as the metric distance between simulated (I_{sim} for simulated current, U_{sim} for simulated voltage) and measured values (I_{meas}: measured current, U_{meas}: measured voltage):

$$\text{acc} = \sqrt{(I_{meas} - I_{sim})^2 + (U_{meas} - U_{sim})^2} \tag{1}$$

The accuracy is supposed to be minimized. By changing the starting parameters, a new acc is calculated. The difference between the new and the starting acc is used to calculate the new parameter sets as described in [8]. The algorithm is stopped as soon the changing of the parameters sets is lower than the stopping criteria.

The a-Si and μc-Si cells are connected in series. Each of the cells has a current source ($I_{ph,a\text{-}Si}$, $I_{ph,\mu c\text{-}Si}$), serial resistance ($R_{s,a\text{-}Si}$, $R_{s,\mu c\text{-}Si}$), parallel resistance ($R_{p,a\text{-}Si}$, $R_{p,\mu c\text{-}Si}$) and at least one diode (saturation current $I_{s1,a\text{-}Si}$, $I_{s1,\mu c\text{-}Si}$, diode factor $n_{1,a\text{-}Si}$, $n_{1,\mu c\text{-}Si}$). As supposed in [9] the one diode-model gives satisfactory results for a-Si cells. The μc-Si layer has a second diode (saturation current $I_{s2,\mu c\text{-}Si}$, diode factor $n_{2,\mu c\text{-}Si}$) [10]. This method is comparable to the method to observe the LID in a-Si cells [11].

The temperature T, the Boltzmann constant k and elementary charge e are also part of the equivalent circuit (figure 1) and its equation (Equation 2-4):

$$I = \min\left(I_{a-Si}, I_{\mu c-Si}\right) \tag{2}$$

with

$$0 = I_{ph,a-Si} - I_{s1,a-Si}\left(\exp\left(e\frac{U - I_{a-Si}R_{s,a-Si}}{n_{1,a-Si}\,kT}\right) - 1\right) - \frac{U + I_{a-Si}R_{s,a-Si}}{R_{p,a-Si}} - I_{a-Si} \tag{3}$$

and

$$0 = I_{ph,\mu c-Si} - I_{s1,\mu c-Si}\left(\exp\left(e\frac{U - I_{\mu c-Si}R_{s,\mu c-Si}}{n_{1,\mu c-Si}\,kT}\right) - 1\right) - I_{s2,\mu c-Si}\left(\exp\left(e\frac{U - I_{\mu c-Si}R_{s,\mu c-Si}}{n_{2,\mu c-Si}\,kT}\right) - 1\right) - \frac{U + I_{\mu c-Si}R_{s,\mu c-Si}}{R_{p,\mu c-Si}} - I_{\mu c-Si} \tag{4}$$

3. Results

As expected, the tandem solar cells are degrading during the first 100h. After annealing, the solar cell does not reach its initial state. The fill factor is still decreased by 1% after annealing the sample at 120°C at several hours. Each parameter of the μc-Si layer shows a variation in the value between the initial and the degraded state. The photocurrent I_{ph} decreases up to 3% of the initial value, also the parallel resistance R_p,I_{S1} and n_1 decreases. The serial resistance R_s increases, attention should be paid on the relation, that the serial resistance of the μc-Si layer is smaller than of the a-Si layer: $R_{s,\mu c\text{-}Si} \approx 0.2 R_{s,a\text{-}Si}$. Also the saturation current I_{S2} and n_2 increases, as seen also in the LID in the one-diode model of a-Si solar cells [11].

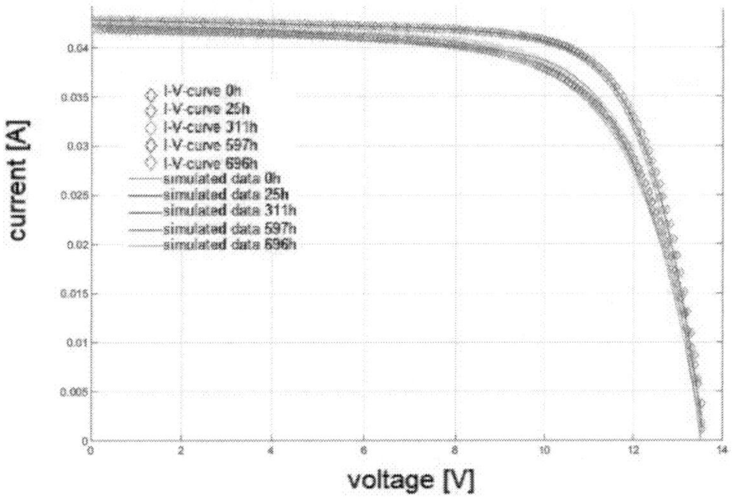

Figure 2. Measured (dots) and simulated (line) I-V-curve of a-Si/μc-Si tandem solar cells during light-induced degradation at an degradation temperature of 50°C.

The changing in the extracted parameters from the experimental data for the degradation can be seen in table 1:

Table 1. One example of the changing of simulating-parameters during LID of different degradation temperatures for the μc-Si layer.

Parameter	μc-Si
$I_{ph}/ I_{ph\_initial}$	0.97
$R_s/ R_{s\_initial}$	1.86
$R_p/ R_{p\_initial}$	0.80
$I_{s1}/ I_{s1\_initial}$	0.01
$n_1/ n_{1\_initial}$	0.93
$I_{s2}/ I_{s2\_initial}$	2.91
$n_2/ n_{2\_initial}$	1.20

4. Conclusions

In our work we observed the degradation in the μc-Si layer of tandem solar cells. This data was analysed by a simulation tool based on Matlab/ Simulink. By this simulation we can extract the parameters of the equivalent circuit for the μc-Si sub cell. By this extraction of these parameters we approach a model for the three-diode-model which was presented in this work.

By this modelling we show that the degradation temperatures have a relevant influence on each parameter of the "three-diode-model". Previous research work observed the tandem structure only as one cell [12] or only observed the LID in the a-Si layer of tandem solar cells [11]. Our method and extracted data will help to improve the innovation and modelling of thin film silicon-based tandem solar cells.

5. Acknowledgments

We acknowledge support from BMBF through Contact No. 03FH008I2 and the support from Ministry for Innovation, Science and Research of North Rhine-Westphalia.

References

[1] Staebler D and Wronski C R 1977 Reversible conductivity changes in discharge produced amorphous Si *Applied Physics Letters* **31** 292 - 294

[2] Weicht J A et al 2013 Light-induced degradation of a-Si/i-Si thin film solar cells under different climate conditions and its influence on performance *Proc. 28th EU PVSEC* (France: Paris) pp. 2542 - 2545

[3] Kleider J P et al 2001 Electronic and topographic properties of amorphous and microcrystalline silicon thin films *Thin Solid Films* **383** 57-60

[4] Chan D S H et al 1986 A comparative study of extraction methods for solar cell model parameters *Solid-State Electronics* **29** 329-337

[5] Shah A V et al 2003 Material and solar cell research in microcrystalline silicon *Solar Energy Materials and Solar Cells* **78** 469-491

[6] Xiangna L and Mingde X 1986"The Staebler-Wronski effect in microcrystalline silicon films *Chinese Physics Letters* **3** 73

[7] Gow J A and Manning C D 1999 Development of a Photovoltaic Array Model for Use in Power-Electronics Simulation Studies *IEEE Proc. of Electric Power Applications* **146** 193–200

[8] Byrd R H et al 2000 A Trust Region Method Based on Interior Point Techniques for Nonlinear Programming *Mathematical Programming* **89** 149–185

[9] Merten J et al 1998 Improved equivalent circuit and analytical model for amorphous silicon solar cells and modules *IEEE Transactions on Electron Devices* **45** 423-429

[10] Meier J et al 1994 Complete microcrystalline p-i-n solar cell—Crystalline or amorphous cell behavior *Applied Physics Letters* **65** 860-862

[11] Weicht J A et al 2015 Changing of the parameters of the "three-diode model" by light-induced degradation at different degradation temperatures of a-Si/μc-Si solar cells *Proc. 31st EU PVSEC* (Germany Hamburg)

[12] Weicht J A et al 2014 Parameter variation of the one-diode model of a-Si and a-Si/μc-Si solar cells for modeling light-induced degradation *J. of Physics: Conference Series IOP Publishing* **559** 12017

Journal of Physics: Conference Series **682** (2016) 012018 doi:10.1088/1742-6596/682/1/012018

Angle dependent light scattering by gold nanospheres

J X Wang[1], A M Nilsson[1], D L A Fernandes[2] and G A Niklasson[1]

[1] Division of Solid State Physics, Department of Engineering Sciences, The Ångström Laboratory, Uppsala University, P. O. Box 534, SE–751 21 Uppsala, Sweden
[2] Division of Physical Chemistry, Department of Chemistry, The Ångström Laboratory, Uppsala University, P. O. Box 534, SE–751 21 Uppsala, Sweden

E–mail: Junxin.Wang@angstrom.uu.se

Abstract. Gold nanocrystals exhibit unique optical properties in enhanced light absorption and scattering owing to their extremely large scattering/absorption cross–sections and large electric field enhancements generated by localized surface plasmon resonance. In this work, the optical properties of gold nanospheres with diameters of 60 nm and 200 nm with remarkable uniformity in size were studied both numerically and experimentally. The total transmittance and reflectance as well as the angle–resolved light scattering intensities of the gold nanospheres were measured. The absorption and scattering coefficients were obtained by fitting the experimental data to the two–flux theory and were in qualitative agreement with single–scattering calculations using the Mie theory.

1. Introduction

Metallic nanocrystals have been receiving extensive attention for applications in optical and optoelectronic devices, and in solar energy technologies because of their extremely large scattering/absorption cross–sections and large electric field enhancements generated by localized surface plasmon resonance (LSPR). LSPR stems from the collective oscillations of conduction–band electrons, endowing noble metal nanocrystals with unique optical properties, *i.e.*, their LSPR wavelengths can be tailored from the visible to infrared regions by synthetically varying their shape and size [1]. Among them, gold nanospheres (NSs) are one of the most attractive plasmonic materials owing to their superior chemical and thermal stability, easy preparation by wet–chemistry approaches, and perfect geometrical symmetry. So far, a number of investigations have been published on plasmonic scattering enhancement on gold nanoparticles [2,3]. These studies focused mainly on the characterization of the absorption and scattering of gold nanoparticles with different shapes and sizes. However, wavelength and angle–resolved light scattering on gold nanoparticles has seldom been reported. In fact, the wavelength and angular dependent light scattering is one of the most critical issues for understanding how light propagates and interacts with the nanoparticles in the scattering layer. A thorough understanding of the spatial light scattering intensity distribution plays an essential role for controlling light scattering, which is highly desirable for further advancing photonic and display techniques such as solar cells, photodetectors, and LEDs [4–6].

In this paper, monodispersed gold nanospheres with diameters of 60 nm (NS60) and 200 nm (NS200) were investigated experimentally in terms of their extinction, total and diffuse reflectance and transmittance. The absorption and scattering cross section of gold nanospheres were numerically explored by Mie theory as well as by fitting the experimental results to two–flux theory. Wavelength and angle–resolved light scattering properties on NS60 and NS200 samples were also investigated.

2. Experimental methods

Gold nanospheres – Gold nanosphere samples of two different sizes were obtained from NanoSeedz Ltd. (Hong Kong SAR, China) as non–standard products. As informed by the manufacturer, the gold nanospheres sample was prepared by a seed–mediated growth approach modified from their previously reported works [7].

Physical techniques – Scanning electron microscopy (SEM) imaging was conducted on an FEI Quanta 400 FEG microscope operating at 20 kV.

Optical measurements – Extinction spectra of gold nanospheres in water solution were measured in the spectral range from 300 nm to 1000 nm using a Perkin-Elmer Lambda 900 ultraviolet/visible/near-infrared spectrophotometer using 1 cm thick quartz cuvettes. The extinction value represents the optical density (OD), which equals the logarithm of (1/transmittance). Total transmittance T_{tot} and reflectance R_{tot}, as well as diffuse transmittance T_{dif} and reflectance R_{dif} were measured in the same Lambda 900 spectrophotometer. The spectrophotometer was equipped with a spectralon coated integrating sphere from LabSphere. Regular transmittance, T_{reg} and specular reflectance R_{spe}, were derived by subtracting the diffuse components T_{dif} and R_{dif} from the total components T_{tot} and R_{tot}. For the transmittance and reflectance measurements, the gold nanospheres sample in water solution was sandwiched between two sealed microscope slides (75 × 25 × 1 mm) with a gap of 50 µm. The refractive index of the glass n_g is close to 1.5.

Wavelength and angle–resolved scattering measurements – The measurement system [14] consists of a 250 W tungsten–halogen lamp with a stabilized *dc* power supply, a monochromator (SP–401), two gratings with blaze wavelength of 0.75 µm and 2.0µm, an off–axis parabolic aluminum mirror from Optical Surfaces Ltd, with an off–axis distance of 80 mm and a focal length of 410 mm. The detector is a silicon diode in the spectral response region from 300 nm to 1100 nm. Lock–in technique is used to monitor the signal from the detector (EG&G model 7260). The sample is centrally mounted and illuminated, the detector with an integrating sphere can be positioned at various angles from 0° to 180° to collect the scattered light that allows angle–resolved measurements to be taken.

3. Theroy

Mie scattering theory – The electrodynamic response of particles of spherical shape can be analytically and effectively solved with Mie theory in a classical electromagnetic framework [8]. In this work, our MATLAB code is based on the program in the appendix of the book by Bohren and Huffman [9]. We compute the scattering (C_{sca}), absorption (C_{abs}) and extinction (C_{ext}) cross sections divided by the volume of single nanosphere. The refractive index of surrounding water is set as 1.33.

Two flux theory – Two–flux models [10] offer a description of the optics of materials exhibiting diffuse light scattering. In detail, light scattering materials are modeled as a slab with thickness *d* in which diffuse light I_d is propagating in the forward direction and diffuse light J_d in the reverse direction [11]. The scattering and absorption coefficients are denoted S and K. The extinction coefficient is the sum of S and K. In the independent scattering approximation, S and K are related to the scattering and absorption cross sections [12] by:

$$S = \frac{fC_{sca}}{V} = \frac{fC_{sca}}{\frac{4}{3}\pi a^3} \tag{1}$$

$$K = \frac{fC_{abs}}{V} = \frac{fC_{abs}}{\frac{4}{3}\pi a^3} \tag{2}$$

where f is the volume fraction and a is the radius of the nanospheres. In two–flux theory, the Kubelka–Munk approximation [13] for the transmittance and reflectance, denoted as T_{KM} and R_{KM}, represent the total transmittance and reflectance for a layer on a transparent background and are given by:

$$T_{KM} = \frac{b(1-R_g)}{(a-R_g)\sinh(bSd) + b\cosh(bSd)} \tag{3}$$

$$R_{KM} = \frac{1-R_g[a-b\coth(bSd)]}{a-R_g+b\coth(bSd)} \tag{4}$$

where R_g is the total reflectance from the background, which is the backside glass/air interface in our case. The parameters a and b in equation (3) and (4) are given in terms of scattering and absorption coefficients, with $a = 1+K/S$ and $b = (a^2-1)^{0.5}$.

This Kubelka-Munk approximation neglects the front surface reflection, and in order to take it into account we obtain [13]:

$$T = \frac{(1-R_c)T_{KM}}{1-R_jR_{KM}} \tag{5}$$

$$R = R_c + \frac{(1-R_c)(1-R_j)R_{KM}}{1-R_jR_{KM}} \tag{6}$$

where R_c is the reflectance of collimated light incident at the front air/glass surface. R_j denotes the reflectance of partially diffuse light from the nanoparticles, impinging onto the front glass/air interface.

4. Results and Discussions

Scanning electron microscopy (SEM) imaging (figure 1a and c) shows that the obtained gold NSs possess spherical shapes and smooth surfaces. The NS60 sample possesses a nearly spherical geometry with an average diameter measured to 60 ± 5 nm. The surface morphology of NS200 is a bit rough, and the average diameter of the spheres was measured to 200 ± 15 nm.

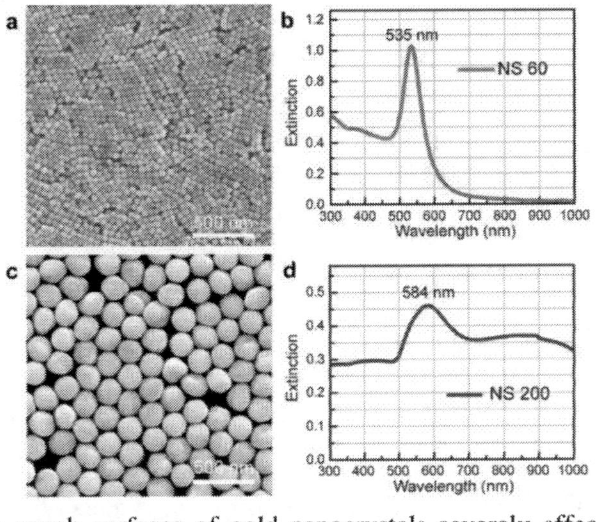

Figure 1. SEM (a,c) and extinction spectra (b,d) of gold nanosphere samples with average diameters of 60 ± 5 nm (a,b) and 200 ± 15 nm (c,d). The particle concentrations and volume fractions are $1.7 \cdot 10^{10}$ particles/mL, $1.9 \cdot 10^{-6}$ and $6.5 \cdot 10^{9}$ particles/mL, $2.8 \cdot 10^{-5}$ for NS60 and NS200, respectively.

The rough surfaces of gold nanocrystals severely affect their scattering intensities and profiles, therefore gold nanocrystals with nearly spherical shapes and smooth surfaces with narrow size distributions are highly sought after. The extinction spectra of NS60 and NS200 samples were acquired in water solution in figure 1b and d. The maximum extinction wavelength of NS60 peaks at 535 nm corresponding to the dipolar plasmon mode. It red shifts to around 800 nm in NS200 sample. A quadrupole plasmon mode appeared around 560 nm for NS200 sample. Peak broadening is clearly observed owing to increasing radiative losses for the gold NSs with larger sizes [15]

In order to obtain the scattering and absorption coefficients from two–flux theory, the gold nanosphere solutions were sandwiched between two sealed glass slides with a gap around 50 μm. The total, diffuse and regular transmittance as well as the total, diffuse and specular reflectance of these samples are shown in figure 2. For the NS60 sample, the total transmittance curve and the regular transmittance curve almost overlap, indicating the main transmittance change comes from the absorption of the nanospheres. The transmittance drops below 350 nm due to the absorption of the glass slides. The diffuse transmittance is almost negligible in NS60, a weak peak centered at 560 nm can be observed owing to the diffuse scattering of the nanoparticles. In regard to the reflectance of NS60 in figure 2c, the specular reflectance mainly corresponds to the reflectance from the glass slide, which is around 8%. The valley near 535 nm corresponds to the strong absorption of the dipolar plasmon mode of NS60. The diffuse reflectance peak at 560 nm indicates the backward scattering of NS60 sample. Compared to NS60 sample, the contribution from the diffuse part is higher in NS200 sample as shown in figure 2b. In figure 2d, the specular reflectance is almost a flat curve at 8%, and the valley around 560 nm can hardly been seen due to the weak quadrupole plasmon absorption of NS200. The diffuse reflectance is significant and originates from backward scattering from the NS200 sample.

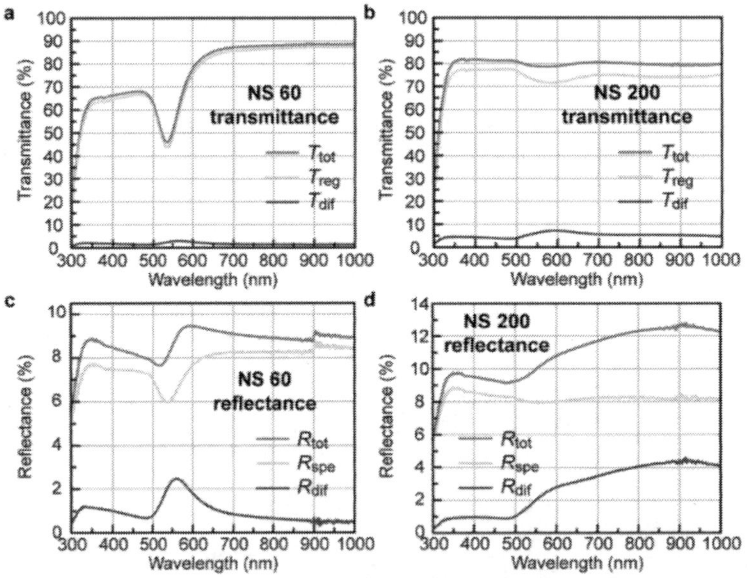

Figure 2. Measured total, regular and diffuse transmittance (upper row) and total, specular and diffuse reflectance (bottom row) of NS60 (a, c) and NS200 (b, d) samples. The particle concentrations and volume fractions are $6.7 \cdot 10^{11}$ particles/mL, $7.6 \cdot 10^{-5}$ and $2.6 \cdot 10^{11}$ particles/mL, $1.1 \cdot 10^{-3}$ for NS60 and NS200, respectively.

Figure 3 compares the scattering, absorption and extinction coefficients per volume when derived from Mie scattering theory or from experimental data using the two–flux theory for NS60 and NS200 samples. The results for the NS60 sample are in good agreement with the Mie theory, indicating a very good morphology and size distribution of the NS60 sample. In addition the low volume fraction of $7.6 \cdot 10^{-5}$ diminishes the interactions between particles so that the results agree well with the Mie theory which deals with a single nanoparticle. The results for the NS200 sample calculated from two–flux theory show discrepancies from the Mie result. We speculate that agglomeration of particles may have occurred and resulted in the red shift of the whole spectrum.

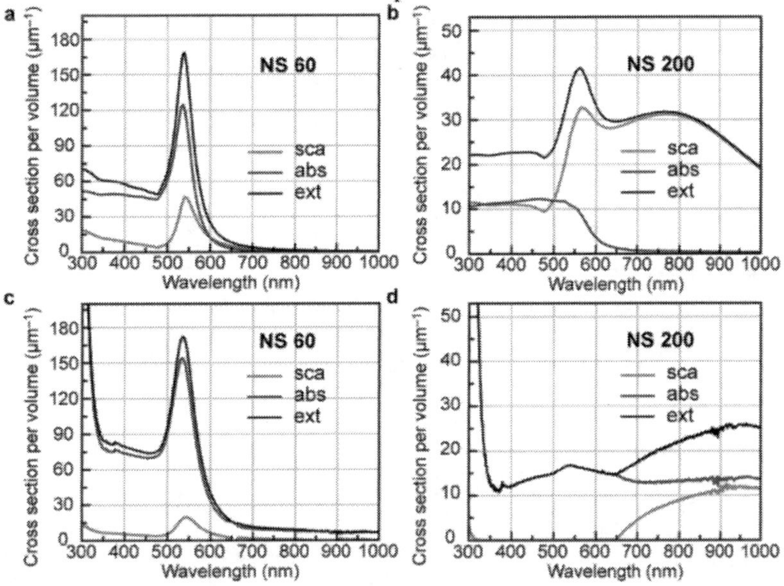

Figure 3. Scattering, absorption and extinction cross section per volume for NS60 (a) and NS200 (b) samples calculated using Mie theory. Scattering, absorption and extinction cross section per volume of NS60 (c) and NS200 (d) calculated from measured *T*, *R* results in figure 2 using two–flux theory.

In figure 4, the scattering intensity is presented as a function of wavelength and angle for the NS60 and NS200 samples. The regular component of the transmitted light dominates for angles from 0° to 4° in our angle–resolved setup. As proved by both numerical simulation and experimental measurement in figure 2 and 3, the diffuse scattering from NS60 sample is relatively low compared to its absorption. A weak diffuse scattering peak at 560 nm can be observed in figure 2a and figure 3c, while the scattering intensities are very low at 500 nm and 600 nm. The large–angle scattering observed at 300 nm and 400 nm may arise from a weak Rayleigh scattering component. In contrast to NS60, NS200 shows stronger scattering intensities, even at wavelengths larger than 500 nm. This is reasonable since the diffuse transmittance on the NS200 is significant at wavelengths larger than 500 nm as shown in figure 2.

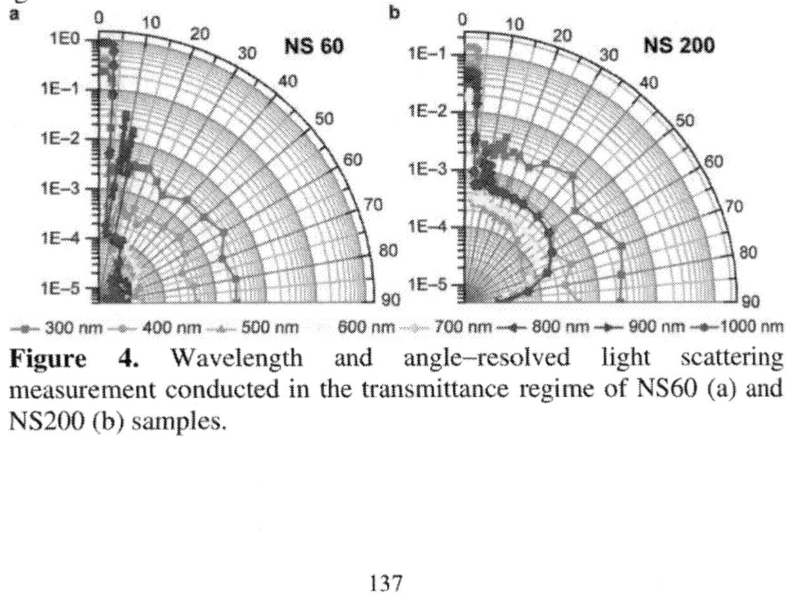

Figure 4. Wavelength and angle–resolved light scattering measurement conducted in the transmittance regime of NS60 (a) and NS200 (b) samples.

5. Conclusions

The extinction, reflectance and transmittance have been measured on suspensions of gold nanospheres with diameters of 60 nm and 200 nm. The scattering and absorption cross sections have been derived based on fitting of experimental data to two–flux theory. For the 60 nm particles the results are consistent with the Mie scattering theory. Wavelength and angle–resolved scattering measurements were carried out, and a strong wavelength and angular dependence of light scattering for the particles was established.

Acknowledgements

This paper was presented at the INERA Conference "Light in Nanoscience and Nanotechnology" in Hissar, Bulgaria, October 19–22, 2015. The conference is part of the INERA REGPOT project of the Institute of Solid State Physics, Bulgarian Academy of Sciences. Chemical gold nanosphere samples of two different sizes were obtained from NanoSeedz Ltd. (Hong Kong SAR, China). Scanning electron microscopy images were performed by Qifeng Ruan.

References

[1] Ni W H, Kou X S, Yang Z and Wang J F 2008 *ACS Nano* 2 677–686
[2] Baek S –W, Park G, Noh J, Cho C, Lee C –H, Seo M –K, Song H and Lee J –Y 2014 *ACS Nano* **8** 3302–3312
[3] Fang C H, Jia H L, Chang S, Ruan Q F, Wang P, Chen T and Wang J F 2014 *Energy Environ. Sci.* **7** 3431–3438
[4] Yang M, Fu Z P, Lin F and Zhu X 2011 *Opt. Express* **19** A763–771
[5] Montgomery G P 1988 *J. Opt. Soc. Am. B* **5** 774–784
[6] David A 2013 *J. Display Technol.* **9** 301–316
[7] Ruan Q F, Shao L, Shu Y W, Wang J F and Wu H K 2014 *Adv. Opt. Mater.* **2** 65–73
[8] Mie G 1908 *Annalen der Physik* **330** 377–445
[9] Bohren C F and Huffman D R 1983 *Absorption and Scattering of Light by Small Particles* (New York: Wiley)
[10] Vargas W E 1998 *Appl. Opt.* **25** 1122–1128
[11] Barrios D, Vergaz R, Sanchez–Pena J M, Granqvist C G and Niklasson G A 2013 *Sol. Energy Mater. Sol. Cells* **111** 115–122
[12] Niklasson G A and Eriksson T S 1988 *Proc. SPIE* **1016** 89–99
[13] Kubelka P 1948 *J. Opt. Soc. Am.* **38** 448–457
[14] Nostell P and Roos A 1999 *Rev. Sci. Inst.* **70** 2481–2494
[15] Rodríguez–Fernández J, Pérez–Juste J, Abajo F J G d and Liz–Marzán L M 2006 *Langmuir* **22** 7007–7010

Journal of Physics: Conference Series **682** (2016) 012019 doi:10.1088/1742-6596/682/1/012019

Properties of SnS thin films grown by physical vapour deposition

M Ganchev*, P Vitanov, M Sendova-Vassileva, G Popkirov, H Dikov

Central Laboratory of Solar Energy & New Energy Sources – Bulgarian Academy of Sciences, 72 Tzarigradsko chaussee, 1784 Sofia

mganchev@abv.bg – corresponding author

Abstract. Thin films of tin sulfide (SnS) were prepared by thermal evaporation technique on glass substrates and on n-type Si substrate and their physical properties were studied. The phase of the obtained thin films before and after thermal treatment was confirmed by X-ray diffraction analysis and Raman spectra. Optical transmission and reflection spectra were measured in the wavelength range 300-1800 nm, and the data were used to determine the direct and indirect optical band gaps. Four-point measurements have revealed that SnS thin film exhibits p-type conduction. Current-voltage characteristics of the SnS/ n-Si structures demonstrate strong photosensitivity and photovoltaic effect. However, in order to be able to evaluate the potential applicability of this heterojunction for photovoltaic or electronic devices, further study and technological optimization has to be conducted.

1. Introduction

Currently, several binary compound semiconductors attracted more and more researchers due to their favorable optoelectronic properties and simple preparation method. One such material is tin monosulphide (SnS), which has some desirable properties, such as high absorption coefficient ($\approx 10^4 cm^{-1}$ near the fundamental band edge) and high hole mobility ($\approx 90\ cm^2V^{-1}s^{-1}$). The elements Sn and S are not toxic and are fairly abundant. It has been shown [1] that SnS is a p-type semiconductor and theoretically it has light conversion efficiency of 24%. A variety of methods for depositing of SnS thin layers were developed as spray- pyrolysis [2], chemical bath deposition and electrodeposition [3,4], chemical vapor deposition [5] and physical vapor deposition [6]. It was shown that the thermal evaporation method yields stoichiometric layers with reasonable thickness of several micrometers free of pinholes, well adherent to the substrate and with good electro-physical properties for solar cells [7]. In this paper, we report on the preparation and characterization of thermally evaporated SnS thin films in order to study their suitability for solar cells application.

2. Experimental

The SnS layers were deposited by thermal evaporation technique under high vacuum (10^{-5} Torr). The distance between source and substrates have been maintained constant at 10 cm and the rate of deposition 2 Å/s, respectively. The source material was obtained by a simple solution precipitation reaction of $SnCl_2$ and thioacetamide (C_2H_5NS) in presence of EDTA at 70 °C (pH 1), thoroughly washed in boiling deionized water and ethanol. Before evaporation the source material was pre-heated in vacuum for 30 min at near 200 °C. The substrates were soda lime glasses sized 26x76 mm^2 or 2 inch n-Si wafers. The evaporation source was a rectangular tantalum boat heated resistively by electric current of 180 – 190 A. The source was covered by quartz wool to avoid damages due to ejected molten lumps on the deposited layers. The source material was subjected to differential scanning calorimetry (DSC) measurements to elucidate the transition temperatures and phase composition of the used material. The DSC unit was a Mettler TC-10-TA with DSC-20 standard cell. The measurement of the optical transmission and reflection spectra was performed using a Shimadzu UV–3600 spectrophotometer in the 300 – 1800 nm range. The Raman spectra were recorded with a Horiba Jobin Yvon LabRam HR 800 micro Raman spectrometer using a HeNe laser. Care was taken to lower the laser intensity to avoid excessive heating. X- ray diffractometry was done with a Brucker D8 advance spectrometer. The thickness of the layers was 1.15 ± 0.05 μm, measured by a high-precision Talystep profilometer (Taylor Hobson). AFM measurements were carried out using a Multimode V (Bruker, Santa Barbara, CA).

3. Results and discussion

3.1. Structural characterization

Figure 1 shows a DSC curve of a probe of the SnS powder used as source material. It can be seen that with the increase of the temperature an endothermic process takes place near 166 °C, with calculated enthalpy of 13.93 J/g. If the same sample is analyzed again or upon reverse scan this thermal absorption feature is not seen any more, indicating that the process is not reversible. The observed endothermic process could be interpreted as a crystallization of amorphous residuals in the precipitated powder and formation of monophase SnS material. In order to have well defined source material for deposition all experiments were performed after annealing of the substrate in the boat before evaporation.

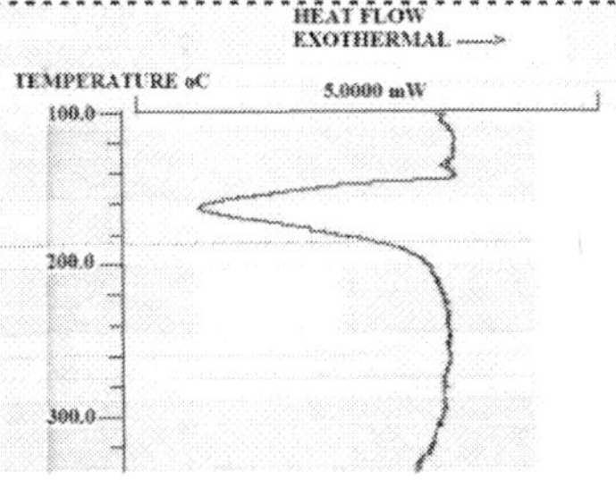

Figure 1. DSC analysis of SnS powder used as source material for physical vapour deposition of thin films.

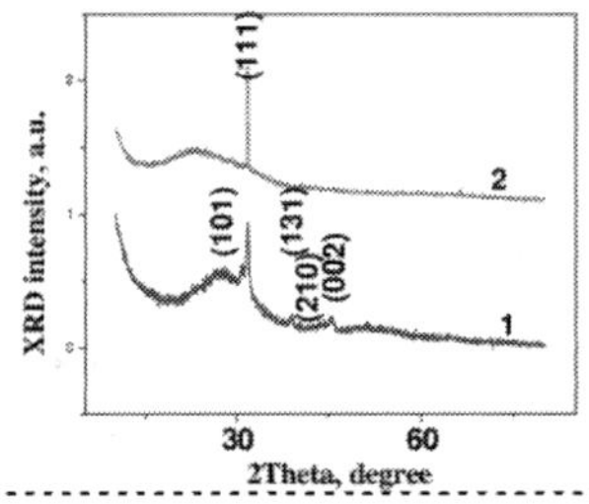

Figure 2. XRD pattern of PVD SnS thin films, as-deposited (1) and annealed in air at 100°C for 1 hour (2).

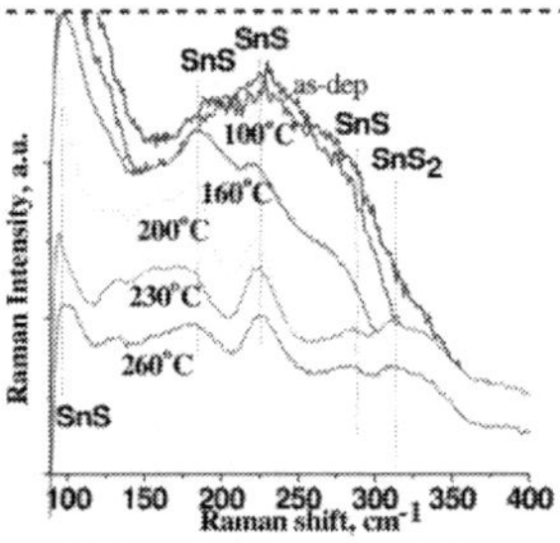

Figure 3. Raman spectra of PVD SnS thin films annealed in air at different temperatures for 1 hour.

Figure 2 presents XRD patterns of SnS thin films as deposited by physical vapour deposition and annealed at 100 °C for 1 hour, respectively. The spectrum of as-deposited SnS thin film presents the main relevant reflexes of the orthorhombic SnS phase from (101), (111), (131) and (002) planes. The peaks are weak with wide base which indicates an arrangement in short range order. The pattern of the sample annealed at 100 °C shows a narrow and tall reflex at the main (111) plane alone, indicating oriented crystallization. At higher temperatures up to 230 °C this specific orientation remains. Figure 3 demonstrates the micro Raman spectra of the as deposited sample and samples annealed in air at temperatures between 100°C and 260°C. The spectra of those two samples are very broad and almost featureless, being characteristic for amorphous material or one with very small nano-crystals. After annealing at higher temperatures bands near the positions characteristic for orthorhombic Herzenbergite SnS phase appear [8]. These bands are quite broad and a bit shifted from the positions observed in monocrystals. This can be linked to the small size of the crystallites comprising the layer.

Journal of Physics: Conference Series **682** (2016) 012019 doi:10.1088/1742-6596/682/1/012019

At annealing temperatures of 230°C and above a new band at about 312 cm^{-1} appears which can be ascribed to SnS_2.

3.2. Optical properties

Optical transmission and reflection spectra measurements were performed at room temperature in the wavelength range of 300 - 1800 nm.

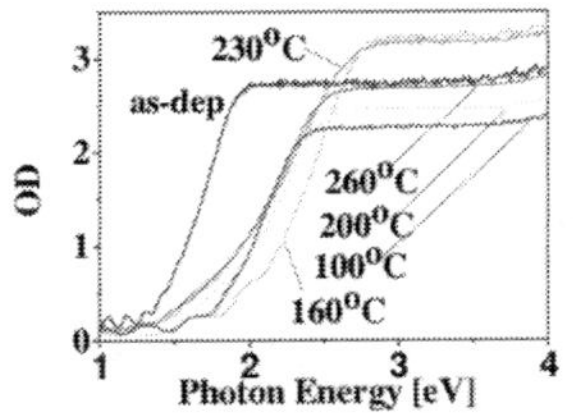

Figure 4. Optical density (absorbance) versus light energy of PVD SnS thin films annealed in air at different temperatures for 1 hour

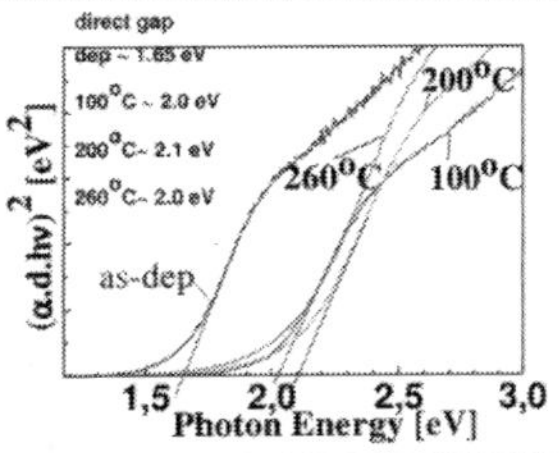

Figure 5. Dependence of $(\alpha dh\nu)^2$ on hν for determination the direct transitions' band gap of PVD SnS thin films, annealed in air for 1 hour.

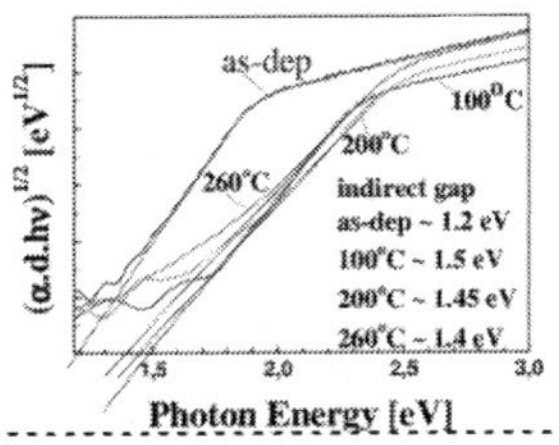

Figure 6. Dependence of $(\alpha dh\nu)^{1/2}$ on hν for determination of the indirect transitions' band gap of SnS films, annealed in air for 1 hour

The optical density (absorbance) versus photon energy of the SnS thin film is shown in Figure 4. Both direct and indirect optical gaps of a crystal or polycrystalline material depend on its structure. In terms of this, while some scientists have classified the optical transition of SnS thin films as direct band gap, others have classified it as indirect band gap [9,10]. The forbidden band value for allowed direct transitions can be determined by the cross point of the linear part of $(\alpha dh\nu)^2$=f($h\nu$) with the $h\nu$ axis On the other hand, the point where the linear part of $(\alpha dh\nu)^{1/2}$=f($h\nu$) graphic intersects the $h\nu$ axis gives the forbidden band value for allowed indirect transitions. Here, α is the absorption coefficient, d is the film thickness and $h\nu$ is the photon energy. In this way we could estimate the allowed direct and indirect energy band values (cf. Figures 5 and 6). It can be observed in Figures 4, 5 and 6 that the optical band gap of the as deposited material is markedly smaller than that of the annealed material.

3.3. Morphological characterization

The surface roughness of SnS layer is determined by AFM (cf. Figure 7). Globular formations with diameter near 200 nm are seen, which are composed of smaller nano-particles with diameter near 15 nm. After annealing up to 200 °C the size of the globules and nano-particles does not change, but at higher temperatures (260 °C) an increase in size of up to 350 nm and 35 nm, respectively, is observed.

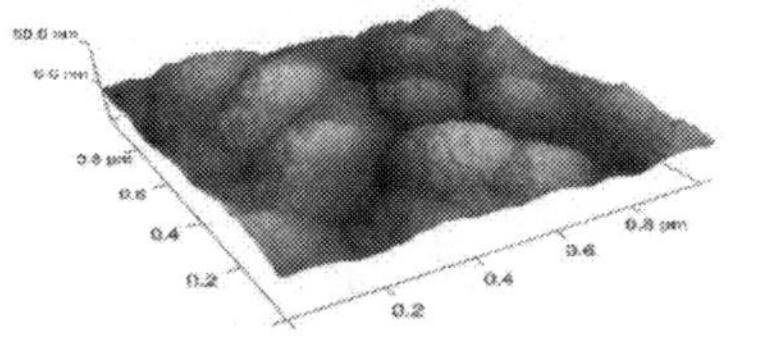

Figure 7. AFM image of the surface of PVD SnS thin film, annealed at 100°C in air for 1 hour.

Journal of Physics: Conference Series **682** (2016) 012019 doi:10.1088/1742-6596/682/1/012019

3.4. Electrical properties

The four point probe measurement of an SnS layer deposited on glass substrate has revealed sheet resistance of 250 Ohm/sq and p-type conductivity. The electrical properties of these films have been studied further by forming test structures as p-n diodes. The SnS thin films were evaporated on n-type Si wafers, FZ, with 7 Ω.cm resistivity and (111) orientation through a shadow mask. The size of the test structures is 2.5 x 2.5 mm^2. The SnS layer was used as deposited. The current–voltage (I-V) characteristic of the SnS/n-Si structure, measured in the dark has a typical diode behaviour of a p+/n junction (see Figure 8). The p+ SnS layer was contacted by a piece of indium, because the SnS layer is soft. The rectification ratio reaches 100 at +/- 2 V and the reverse current is 2.10^{-8} A at V= -2V, as can be seen from Figure 9.

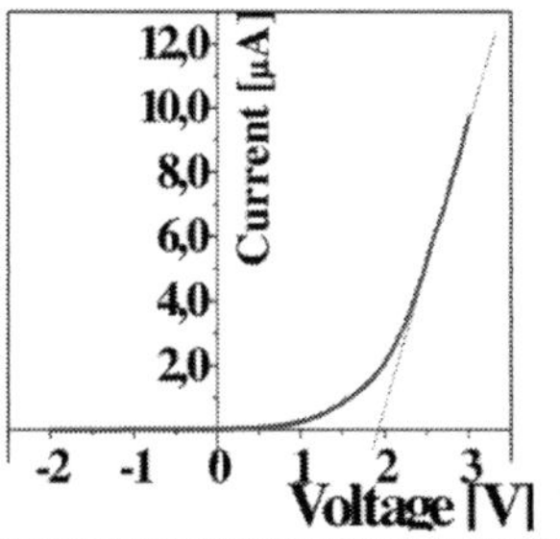

Figure 8. I-V curve with diode characteristics of hetero-structure SnS / n-Si

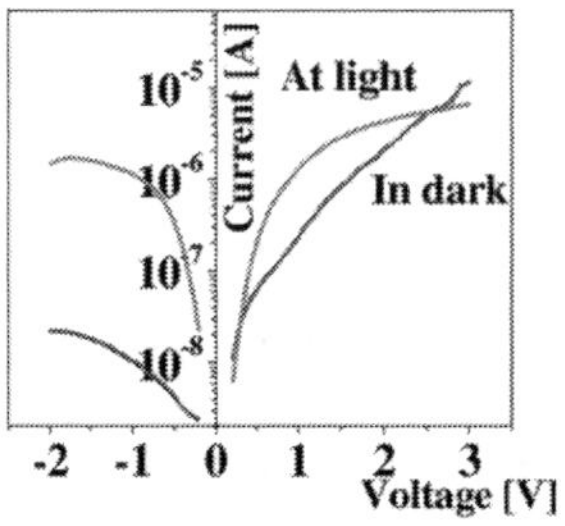

Figure 9. The semi-log plot of I-V characteristic of SnS/n-Si structure measured in the dark and upon illumination

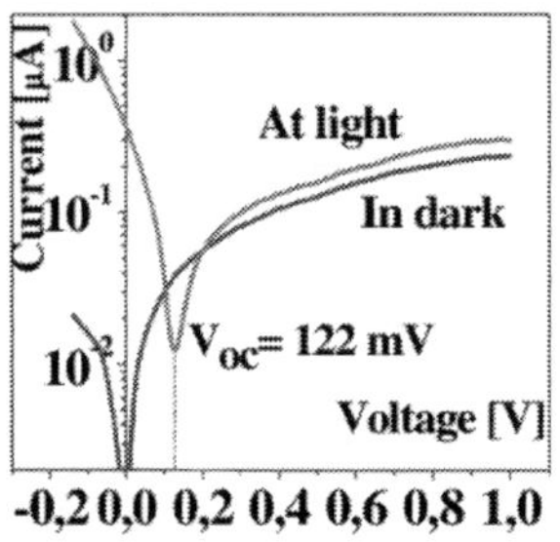

Figure 10. The semi-log plot of I-V characteristic of TCO/SnS/n-Si structure measured in the dark and upon illumination

Using the simple piecewise-linear model of a p/n diode, the on-state voltage, V_o, can be defined as intersection of the linear approximation of the current in forward direction with the V-axis of the I-V curve. As can be seen from Figure 8, V_o = 1.9 V, confirming that the diode structure is a heterojunction. Upon illumination (100 mW/cm^2), the reverse current of the p+/n heterojunction increases up to 2 orders of magnitude (Figure 9), demonstrating considerable illumination sensitivity. A second type of SnS/n-Si structure with transparent conductive layer as an electrode was used for test of the photovoltaic effect. The area of the device was 4.10^{-2} cm^2. The semi-log plot of I-V characteristic of this structure measured in the dark and upon illumination 100 mW/cm^2 is shown in Figure 10. A photovoltaic effect is obtained, with an open circuit voltage of V_{oc} = 0.112 V. The low photovoltaic performance of this structure is maybe due to the high series resistance of the test device. The carrier transport mechanism in SnS layers is not yet clear and should depend on the polycrystalline nature of the material.

Conclusions

Thin films of tin sulphide (SnS) were prepared by thermal evaporation technique on glass substrates and on n-type Si substrate, and their physical properties were studied. The used source material SnS powder undergoes a transition to full crystallization at temperature of 160 °C. During annealing the evaporated layers crystallize in preferred orientation (111) of the orthorhombic SnS phase (Herzenbergite). The Raman spectra have shown the vibrations characteristic for orthorhombic SnS and an additional one for SnS_2 for samples obtained after annealing above 200°C. AFM analysis has shown globular surface morphology with nano-crystalline particles with diameter of 15 nm which have grown up to 35 nm after annealing. Optical investigations have presented indirect band gap for annealed samples of 1.4 eV which coincides with previous investigations.

The rectifying behaviour of the presented SnS/n-Si structures is typical for heterojunctions, having rectification ratio of 100 at +/- 2 V and reverse current of 2.10^{-8} A at -2V. Strong photosensitivity and a photovoltaic effect could be observed. However, in order to be able to evaluate the potential for photovoltaic or electron device applications, further study and technological optimization have to be conducted.

Acknowledgements
This work was supported by Bulgarian National Science Foundation under contract TO2 – 12/ 12.12.2014.

References

[1] Loferski J 1956 *J. Appl. Phys.* vol 27 **7** p 777
[2] Koteswara R N and Ramakrishna R K 1998 *Thin solid films* **325** 1-2 p 4
[3] Safonova M, Nair P K, Mellikov E, Garcia A R, Kerm K, Revathi N, Romman T, Mikli V and Volobueva O 2014 *J Mater Sci: Mater Electron* **25** 3160
[4] Mishra K, Rajeshwar K, Weiss A, Murley M, Engelken R D, Slayton M and McCloud H E 1989 *Journal of the Electrochemical society* **136** p1915
[5] Engelken R D, McCloud H E, Lee C, Slayton M and Goreishi H 1987 *Journal of the Electrochemical Society* **134** p 2696
[6] Goswami A and Mitra A 1975 *Indian Journal of Pure and Applied Physics* **13** p 508
[7] Ogah O E, Zoppi G, Forbes I and Miles R W 2008 Proc. 23 EUPVSEC, 1-5 Sept., Valencia, Spain
[8] Sinsermsuksakul P, Heo J, Noh W, Hock A S and Gordon R G. 2011 *Adv. Energy Mater.* **1** p 1116
[9] Burton L A, Colombara D, Abellon R D, Grozema F C, Peter L M, Savenije T J, Dennler G and Walsh A 2013 *Chem. Mater.* **25** 4908
[10] Guneri E, Ulutas C, Kirmizigul F, Altindemir G, Code F and Gumus C 2010 *Applied Surface Science* **257** 1189

Vibrational spectroscopy of Ga^+ ion implanted ta-C films

M Berova[1], M Sandulov[1], T Tsvetkova[1], L Bischoff[2], R Boettger[2] and M Abrashev[3]

[1]Institute of Solid State Physics, Bulgarian Academy of Sciences, 72 Tzarigradsko Chaussee, 1784 Sofia, Bulgaria
[2]Institute of Ion Beam Physics and Materials Research, Helmholtz-Zentrum Dresden-Rossendorf, P.O.B. 51 01 19, 01314 Dresden, Germany
[3]Faculty of Physics, Sofia University, 1164 Sofia, Bulgaria

*E-mail: tania_tsvetkova@yahoo.co.uk

Abstract. In the present work, low energy Ga^+ ion beam implantation was used for the structural and optical properties modification of tetrahedral amorphous carbon (ta-C) thin films, using gallium (Ga^+) as the ion species. Thin film samples (d~40nm) of ta-C, deposited by filtered cathodic vacuum arc (FCVA), have been implanted with Ga^+ at ion energy E = 20 keV and ion doses $D=3.10^{14}\div3.10^{15}$ cm^{-2}. The Ga^+ ion beam induced structural modification of the implanted material results in a considerable change of its optical properties, displayed in a significant shift of the optical absorption edge to lower photon energies as obtained from optical transmission measurements. This shift is accompanied by a considerable increase of the absorption coefficient (photo-darkening effect) in the measured photon energy range (0.5÷3.0 eV). These effects could be attributed both to additional defect introduction and increased graphitisation, as well as to accompanying formation of bonds between the implanted ions and the host atoms of the target, as confirmed by infra-red (IR) and Raman measurements. The optical contrast thus obtained (between implanted and unimplanted film material) could be made use of for information archiving, in the area of high-density optical data storage, while using focused Ga^+ ion beams.

1. Introduction

Thin film materials of tetrahedral amorphous carbon (ta-C) have attracted great interest for a long time from both scientific and industrial perspectives. The term tetrahedral is used to describe amorphous carbon films with a large percentage of *sp*3 bonding (up to 87%). The films are manufactured using a variety of techniques, including filtered cathodic vacuum arc (FCVA), pulsed laser deposition (PLD) and mass selected ion beam deposition (MSIBD) [1-4]. The high *sp*3 content in the films results in unique properties that include extreme hardness (~70 GPa), chemical inertness, high electrical resistivity, and wide optical band gap [5–7]. Other important factors which make the films an attractive material for coatings include a smooth surface and low friction, thermal stability and transparency over a wide spectral range. These properties also offer advantages as compared to another wide optical bandgap material – silicon carbide (SiC) – for uses in nano-scale optical data recording for archival information storage using focused ion beams (FIB) techniques, where SiC thin films have found useful applications recently [8-15].

Journal of Physics: Conference Series **682** (2016) 012020 doi:10.1088/1742-6596/682/1/012020

In the case of polycrystalline silicon carbide (pc-SiC) thin films, ion bombardment is used to amorphise areas of the films by computer operated FIB systems, thus creating useful optical contrast between non-irradiated polycrystalline areas and the irradiated amorphous areas, which can be further used for nano-scale optical data recording for archival information storage [8-10]. In the case of hydrogenated amorphous silicon carbide (a-SiC:H) films, computer operated FIB systems are used to both introduce irradiation defects and additionally chemically modify the amorphous structure of the films, thus reducing their optical bandgap in even a more effective manner for the useful creation of optical contrast between implanted and non-implanted areas of the film material for applications in nano-scale optical data recording [11-15].

In both polycrystalline and amorphous SiC film materials, a considerable part in the creation of useful optical contrast between irradiated and non-irradiated areas of the films is played by the transformation of substantial part of the present diamond-like (sp3) carbon bonds, before the irradiation, into graphite-like (sp2) carbon bonds, as a result of it [10,15]. It is expected, that a similar mechanism of the carbon bonds transformation would result when applying ion bombardment with different ions, e.g. nitrogen (N^+) and gallium (Ga^+) ions, in the case of ta-C films, so that to achieve useful optical contrast between irradiated and non-irradiated areas of the films. The use of gallium as the ion implanted species is particularly attractive since it is available in standard focused ion beam (FIB) machines, and in addition has been shown to be capable of generating large optical contrasts [16]. The underlying structural modification, induced by the N^+ and Ga^+ ion bombardment, has been investigated by Raman and ifra-red (IR) spectroscopy measurements.

2. Experimental

Thin ta-C films (d~40 nm) were deposited on c-Si (100) substrates using a commercial FCVA system (Commonwealth Scientific Corporation). Carbon plasma is produced from the arc spot on the cathode, 99.999% pure graphite in high vacuum. Cathodic arcs are prolific generators of highly ionized carbon plasmas. With the FCVA technique, the plasma stream is steered through a magnetic filter to eliminate neutral particles generated at the cathode. At the filter exit, the fully ionized plasma, consisting of carbon ions and electrons, streams towards the substrate. The films were deposited at room temperature with an arc current of 120 A under floating conditions.

Ion implantation of N^+ and Ga^+ was carried out at room temperature (RT) using a commercial broad-beam ion implanter. The ion-beam intensity was I~2 μA/cm^2, the ion energy was E=20 keV, and the ion doses used were $D=3.10^{14} \div D=3.10^{15}$ cm^{-2}. SRIM simulation program [17] was used to determine the projected range Rp ~ 29 nm and the struggle ΔRp ~ 10 nm, for N^+, and Rp ~ 17 nm and ΔRp ~ 4 nm, for Ga^+, implanted ions into the ta-C film samples (d ~ 40 nm).

The structural modifications of the ta-C films by the N^+ and Ga^+ ion implantation were monitored by Raman and Infrared (IR) spectroscopy. The Raman spectra were obtained using a LabRAM HR Visible (Horiba Jobin- Yvon) Raman spectrometer. An objective X50 was used both to focus the incident laser beam onto the pellets surface into a spot with a diameter about 2μm and to collect the scattered light. The used excitation was He-Ne 633 nm laser line. To avoid overheating the measurements were performed with a reduced laser power of 0.6 mW.

Fourier Transform Infrared (FTIR) spectroscopy measurements were done by Shimadzu FTIR Spectrophotometer IRPrestige-21 in the spectral range 350-4000 cm^{-1} in reflectance mode (specular reflectance attachment), Al mirror. The samples were measured in absorbance

3. Results and discussion

3.1. Raman measurements

We have obtained and compared Raman measurement results for ta-C films implanted with Ga^+ ions at doses $D_1=3x10^{14}$ions/cm^2 and $D_2=3x10^{15}$ions/cm^2, and an unimplanted film was used like a reference sample. The analyzed spectrum range was from 100 cm-1 to 3500 cm^{-1} and the part 1000 ÷2200 cm^{-1} was studied as the spectrum responsible for the film's scattering.

Journal of Physics: Conference Series **682** (2016) 012020 doi:10.1088/1742-6596/682/1/012020

In figure 1 the comparison is depicted between Raman band intensities of three samples. It is well manifested that the spectrum band of the unimplanted sample is rather less in intensity than the band's intensities of the implanted samples. The Raman shift is calculated so that to be interpreted and compared with other science sources.

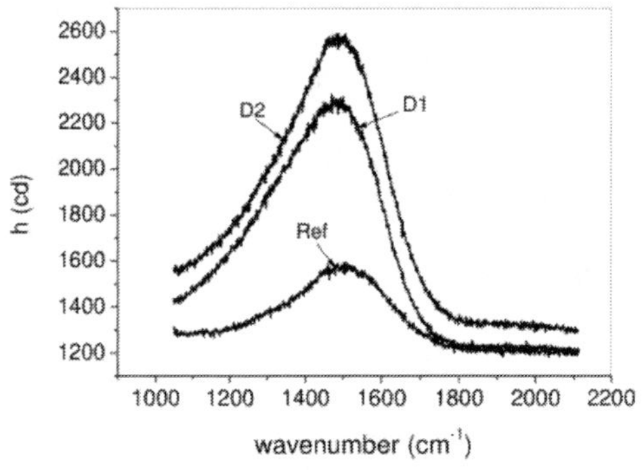

Figure 1. Raman spectra of ta-C films: unimplanted (Ref.), Ga^+ implanted with a dose D1=3.10^{14} cm^{-2} (D1) and Ga^+ implanted with a dose D2=3.10^{15} cm^{-2} (D2).

Direct information from the intensity measurements could not be given but changes in the ratio between I_D/I_G, the intensities of the D and G peaks in the Raman spectra [18], could bring same useful information about changed graphite-like bonded structures in the films. For that reason, in order to determine the values of I_D and I_G bands and to compare their intensities, the options of Origin 8.1 programme were used to deconvolute and to separate D and G peak. With the further option to calculate more precisely the intensity values, the most common Raman spectra fitting method was used, which is employing two Gaussian peaks with linear background subtraction.

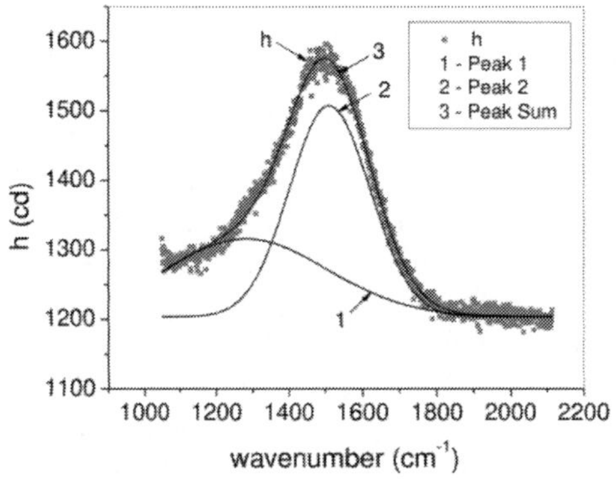

Figure 2. Deconvoluted Raman spectrum for reference sample.

The G and D peaks are due to sp^2 sites only. The G peak is due to the bond stretching of all pairs of sp2 atoms in both rings and chains. The D peak is due to the breathing modes of sp2 rings. In figures 2, 3 and 4 are represented deconvoluted Raman spectra of the unimplanted ta-C film, and those for the implanted with Ga^+ ions films, with doses D_1 and D_2, respectively, so that to reveal more information about the structure of the films. For the ta-C film implanted with D_1, I_D=525 cd and I_G=954,4 cd while for the sample implanted with the higher D_2 dose I_D=561 cd and I_G=756 cd. The ratios were calculated as follows: I_D/I_G=0.74, I_D/I_G=0.56 and I_D/I_G=0.37, for the implanted samples with doses D_2 and D_1 and the unimplanted sample, respectively. The tendency for growth of this ratio in ta-C films is in accordance with other scientific literary sources and it is believed that the arising of this effect is due to increasing of sp^2 sites in chains and their decreasing in rings while disordering increases. The lower ratio I_D/I_G in combination with high G peak is a sufficient condition to confirm higher sp^3 in the film.

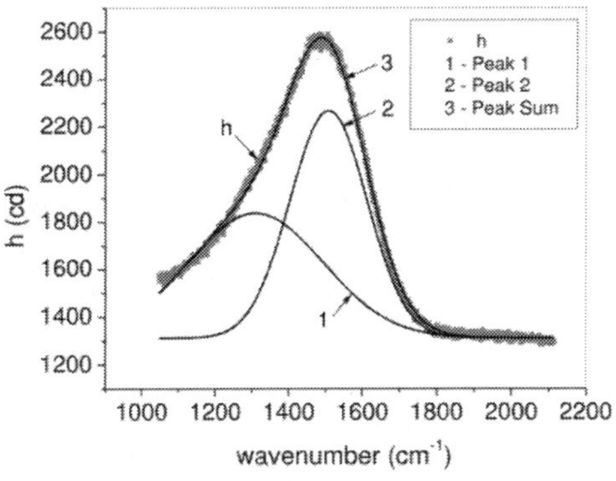

Figure 3. Deconvoluted Raman spectrum for Ga^+ implanted ta-C sample, D1=3.10^{14} cm^{-2}.

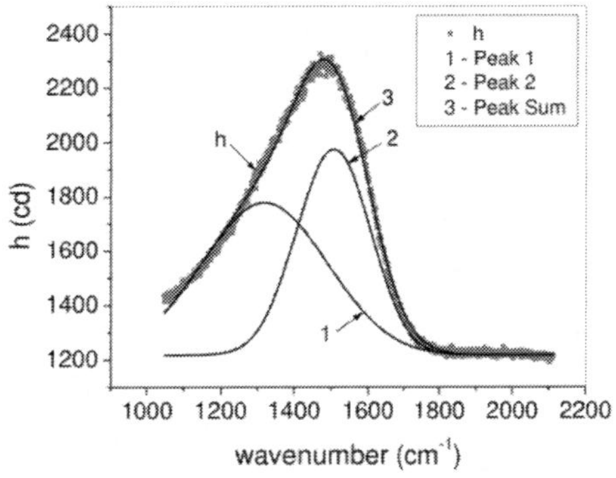

Figure 4. Deconvoluted Raman spectrum for Ga^+ implanted ta-C sample, D2=3.10^{15} cm^{-2}.

The values of the FWHM (full width at half maximum) for the G and D peak, $FWHM_G$ and $FWHM_D$, were obtained for the three kinds of samples. The values of $FWHM_G$ for the unimplanted and the implanted samples with D1 and D2 are 219,35 cm^{-1}, 209,45 cm^{-1} and 199,92 cm^{-1}, respectively. A tendency is observed for decreasing of $FWHM_G$ values while applying ion implantation and its influence is increasing with the dose. This implies a decrease of the sp^3 content and an increase of sp^2 clustering areas, while disordering increases. The ion beam induced structural modification is also related to a slight shift in the G peak frequency, demonstrated as a position shift of the G band towards lower wavenumbers with the increase of the implantation dose (figures 2,3 and 4).

3.2. Infrared Spectra

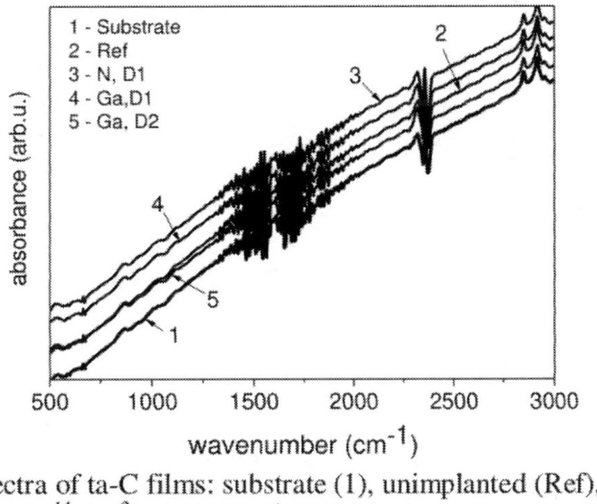

Figure 5. IR spectra of ta-C films: substrate (1), unimplanted (Ref), N^+ implanted with a dose D1=3.10^{14} cm^{-2} (N, D1) Ga^+ implanted with a dose D1=3.10^{14} cm^{-2} (Ga, D1) and Ga^+ implanted with a dose D2=3.10^{15} cm^{-2} (Ga, D2).

We have measured the infrared spectra in reflectance mode in the spectral range 350-3000 cm^{-1} for ta-C film samples without implantation (reference), ta-C implanted with N^+ ions with dose 3.10^{14}ions/cm^2, ta-C implanted with Ga^+ ions with dose 3.10^{14}ions/cm^2 and ta-C implanted with Ga^+ dose 3.10^{15}ions/cm^2 (figure 5).

In figure 5 are seen the following lines: 1490 cm^{-1}, 1550 cm^{-1}, 1630 cm^{-1}, 1560 cm^{-1}, 2375 cm^{-1}. The lines are due at: 1490 cm^{-1}, 1550 cm^{-1}, 1630 cm^{-1}- C-C stretch (in ring), 2375 cm^{-1} –C(triple bond)C– stretch or C(triple bond)N– stretch, 2800-2900 cm^{-1} is due to air contamination and is a CO_2 line. We observe that triple bonds have higher absorption than aromatic bonds. The film is near the standard for diamond-like carbon (DLC) films: 80% sp^3 and 20% sp^2. There is small growth of the absorption as a result of the implantation.

The graphics of figure 6 shows the increase of absorption of the ta-C films after N^+ and Ga^+ ion implantation as compared with the Si substrate. In figure 7 there is a line for which the wavenumber corresponds to that of observed gallium compounds - around 660 cm^{-1} [19, 20]. In figure 8 there are lines around 1100 cm^{-1} and 1250 cm^{-1} which are due C–H or C–N stretching bonds [21]. The growth of the absorption is more visible in figures 7 and 8 than in figure 5, and we can conclude that absorption of the N^+ implanted sample is higher than the reference sample, while the decrease of the absorption of the Ga^+ ion implanted ta-C films (which is even greater with the Ga^+ dose increase), as compared with the N^+ implanted one, is presumably due to the gallium colloids formation and their growth resulting in increased light scattering [22].

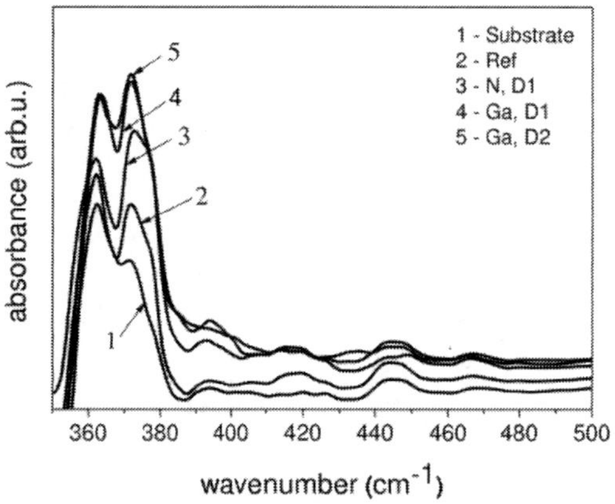

Figure 6. IR spectra of ta-C films: substrate (1), unimplanted (Ref), N^+ implanted with a dose $D1=3.10^{14}$ cm^{-2} (N, D1) Ga^+ implanted with a dose $D1=3.10^{14}$ cm^{-2} (Ga, D1) and Ga^+ implanted with a dose $D2=3.10^{15}$ cm^{-2} (Ga, D2).

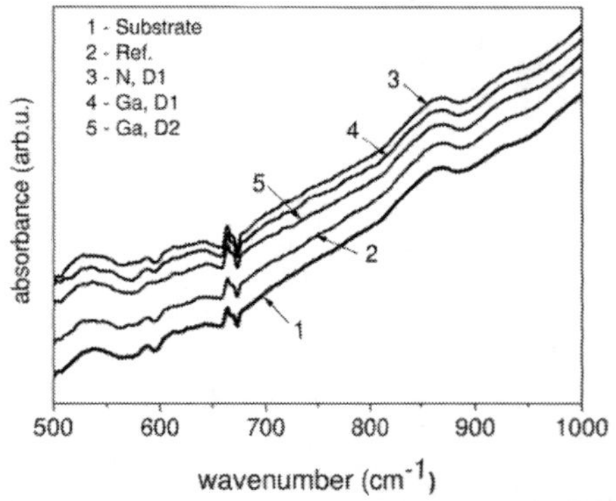

Figure 7. IR spectra of ta-C films: substrate (1), unimplanted (Ref), N^+ implanted with a dose $D1=3.10^{14}$ cm^{-2} (N, D1) Ga^+ implanted with a dose $D1=3.10^{14}$ cm^{-2} (Ga, D1) and Ga^+ implanted with a dose $D2=3.10^{15}$ cm^{-2} (Ga, D2).

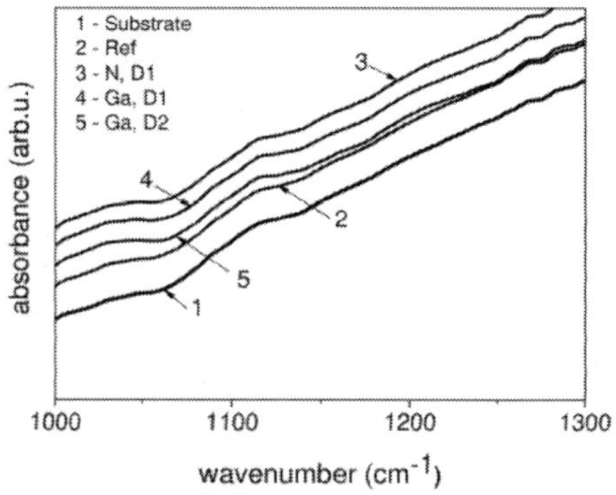

Figure 8. IR spectra of ta-C films: substrate (1), unimplanted (Ref), N^+ implanted with a dose $D1=3.10^{14}$ cm^{-2} (N, D1) Ga^+ implanted with a dose $D1=3.10^{14}$ cm^{-2} (Ga, D1) and Ga^+ implanted with a dose $D2=3.10^{15}$ cm^{-2} (Ga, D2).

4. Conclusion

The obtained results in the present work have shown that choosing relatively low N^+ and Ga^+ ion doses, $D=3.10^{14} \div D=3.10^{15}$ cm^{-2}, and ion energy, E=20 keV, results in considerable modification of the structural and optical properties of the implanted ta-C films, indicating some increased graphitisation and assumed electronic properties modification, as shown by the Raman and IR spectroscopy results. The related increase of the absorption coefficient (α) in the visible wavelength range is sufficient in magnitude to fulfill the requirements for optical contrast needed for applications in optical data recording and archival information storage.

Acknowledgements
The authors would like to thank the group of Prof. S. Kitova for the samples preparation. The authors would also like to thank the staff of the ion beam center at the Helmholtz-Zentrum Dresden-Rossendorf e.V., a member of the Helmholtz Association, for performing the ion implantation.

References

[1] McKenzie D, Muller D and Pailthorpe BA 1991 *Phys. Rev. Lett.* **67** 773
[2] Fallon PJ, Veerasamy VS, Davis CA, Robertson J, Amaratunga GAJ, Milne WI and Koskinen J 1993 *Phys. Rev. B* **48**, 4777
[3] Friedmann TA, McCarty KF, Barbour JC, Siegal MP and Dibble DC 1996 *Appl. Phys. Lett.* **68** 1643
[4] Lifshitz Y, Kasi SR and Rabalais JW 1989 *Phys. Rev. Lett.* **62** 1290
[5] Silva SRP, Xu S, Tay BK, Tan HS, Scheibe HJ, Chhowalla M and Milne WI 1996, *Thin Solid Films* **290–291** 317
[6] Veerasamy VS, Yuan J, Amaratunga G, Milne WI, Gilkes KWR, Weiler M and Brown L *Phys. Rev. B* 1993 **48** 17954
[7] Robertson J 1991 *Solid State Chem.* **21** 199
[8] Ruttensperger B, Krötz G, Müller G, Derst G and Kalbitzer S 1991 *J. Non-Cryst. Solids* **137-138** 635
[9] Müller G 1993 *Nucl Instrum Methods B* **80-81** 957
[10] Kalbitzer S 2004 *Nucl Instrum Methods B* **218** 343
[11] Tsvetkova T 1996 Ion Beam Modification of Amorphous Silicon-Carbon Alloys *Beam Processing of Advanced Materials*, ed J Singh, S Copley and J Mazumder (ASM International, Materials Park) pp 207-2018
[12] Tsvetkova T, Takahashi S, Zayats A, Dawson P, Turner R, Bischoff L, Angelov O and Dimova-Malinovska D 2005 *Vacuum* **79** 94
[13] Tsvetkova T, Takahashi S, Zayats A, Dawson P, Turner R, Bischoff L, Angelov O and Dimova-Malinovska D 2005 *Vacuum* **79** 100
[14] Takahashi S, Dawson P, Zayats AV, Bischoff L, Angelov O, Dimova-Malinovska D, Tsvetkova T and Townsend P 2007 *J. Phys. D: Appl Phys.* **40** 7492
[15] Tsvetkova T, Sellin P, Carius R, Dimova-Malinovska D and Angelov O 2007 *J. Optoel. Adv. Mater.* **9** 375
[16] Bischoff L, Teichert J, Kitova S and Tsvetkova T 2003 *Vacuum* **69** 73
[17] Ziegler JF, Biersack JP and Littmark U 1985 *The Stopping and Range of Ions in Matter* vol 1 (New York: Pergamon)
[18] Ferrari A C and Robertson J 2000 *Physical Review B* **61** 14095
[19] Ferrari A C, Rodil S E and Robertson J 2003 *Physical Review B* **67** 155306
[20] Zhao Y, Frost R L, Yang J and Martens W N 2008 *J.Phys.Chem. C* **112** 3568.
[21] Dwivedi N, Kumara S, Tripathia R K, Careyc J D, Malikb H. K. and Dalaia M K 2012 *Appl. Mater. Interfaces* **4** 5309
[22] Berova M, Sandulov M, Tsvetkova T, Karashanova D, Boettger R and Bischoff L 2015 *J. Phys.: Conf. Ser.* in press

Sputter deposited $W_{1-x-y}Ni_xTi_yO_3$ thin films: Electrochromic properties and durability

M Morales-Luna[1,2], M A Arvizu[1], C G Granqvist[1] and G A Niklasson[1]
[1] Department of Engineering Sciences, The Ångström Laboratory, Uppsala University, P. O. Box 534, SE-751 21 Uppsala, Sweden
[2] Departamento de Nanociencias y Nanotecnología, Centro de Investigación y de Estudios Avanzados del I.P.N., A.P. 14-740, 07360 México D.F., Mexico

E-mail: mmorales@fis.cinvestav.mx

Abstract. Previous research demonstrated that a small amount of nickel enhances the coloration efficiency of tungsten–nickel oxide electrochromic (EC) thin films with respect to that of pure tungsten oxide (WO_3) films. Furthermore the incorporation of titanium gives an improvement in the durability of tungsten–titanium oxide EC thin films. In this work we investigated the EC performance of tungsten–nickel–titanium oxide ($W_{1-x-y}Ni_xTi_yO_3$) EC thin films with emphasis on durability. The films were deposited on indium tin oxide covered glass by reactive dc sputtering from tungsten, tungsten–titanium alloy and nickel targets. Cyclic voltammetry was performed using 1 M $LiClO_4$ in propylene carbonate as electrolyte. The voltage window was chosen to induce fast degradation of the samples within 80 cycles. Elemental compositions were obtained by Rutherford Backscattering Spectroscopy.

1. Introduction

Global warming is a widely accepted phenomenon with many adverse effects for the earth's ecosystems and for human activities. It is caused by excessive use of certain energy sources, such as fossil fuels, that lead to high carbon dioxide emissions. Hence research aimed at the generation of clean energy, as well as energy saving technologies, is essential. The built environment is particularly important since it accounts for 30–40% of the primary energy consumption. A lot of this energy is used to provide artificial illumination and for cooling or heating purposes. Windows are of great importance in this context since they are weak links in the buildings' energy system and frequently let in or out too much energy [1]. There are numerous ways to improve windows; one of them is to go to the so called "smart window" which allows modulation of the transmittance of solar energy and visible light in order to create energy efficiency [1].

Smart windows can be made with electrochromic (EC) materials and allow transparency modulation upon the application of a small potential difference [2]. An EC device is composed of several parts and has a pure ion conductor (electrolyte) in the middle. This part is sandwiched between two layers, one of them being the main EC thin film and the other an ion storage film that could have EC properties or not. The layers are coated on transparent conductors to which the voltage source is connected, and the entire system can be laminated between transparent glass or PET sheets [3]. The most studied EC material is amorphous tungsten oxide, which allows high optical modulation. It is well known that tungsten oxide (WO_3) is a cathodic EC material (colors under charge insertion) with a perovskite-type structure based on corner-sharing WO_6 octahedra [3]. The ion storage film should be either transparent during charge insertion and extraction or possess EC properties opposite in nature with respect to the main EC film,

Journal of Physics: Conference Series **682** (2016) 012021 doi:10.1088/1742-6596/682/1/012021

i.e., it should color anodically (under charge extraction) if the main EC film has cathodic properties, as for the case of WO_3.

Extensive color/bleach cycling of WO_3 films leads to a loss of the EC properties, and previous research has been focused on ways to improve their durability as EC films. A good option is the addition of a certain concentration of titanium oxide (TiO_2) into the WO_3 film, and a recent study of Ti doping of WO_3 films showed enhanced durability even when the voltage span was increased so as to subject the samples to stringent conditions [4].

Nickel oxide (NiO) is an excellent choice as ion storage film in EC devices. Several recent studies [5,6,7,8] have been devoted to its compatibility with electrolytes appropriate for WO_3, charge density exchange, durability and coloration efficiency. In particular, we note recent work on composites of WO_3 and NiO which produced interesting results regarding their optical, structural and electrochromic properties [9,10].

The principal objective of our present work is to investigate the effects of Ti addition to a Ni–W oxide matrix. The study is focused on the durability of EC films under harsh conditions obtained by applying a wide voltage window to accomplish electrochemical charge insertion/extraction, and we also study changes in the optical properties.

2. Film preparation and characterization

Thin films of W oxide, W–Ni oxide and W–Ni–Ti oxide were prepared by reactive DC magnetron sputtering in a deposition system based on a Balzers UTT 400 unit. Targets were 5-cm-diameter plates of W, Ni, W(95wt%) + Ti(5wt%) and W(90wt%) + Ti(10wt%) (Plasmaterials), all with 99.99% purity. The sputter system was first evacuated to 2×10^{-4} mTorr. Argon and oxygen, both of 99.997% purity, were then introduced through mass-flow controlled gas inlets. The O_2/Ar ratio was set to 0.15, and the pressure was 30 mTorr; these parameters were chosen based on previous studies [9]. To obtain different concentration of W–Ni–Ti oxide, the discharge power of the targets were maintained at $P_{WTi} + P_{Ni} =$ 230 W; some more details on sample preparation are given elsewhere [4].

For studying electrochemical and optical properties, the films were deposited onto unheated 5×5 cm^2 glass plates coated with transparent and electrically conducting layers of In_2O_3:Sn (ITO) with an electrical sheet resistance of 40 ohms per square. Carbon substrates were used for films for Rutherford Backscattering Spectrometry (RBS) analysis, which was performed at the Tandem Laboratory of Uppsala University using 2 MeV ^{4}He ions backscattered at an angle of 170°. RBS data were fitted to a model of a film–substrate system by the use of the SIMNRA program [11]. Optical transmittance was recorded in situ during electrochemical cycling in the 380–800-nm wavelength range using a fiber optical instrument from Ocean Optics. Film thicknesses were around 300 nm as measured by surface profilometry using a Bruker DektakXT instrument. X-ray diffraction analyses were made using a Siemens D5000 instrument and the diffractograms showed only peaks due to the ITO underlayer and no evidence of peaks associated to W, W–Ni or W–Ni–Ti oxides. Hence all of the as-deposited films were amorphous.

3. Electrochemical and optical data

Ternary Ni–Ti–W oxide EC films were investigated by cyclic voltammetry (CV) using a Solartron 1286 electrochemical interface in a three-electrode electrochemical cell with the sample as working electrode and Li foils serving as reference and counter electrodes. The electrolyte was 1M $LiClO_4$ in propylene carbonate (Li–PC). All measurements took place inside a glovebox with inert argon atmosphere. The voltage range and sweep rate were 1.7–4V *vs.* Li and 10 mV/s, respectively. Figure 1 shows voltammograms for EC films with various compositions at selected CV cycles. The films without titanium show poor stability within 80 CV cycles, the same behaviour is found for the $W_{0.78}Ni_{0.12}Ti_{0.1}O_3$ film. However the $W_{0.67}Ni_{0.16}Ti_{0.17}O_3$ sample is clearly more stable.

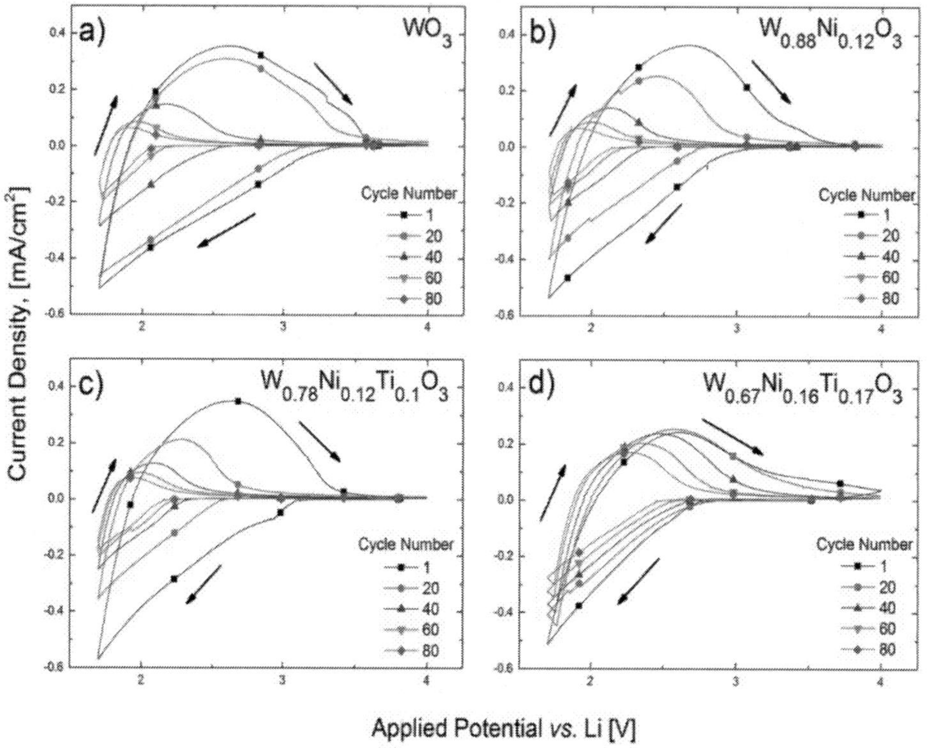

Figure 1. Cyclic voltammograms for (a) W oxide, (b) W–Ni oxide, and (c)–(d) W–Ni–Ti oxide films of the shown compositions immersed in 1M Li–PC. Data were taken after the indicated numbers of cycles for the voltage sweep range 1.7–4.0 V *vs.* Li. Arrows indicate the direction of insertion and extraction of charge.

It is important to study the extracted and inserted charge density Q as a function of the number of voltammetry cycles to show the electrochemical activity of the films. Such evolution is illustrated in figure 2(a), where we confirm that only the $W_{0.67}Ni_{0.16}Ti_{0.17}O_3$ sample is able of maintain its charge capacity, whereas all of the other films show drastic drops of their charge capacity within the first 40 cycles. One way to emphasize the modifications in the charge capacity is to study the difference between the extracted and inserted charge density $\delta Q = \left| Q_{\text{inserted}} + Q_{\text{extracted}} \right|$, with $Q_{\text{inserted}} < 0$ and $Q_{\text{extracted}} > 0$, as shown in figure 2(b). Some samples seem to present a saturation of the amount of charge that could not be extracted, which is the case for films without titanium. The $W_{0.78}Ni_{0.12}Ti_{0.1}O_3$ sample, on the other hand, displays a high difference between the inserted and extracted charge in the first cycle, and then the difference tends to decrease. The $W_{0.67}Ni_{0.16}Ti_{0.17}O_3$ sample starts with a low difference and maintains this property during the 80 cycles.

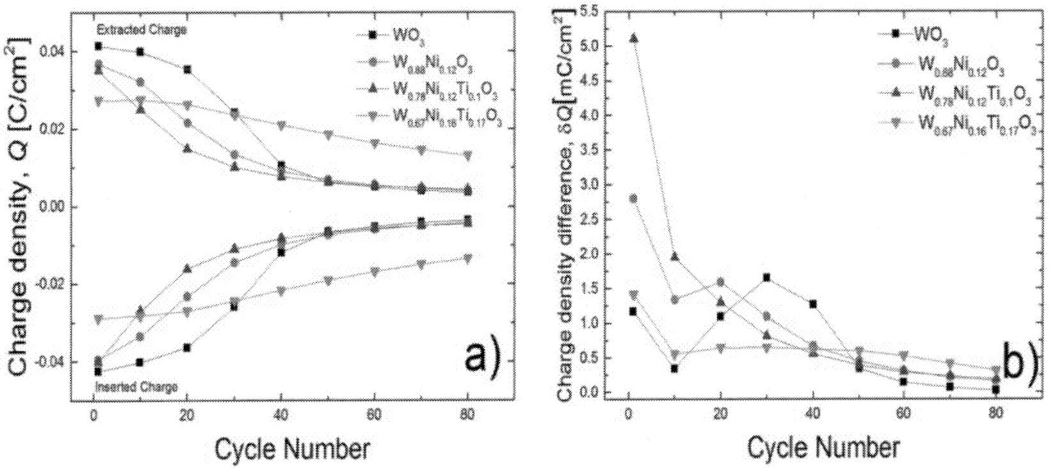

Figure 2. (a) Extracted and inserted charge density during voltammetric cycling of films with the shown compositions, and (b) difference in charge density for successive charge extractions and insertions. Symbols indicate data and lines were drawn for convenience.

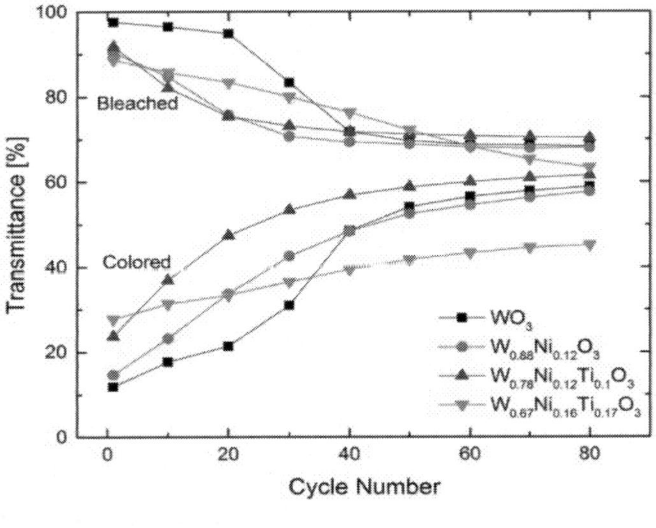

Figure 3. Evolution of transmittance at the wavelength 550 nm in the bleached and colored states for films with the shown compositions. The films were immersed in 1M Li–PC and studied as a function of cycle number.

4. Optical transmittance

Optical transmittance T was recorded in situ during CV cycling, using an Ocean Optics fiber-optic instrument, at the wavelength 550 nm. Figure 3 shows transmittance in bleached and colored states for the same selected CV cycles as in figure 2. The pure tungsten oxide film lost its optical modulation during the charge insertion/extraction process. This is different from the case of films containing nickel and titanium that presented significant changes between the bleached and the colored states even after 40 cycles. An important aspect to highlight is the color of the films after cycling; we found it to be brownish–yellowish, which is typical for samples that have undergone several cycles using a large voltage window and a Li–PC electrolyte [4]. However, and in accordance with the electrochemical results, this yellowish tint was much less marked for the $W_{0.67}Ni_{0.16}Ti_{0.17}O_3$ sample, indicating a lower degradation of the films.

As a complement to the optical and electrochemical characterization, we calculated the coloration efficiency (CE) by CE = $ln(T_b/T_c)/\Delta Q$, where ΔQ is the inserted charge density and b and c denote bleached and colored states, respectively. Data on CE as a function of cycle number are shown in figure 4, where it can be observed that W–Ni oxide shows the best response in CE, confirming results in previous research [10]. However the incorporation of titanium in W–Ni oxide produces a decrease in CE, as can be seen in figure 4. It is important to note here that the high values of CE for pure W oxide and for W–Ni oxide should not be overrated because of the mentioned yellowish coloration of these films.

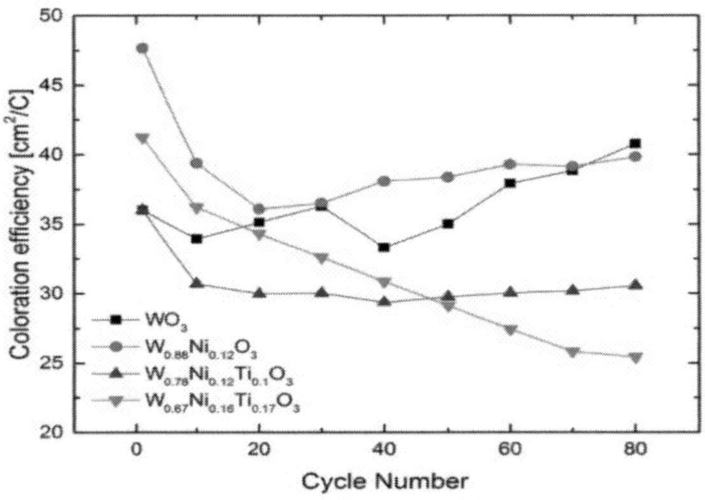

Figure 4. Evolution of coloration efficiency at the wavelength 550 nm for films with the shown compositions. The films were immersed in 1M Li–PC and studied as a function of cycle number.

5. Conclusions

W oxide and W–Ni oxide films suffered clear deterioration in their electrochromic properties during less than 50 voltammetric cycles in a wide voltage range (1.7–4 V *vs.* Li) but an addition of ~17at.% of Ti improves the stability considerably. On the other hand the Ti content diminished the coloration efficiency. Future studies should investigate whether different concentrations of Ni could improve the coloration efficiency and eventually lead to electrochromic films with both good coloration and high durability.

Acknowledgments

This paper was presented at the INERA Conference "Light in Nanoscience and Nanotechnology" in Hissar, Bulgaria, October 19–22, 2015. The conference is part of the INERA REGPOT project of the Institute of Solid State Physics, Bulgarian Academy of Sciences. We are grateful to Daniel Primetzhofer and the staff of the Tandem Accelerator Laboratory at Uppsala University for support with RBS measurements. MML and MAA thank the Mexican Council for Science and Technology (CONACyT) for financial assistance to carry out research as visiting PhD student and as postdoctoral researcher at Uppsala University, respectively. Our work was also supported by the European Research Council under the European Community's Seventh Framework Program (FP7/2007–2013)/ERC Grant Agreement No. 267234 ("GRINDOOR").

References

[1] Granqvist C G, Green S, Niklasson G A, Mlyuka N R, Kræmer S von and Georén P 2010 *Thin Solid Films* **518** 3046-53
[2] Niklasson G A and Granqvist C G 2007, *J. Mater. Chem.* **17** 127–56
[3] Granqvist C G 1995 *Handbook of Inorganic Electrochromic Materials* (Amsterdam: Elsevier)
[4] Arvizu M A, Triana C A, Stefanov B I, Granqvist C G and Niklasson G A 2014 *Sol. Energy Mater. Sol. Cells* **125** 184–89
[5] Wen R-T, Niklasson G A and Granqvist C G 2014 *Thin Solid Films* **565** 128–35
[6] Wen R-T, Granqvist C G and Niklasson G A, 2014 *Appl. Phys. Lett.* **105** 163502
[7] Yuan Y F, Xia X H, Wu J B, Yang J L, Chen Y B and Guo S Y 2011 *Electrochim. Acta* **56** 2627–32
[8] Green S, Backholm J, Georén P, Granqvist C G and Niklasson G A 2009 *Sol. Energy Mater. Sol. Cells* **93** 2050-55
[9] Green S V, Kuzmin A, Purans J, Granqvist C G and Niklasson G A 2011 *Thin Solid Films* **519** 2062-66
[10] Green S V, Pehlivan E, Granqvist C G and Niklasson G A 2012 *Sol. Energy Mater. Sol. Cells* **99** 339-44
[11] Mayer M, 1999 *AIP Conf. Proc.* **475** 541–44

Journal of Physics: Conference Series **682** (2016) 012022 doi:10.1088/1742-6596/682/1/012022

Photocatalytic degradation of Reactive Black 5 and Malachite Green with ZnO and lanthanum doped nanoparticles

N Kaneva, A Bojinova and K Papazova

Laboratory of Nanoparticle Science and Technology, Department of General and Inorganic Chemistry, Faculty of Chemistry and Pharmacy, University of Sofia, 1164 Sofia, Bulgaria

E-mail: nina_k@abv.bg

Abstract. Here we report the preparation of ZnO particles with different concentrations of La^{3+} doping (0, 0.5 and 1 wt%) via sol-gel method. The nanoparticles are synthesized directly from $Zn(CH_3COO)_2.2H_2O$ in the presence of 1-propanol and triethylamine at 80°C. The conditions are optimized to obtain particles of uniform size, easy to isolate and purify. The nanoparticles are characterized by SEM, XRD and UV-Vis analysis. The photocatalytic properties of pure and La-doped ZnO are studied in the photobleaching of Malachite Green (MG) and Reactive Black 5 (RB5) dyes in aqueous solutions upon UV illumination. It is observed that the rate constant increases with the La loading up to 1 wt%. The doping helps to achieve complete mineralization of MG within a short irradiation time. 1 wt% La-doped ZnO nanoparticles show highest photocatalytic activity. The La^{3+} doped ZnO particles degrade faster RB5 than MG. The reason is weaker N=N bond in comparison with the C–C bond between the central carbon atom and N,N-dimethylaminobenzyl in MG. The as-prepared ZnO particles can find practical application in photocatalytic purification of textile wastewaters.

1. Introduction

Textile wastewaters are a large problem for conventional treatment plants in the entire world. The release of these wastewaters in natural environments is very problematic to aquatic life [1] and mutagenic to human [2]. The traditional treatment techniques applied in textile wastewaters are not a complete solution to the problem due to the biological resistance of some dyes. There are many kinds of dyes available in the market. Based on the chemical structure of chromophore, 20-30 different groups of dyes can be discerned. Anthraquinone, phthalocyanine, triarylmethane and azo dyes are quantitatively the most important groups. The azo dyes, characterized by having an azo group consisting of two nitrogen atoms (-N=N-), are the largest class of dyes used in textile industry. Additionally, these dyes are the most problematic pollutants of textile wastewaters. Therefore, in this work, we have selected the dye Reactive Black 5 (RB5) as one of the most used reactive dyes for textile finishing. The other dye Malachite Green oxalate (MG) is used as a model pollutant in order to compare the particles behavior.

Many methods have been used to synthesize ZnO particles, for example, sol–gel [3, 4], hydrothermal [5] and co-precipitation [6]. Nowadays, various wet chemical methods have been consolidated and are now fashionable for synthesizing ZnO nanoparticles because they provide a simple and economic route as well as low temperature requirement. The addition of appropriate

Journal of Physics: Conference Series **682** (2016) 012022 doi:10.1088/1742-6596/682/1/012022

surfactants can help to control the ZnO morphology. The morphology/surface area plays an important role in the photocatalytic activities of metal oxides. The reason is that doping of metal oxide and/or rare earth metals (like La) increases the surface defects [7, 8].

In our previous studies we have tested the photocatalytic degradation of Malachite Green and Reactive Black 5 using nanostructured films [9-12]. Thin films of zinc oxide were obtained by three different syntheses using different precursors: (*i*) zinc acetate dihydrate, 2-methoxyethanol and monoethanolamine [9], zinc oxide powder, polyethyleneglycol, ferrous (III) chloride, $ZnCl_2$ [10, 11], zinc acetate dihydrate, n-propanol and triethylamine [12]. Based on these results, we use the latter sol-gel method for synthesis of zinc oxide nanoparticles.

Recently, La-doped ZnO materials (nanoparticles, nanorods, nanowires etc.) have attracted much attention in the photocatalytic processes owing to their high photocatalytic activity in the degradation of organic contaminants. The reasons for this effect are the suppression of electron–hole recombination, large content of oxygen vacancies, and strong absorption of OH− ions on the catalysts surface [13–15].

The zinc oxide particles are usually synthesized by sol–gel method, allowing good control of the size and morphology of the submicron particles. For this purpose, surfactants species, such as monoethanolamine and diethanolamine are applied in the preparation of ZnO particles, described in the literature [16, 17]. Recent reports on the preparation of pure and rare earth doped ZnO nanoparticles demonstrate the potentials for their applications in biomedical research [18, 19]. Here we investigate the application of such particles (undoped and doped) as a photocatalyst.

Here we report for the first time on the synthesis and characterization of La-doped ZnO catalysts and their photocatalytic activity in the degradation of MG and RB5 at different initial concentrations: 3, 5, 7 and 10 ppm. All the materials have been characterized by SEM, XRD and UV–VIS analysis.

2. Experimental

2.1 Sample preparation

Zinc acetate dihydrate (> 99.5 %), 1-propanol (> 99.0 %) and triethylamine (> 99.5 %) were from Fluka. Lanthanum oxide (> 99.9 %) and acetic acid were from Sigma Aldrich. Malachite Green ($C_{52}H_{54}N_4O_{12}$, λ_{max} = 615 nm, dye content ≥90%, Croma GmbH) oxalate and Reactive Black 5 ($C_{26}H_{21}N_5Na_4O_{19}S_6$, λ_{max} = 585 nm, dye content 55%, Sigma-Aldrich) were used in the photocatalytic experiments. Distilled water was used in all experiments. All other chemicals and reagents were analytical reagent grade.

The synthesis of lanthanum acetate (III) was carried out following the reactions:

$$La_2O_3 + 6CH_3COOH \rightarrow 2La(CH_3COO)_3 + 3H_2O$$

For this purpose, La_2O_3 powder was dissolved in a solution of CH_3COOH. The materials are mixed together in a round-bottomed flask and stirred at a room temperature. The resultant solution was heated at 80°C for 30 min and washed by repeated decantation with 25 ml portions of H_2O. The suspension was filtered and dried at 80°C for 2 h. The obtained product consisted of almost white, highly lustrous, brittle pieces, which were stirred with a mortar into fine particles and dried again at 80°C for 2 hours.

Zinc oxide particles with different concentrations of La^{3+} doping (0, 0.5 and 1 wt%) were synthesized by sol-gel method. Zinc acetate ($Zn(CH_3COO)_2.2H_2O$), lanthanum acetate ($La(CH_3COO)_3$) and ethylene glycol ($C_2H_4(OH)_2$) were dissolved in 1-propanol, and then triethylamine (TEA) was added to the solution as a stabilizer. The substances were mixed together in a round-bottomed flask with reversed condenser and stirred at a room temperature for 15 minutes. The obtained clear solution was heated at 80°C upon magnetic stirring for 30 min to obtain particles of average size of 150-230 nm. The resultant solution was clear and homogenous. The obtained submicron particles were separated by centrifugation at 6000 rpm for 30 min, twice washed with pure 1-propanol and dried at open air to obtain pure and doped ZnO particles.

2.2. Sample characterization

The as-obtained pure ZnO particles and doped with different percent La^{3+} were first imaged by a scanning electron microscope (SEM) JSM-5510 (JEOL), operated at 10 kV of acceleration voltage. The investigated samples were coated with gold by JFC-1200 fine coater (JEOL) before observation.

X-Ray diffraction (XRD) spectra were taken at room temperature by powder diffractometer (Siemens D500 with CuK_α radiation within 2θ range of 30-70 deg at a step of 0.05 deg 2θ and counting time 2 s/step). This method was used for determination of samples crystallinity phase and composition.

The photocatalytic activities of ZnO and ZnO/La^{3+} particles were investigated and compared by the photoinitiated decolorization of the dyes Malachite Green and Reactive Black 5 in aqueous solution (upon UV-light illumination). The concentration of ZnO particles was 0.12 mg/ml. The measurement was conducted in a cylindrical glass vessel of volume 150 ml, equipped with a magnetic stirrer and UV-lamp (main emission at 370 nm) [20]. The light power density at the sample position was 0.66 mW/cm^2. The light irradiation was measured with research radiometer of Ealing Electro-optics, Inc. The lamp (Sylvania BLB 18W) was fixed ~15 cm above the treated solution. The degradation process of MG and RB5 were monitored by absorbance spectroscopy after aliquot sampling at regular time intervals. Each aliquot sample (2 ml) returned back to the reaction vessel immediately after the spectrophotometrical measurement. The particles (ZnO and ZnO/La^{3+}) were then removed from the aliquot by centrifugation (6000 rpm, 3 min) before the UV-VIS spectrophotometric measurement of dyes concentration. After that, the aliquot, together with the photocatalyst particles were returned back to the glass reactor. In all experiments the optical absorbance spectra were measured by spectrophotometer Jenway 6400 in the wavelength range from 400 to 800 nm.

Five series of experiments were carried out with: pure ZnO particles, doped with different percent La^{3+} (0.5 and 1%). The effect of initial concentration of MG and RB5 (for 3, 5, 7 and 10 ppm) was also evaluated. All photocatalytic tests were performed at constant stirring rate (500 rpm) at room temperature (25°C).

3. Results and discussion

3.1. Structural characterization of ZnO and ZnO/La^{3+} particles

The SEM images of the as-prepared ZnO and $ZnO:La^{3+}$ particles are shown in figure 1. As seen, the morphology is well ordered and spherical colloidal particles with average diameter size of about 150-230 nm are formed. The SEM investigation reveals that after lanthanum doping, the average size of ZnO colloidal particles is increased (figures 1b, c). The observed influence is expected to prove the photocatalytic efficiency of the ZnO. Particle size and surface area have key effects on the

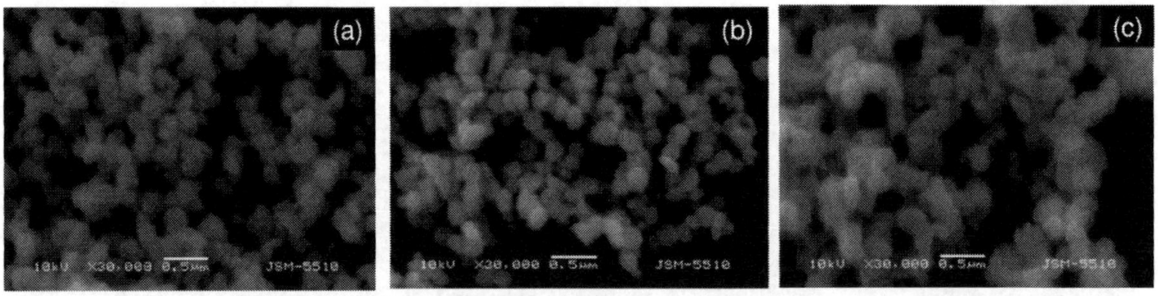

Figure 1. SEM images of ZnO particles at different concentrations of La^{3+} doping : (a) 0%; (b) 0.5 wt% and (c) 1 wt%.

photocatalytic activities of photocatalysts. The particle size distribution, determined from the SEM images is shown in figure 2. Average grain sizes of about 150, 210 and 225 nm are determined for the ZnO samples, doped with 0, 0.5 and 1 wt% respectively. There is a tendency of increasing the particle size with the rise of dopants concentration.

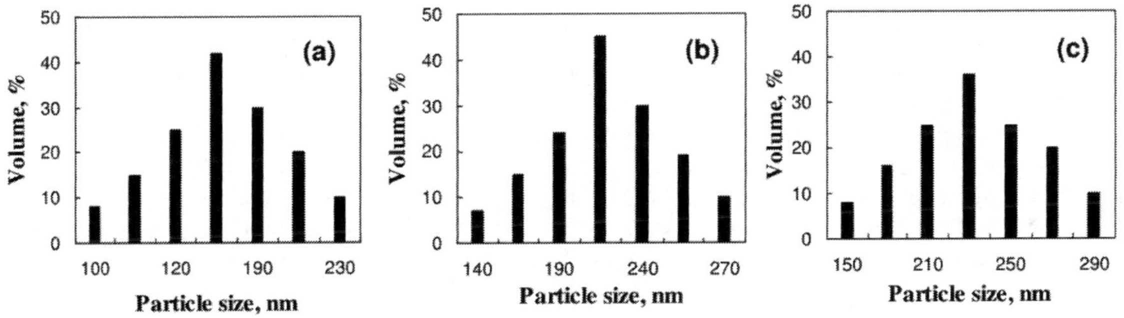

Figure 2. Statistical volume % distribution of ZnO particles doped at different concentrations of La^{3+}: (a) 0%; (b) 0.5 wt% and (c) 1 wt%.

X-ray diffractogram of the particles has been analyzed to obtain information about various crystalline aspects. Figure 3a shows XRD patterns of synthesized undoped ZnO nanopowders. The sharp and intense peaks indicate that the samples are of high crystallinity and ZnO nanoparticles have polycrystalline structure. The XRD peaks for (100), (002) and (101) planes indicate the formation ZnO of phase pure hexagonal wurtzite structure. The high intensity of (101) peak suggests that the growth of nanoparticles has taken place along this direction of crystallization of ZnO. No peaks due to impurities are observed. Figure 3b shows the XRD patterns of 1wt% La doped ZnO catalysts are similar to that of ZnO [21], suggesting that there is no change in the crystal structure upon La loading. This also indicates that La^{3+} is uniformly dispersed onto ZnO nanoparticles [22].

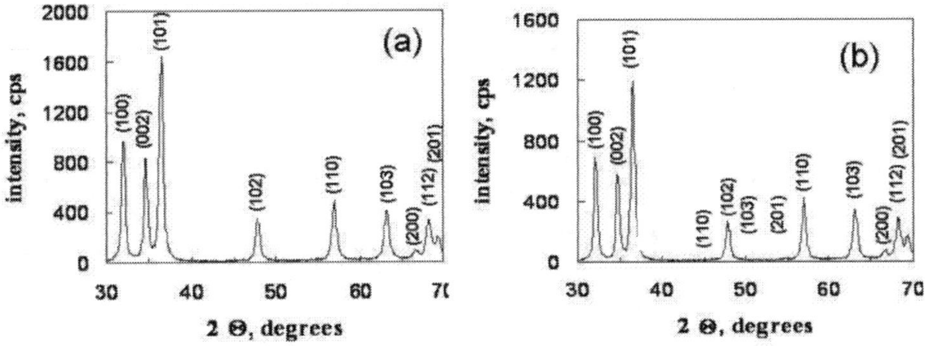

Figure 3. XRD spectra of ZnO particles prepared by sol-gel and doped at: (a) 0 wt% and (b) 1 wt% La^{3+}.

3.2. Photocatalytic activity

In order to determine the photocatalytic activity of ZnO and La-doped ZnO, a series of experiments are carried out with Malachite Green (MG) and Reactive Black 5 (RB5) in aqueous suspensions. The experiments under UV light illumination exhibit a much higher decolorization rate of RB5 in comparison with MG solutions. The results show that the photocatalytic decolorization of MG and RB5 by ZnO colloids under UV illumination follows pseudo first-order kinetics.

The concentrations ratio (C/C_0, where C_0 is the initial concentration of dye and C is the concentration of dye after irradiation at selected time interval) is calculated to evaluate the degradation of the dyes. Curves of MG and RB5 degraded by ZnO colloids with different La^{3+} doping concentrations are shown in Fig. 4 and 5. As it represented in the figures at different concentration of the dye (3, 5, 7 and 10 ppm), undoped ZnO has a smaller ability to mineralize the dyes (about 81-89% in 180 min, calculated using the relation $D = [(C_o\text{-}C)/C_o]\times 100)$ under UV-light irradiation. However, all of the La-doped ZnO photocatalysts exhibit higher photocatalytic activity than the pure ZnO. This may be attributed both to the better absorption in UV-light range and to the larger number of oxygen vacancies or defects, produced by La^{3+} doping. During the 180 min irradiation, the order of the photocatalytic activities of the La-doped ZnO with different La^{3+} doping concentrations is as follows: 1>0.5>0 wt%. This dependence of photocatalytic activity is observed for both dyes. The 1 wt% La-doped ZnO photocatalyst exhibits the highest photocatalytic efficiency for decolorization of MG and RB5 (94.22% and 99.43% after 180 min). That suggests that the doping of La^{3+} may enhance the photocatalytic activity of ZnO particles and there is an optimal doping concentration of La^{3+} ions in ZnO. The pure particles have the smallest photocatalytic activity and slowest degradation rate of the dye.

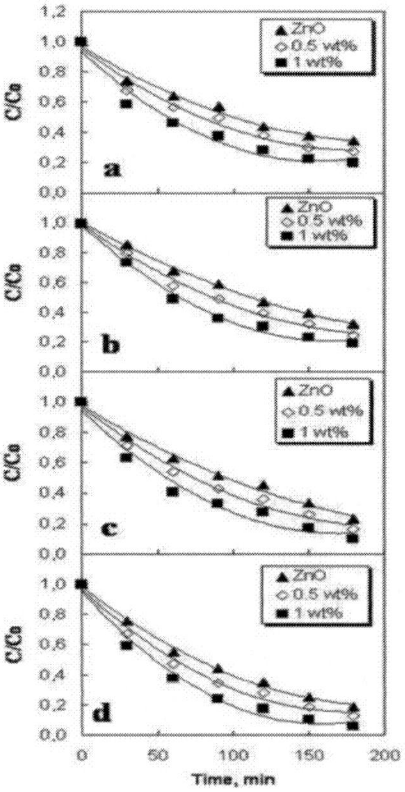

Figure 4. Photocatalytic activity of the Malachite Green at initial concentration: (a) 3 pppm, (b) 5 ppm, (c) 7 ppm and (d) 10 ppm in aqueous solution (under UV-illumination) with ZnO particles with different amounts of La^{3+}doping.

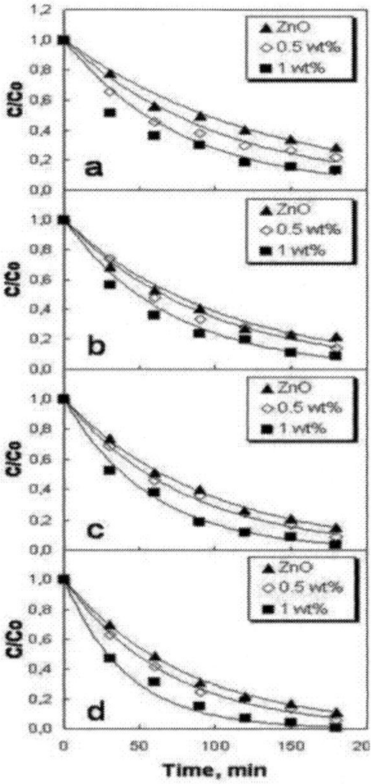

Figure 5. Photocatalytic efficiency of pure and doped ZnO particles for degradation of Reactive Black 5 at initial concentration of: (a) 3 pppm, (b) 5 ppm, (c) 7 ppm and (d) 10 ppm in water solution under UV-light illumiination.

The ZnO particles obtained by sol-gel method possess higher activity and faster degrade RB5 in comparison to MG. The reason for decomposition of RB5 is OH radicals that attack first the azo groups and thus destructing the long conjugated p-systems [23]. As the N=N bonds are easier for destruction than the aromatic ring structures. This explains the higher photoactivitity of ZnO films in the decomposition of RB5, compared to MG, which does not contain N=N bond. The N=N bond is weaker than the C–C bond between the central carbon atom and N,N-dimethylaminobenzyl in MG, which results in the formation of 4-(N,N-dimethylamino) methylbenzylone as most common intermediate [24].

The kinetics of the reaction is obtained by plotting the natural logarithm of the concentration ratios versus the irradiation time. Straight lines are obtained, indicating that the reactions are of first order kinetics. The particles with the lowest rate of reaction are the one that contains ZnO for degradation of Malachite Green (k = 0.0046 min^{-1}) and Reactive Black 5 (k = 0.0113 min^{-1}) under UV-light illumination. The values of the rate constants increases with the dopant concentration increase (0.5wt% La^{3+} - k_{MG} = 0.0091 min^{-1} and k_{RB5} = 0.0122 min^{-1}, 1 wt% La^{3+} - k_{MG} = 0.0154 min^{-1} and k_{RB5} = 0.0243 min^{-1}). This means that the incorporation of La-doped ZnO in the particles increases the photocatalytic efficiency. These values are presented in figure 6.

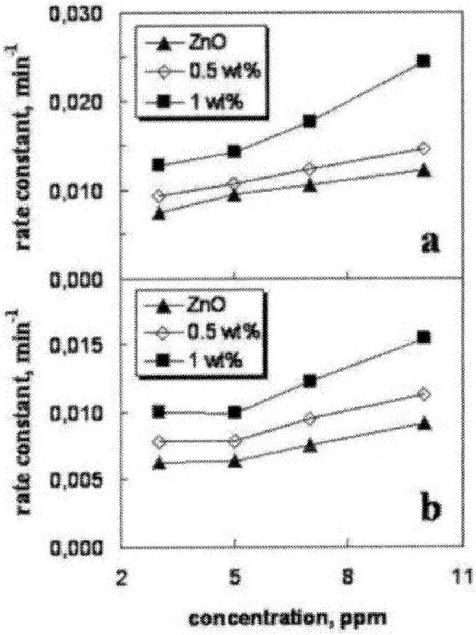

Figure 6. Comparison of the apparent rate constants of photocatalysis of for doped ZnO particles at different initial concentration under UV-light illumination of: (a) RB5 and (b) MG.

Among the catalysts studied, 1.0 wt% La-doped ZnO is found to be more active, showing higher photocatalytic activity for the degradation of RB5 in comparison with MG. Thus, we can conclude that the optimum loading of La is 1 wt% which may be more efficient for separating photoinduced electron–hole pairs and enhance the photocatalytic activity.

4. Conclusions

A series of La-doped ZnO particles with different La doping content (0, 0.5 and 1 wt%) are prepared by sol-gel method and characterized by Scanning Electron Microscopy, X-ray diffraction and UV-VIS analyses. The colloidal particles (pure and doped) are spherical in shape with hexagonal wurtzite structures, which demonstrates that the dopant La^{3+} ions have no effect on the crystal structure. The kinetics of photocatalytic reaction is systematically studied under UV at different variable system parameters – amount of lanthanum doping and initial concentration of Malachite Green and Reactive Black 5 in the reaction solution. The 1 wt% La-doped ZnO photocatalyst exhibits the highest photocatalytic decolorization efficiency, leading to as much as 99.43% photocatalytic degradation of

RB5 in 180 min. The photocatalytic efficiency of pure ZnO particles is lower than the activity achieved by La-doped sample, giving a thickness-based rate constant of pure ZnO << ZnO:La^{3+}(1%) under UV-light illumination. This fact is observed for both dyes. ZnO and ZnO:La particles are reproducible and stable. The doped particles have the highest photocatalytic activity in comparison to the undoped ones. All these observations prove that La-doped ZnO is a potential candidate for the practical application in photocatalytic degradation of organic contaminants from water solutions.

Acknowledgements
This research is financially supported by the project DFNI-T02/16 (№2742) and Russian Ministry of Education as part of the base portion of the state task № 2014/151 (Project Code 117).

5. References

[1] Chung T and Stevens J 1993 *Environ. Tox. Chem.* **54** 435
[2] Chung T, Stevens J and Cerniglia E 1992 *Crit. Rev. Microbiol.* **18** 175
[3] Widiyastuti W, Maula I, Nurtono T, Taufany F, Machmudah S, Winardi S and Panatarani C 2014 *Chem. Eng. J.* **52** 254
[4] Ba-Abbad M, Kadhum A, Mohamad A, Takriff M and Sopian K 2013 *J. All. Comp.* **550** 63
[5] Maryanti E, Damayanti D, Gustian I and Yudha S 2014 *Mat. Lett.* **118** 96
[6] Raoufi D 2013 Renewable Ener. **50** 932
[7] Khatamiana M, Khandara A., Divbanda B, Haghighib M and Ebrahimiasl S 2012 *J. Mol. Catal. A: Chem.* **365** 120
[8] Sin J, Lam S, Lee K and Mohamed A 2014 *Ceram. Intern.* **40** 5431
[9] Kaneva N, Bojinova A, Papazova K and Dimitrov D 2014 *J. Univ. Chem. Technol. Metall.* **49** 149
[10] Kaneva N and Dushkin C 2011 *Coll. Surf. A* **382** 211
[11] Kaneva N and Bojinova A 2012 *Compt. rend. de l'Acad. bul. des Scie.* **65** 1349
[12] Kaneva N, Papazova K, Bojinova A and Dimitrov D 2013 *Nanoscience&Nanotechn*ology, eds E Balabanova, E Mileva **13** 110
[13] Korake P, Dhabbe R, Kadam A, Gaikwad Y and Garadkar K 2014 *J. Photochem. Photobiol. B: Biol.* **130** 11
[14] Suwanboona S, Amornpitoksuk P, Sukolrat A and Muensit N 2013 *Ceram. Intern.* **39** 2811
[15] Jia T, Wang W, Long F, Fu Z, Wang H and Zhang Q 2009 *J. All. Comp.* **484** 410
[16] Li H, Wang J, Liu H, Zhang H and Li X. 2005 *J. Cryst. Growth* **275** 943
[17] Cheng X, Zhao H, Huo L, Gao S and Zhao J 2004 *Sens. Actuators B* **102** 248
[18] Tripathy N, Hong T, Ha K, Jeong H and Hahn Y 2014 *J. Hazard. Mater.* **270** 110
[19] Wolska E, Kaszewski J, Kiełbik P, Grzyb J, Godlewski M and Godlewski M 2014 *Opt. Mater.* **36** 1655
[20] Dushkin C, Stoianov S, Bojinova A and Russev S 2006 *Ann. de L'Univ. Sofia „St. Kl. Ohridski" fac. de chim.* **98/99** 73
[21] Wang J and Gao L 2005 *J. Am. Ceram. Soc.* **88** 1637
[22] Lin J, Yu J, Lo D and Lam S 1999 *J. Catal.* **183** 368
[23] Lucas M and Peres J 2006 *Dyes & Pigments* **71** 236
[24] Guandao G, Aiyong Z, Meng Z, Jinlong C and Quanxing Z 2008 *Chin. J. Catal.* **29** 426

Investigation of sol-gel yttrium doped ZnO thin films: structural and optical properties

T Ivanova[1], A Harizanova[1], T Koutzarova[2] and B Vertruyen[3]
[1]Central Laboratory of Solar Energy and New Energy Sources, Bulgarian Academy of Sciences, blvd. Tzarigradsko chaussee 72, Sofia, Bulgaria
[2]Institute of Electronics, Bulgarian Academy of Sciences, blvd. Tzarigradsko chaussee 72, Sofia, Bulgaria
[3]LCIS/SUPRATECS, Institute of Chemistry B6, University of Liege, Sart-Tilman, B-4000 Liege, Belgium

E-mail: tativan@phys.bas.bg

Abstract. Nanostructured metal oxide films are extensively studied due to their numerous applications such as optoelectronic devices, sensors. In this work, we report the Y–Zn–O nanostructured films prepared by sol-gel technology from sols with different concentration of yttrium precursor, followed by post-annealing treatment. The Y doped ZnO thin films have been deposited on Si and quartz substrates by spin coating method, then treated at temperatures ranging from 300-800°C. XRD analysis reveals modification of the film structure and phases in the doped ZnO films.

1. Introduction

ZnO gains a great scientific research in respect to its interesting properties such as wide band gap (3.37 eV at room temperature) with large exciton binding energy (60 meV), high chemical stability, low dielectric constant, large electrochemical coupling coefficient, high thermal conductivity, binding, antibacterial and UV protection [1,2]. ZnO films possess high transmittance in the infrared and visible spectral regions and high refractive index, which extend their application range. They are used as chemical and biological gas sensors, UV light emitters, thin film transistors, solar cell windows [3].

The doping of ZnO with rare earth impurities such as Y has also been reported as its ionic radius is very close to that of Zn [4]. ZnO:Y nanostructures are studied in respect to improved photocatalytic properties, enhance intensity of UV emission in photoluminescence, conducting transparent coatings etc [5, 6].

In this work, we report the Y–Zn–O nanostructured films prepared by sol-gel technology from sols with different concentration of yttrium precursor, followed by post-annealing treatment. Sol-gel technology has been proved to be very successfully for obtaining doped ZnO films. The sol-gel approach offers possibility to control the film stoichiometry, composition modification (mixing on a molecular level), cost effective process, inexpensive equipments resulting in homogeneous and smooth thin films [7].

[1] To whom any correspondence should be addressed.

The sol-gel Y doped ZnO thin films have been deposited on Si and quartz substrates by spin coating method, then treated at the temperatures ranging from 300-800°C. XRD analysis reveals the modification of the film structure and phases in the doped ZnO films. Vibrational properties are investigated by FTIR spectroscopy and optical behavior has been studied by UV-VIS spectrophotometry (transmittance and reflectance spectra). Optical band gaps are estimated as a function of the annealing temperatures.

2. Experimental details

The sol solution for ZnO deposition has been previously reported [8]. The precursor used is zinc acetate, dissolved in an absolute ethanol (0.4 M concentration). The complexing agent and stabilizer used is monoethanolamine (MEA). The molar ratio MEA/Zn is fixed to 1. The yttrium has been involved by adding $Y(NO_3)_3$ monohydrate in the following weight percents: 0,1 and 1. The corresponding sols and thin films obtained from them are labeled ZnO:Y 0.1 and ZnO:Y 1 and they are referred with these labels throughout the text. For depositing ZnO:Y thin films, the Zn solutions with Y additive were homogenized by ultrasonic treatment for few hours. The sol solutions are found to be very stable retaining their film forming properties for more than three months.

The films were deposited by spin coating method at 4000 rpm on Si wafers and quartz substrates, the samples undergo five layer depositions. The preheating temperature used (heating between layers) is 300°C for 10 minutes. After repeating the coating procedures five times, the films have been subjected to annealing at temperatures of 300 to 800°C in air ambient for 1 hour. The heating rate for the preheating and annealing temperatures has been kept constant 10°C/minute.

XRD spectra of the sol-gel films have been recorded by means of XRD diffractometer Bruker D8, at the grazing angle 2° and step time of 8 s and step a of 0.1° FTIR measurements are performed in the spectral region 350-4000 cm^{-1} by Shimadzu FTIR Spectrophotometer IRPrestige-21. Optical measurements have been were done by using UV-VIS-NIR Shimadzu 3600 spectrophotometer.

3. Results and discussions

The ZnO and ZnO:Y films, annealed at 600 and 800°C have been subjected to XRD measurements and the recorded diffraction patterns are presented in figure 1.

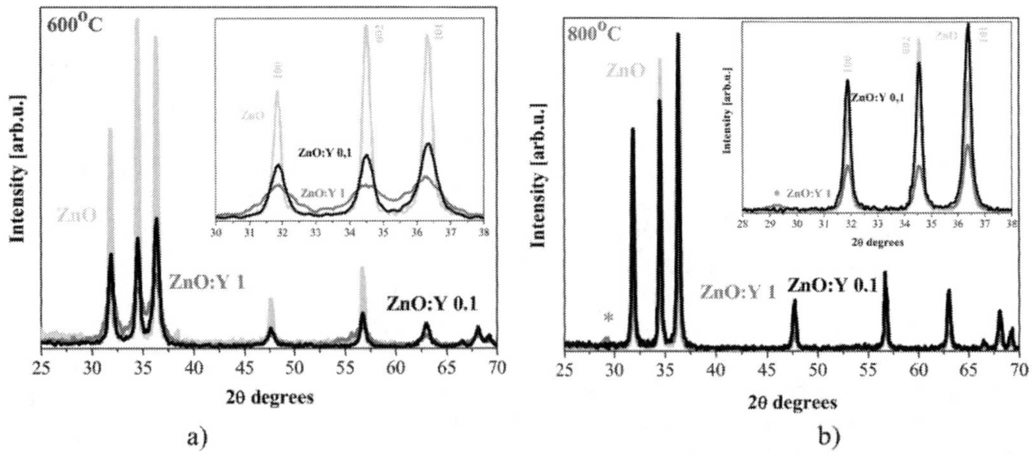

Figure 1. XRD spectra of ZnO:Y films compared to ZnO and annealed at 600 (a) and 800°C (b). The inset figures represent enlarged diffraction patterns in the 2θ range 25 - 39°.

The yttrium doping leads to suppress film crystallization as it is clearly observed for the films, annealed at 600°C. At the highest annealing temperature (800°C) only the film with the lower Y addition reveals a degree of crystallization close to that of ZnO films. The XRD patterns of ZnO,

ZnO:Y 0.1 and ZnO:Y 1 show predominantly diffraction peaks due to wurtzite ZnO phase (JSPDS 01-070-8070) with weak lines related to ZnO_2 phase (JSPDS 04-005-4315). Interesting features appear in the XRD spectra of ZnO:Y1 (very weak line) (figure 1b). The extra peak at 2θ=29,1° (marked with asterisk) can be assigned to (222) reflection of cubic Y_2O_3 phase [JCPDS 01-073-1334, 9]. Additional experiments with higher yttrium concentrations reveal that cubic yttrium oxide appeared as separate phase after high temperature annealing.

The average crystallite sizes can be estimated from the XRD data according to Scherrer's formula (using XRD peaks: 100, 002 and 101) and the results obtained are given in table 1. Table 1 presents also the calculated lattice parameters of the wurtzite ZnO phase and the dislocation density of the films. Dislocations are an imperfection in a crystal associated with misregistry of the lattice in one part of the crystal with respect to another part. The dislocation density of the films is given by the Williamson and Smallman's relation [10]: $\delta = n/d^2$ (1), where n is a factor, which equals unity giving minimum dislocation density and *d* is the grain size. Dislocation densities manifest a decrease with increasing annealing temperatures, which indicates a lower concentration of lattice imperfections.

Table 1. Crystallite sizes (d), dislocation density and lattice parameters of ZnO and ZnO:Y films.

	ZnO	**ZnO:Y 0.1**	**ZnO:Y 1**
600°C			
d, (nm)	30.9	15.8	8.9
δx10^{-4}, (1/nm^2)	10.5	40.1	126.2
a (Å)	3.243	3.246	3.243
c (Å)	5.193	5.197	5.191
800°C			
d, (nm)	36.2	31.6	24.0
δx10^{-4}, (1/nm^2)	7.6	10.0	17.4
a (Å)	3.241	3.245	3.243
c (Å)	5.185	5.190	5.195

The yttrium component influences the crystallization of the films and respectively the crystallite sizes are found to be smaller than those of ZnO films for the two annealing temperatures. XRD study reveals that yttrium doping content affects considerably the crystalline structure of sol-gel ZnO:Y thin films and formation of Y_2O_3 phase begins at higher Y concentrations.

FTIR spectroscopy has been used for studying the vibrational properties of ZnO:Y films as it is well known that FTIR spectroscopy is a very sensitive characterization method and can contribute to XRD conclusions. Figure 2 shows FTIR spectra of ZnO and ZnO:Y films, treated at temperatures of 300°C and 800°C.

The lowest annealing temperature of 300°C results in FTIR spectra revealing a weak band at 668 cm^{-1} (Zn-O stretching mode). This band can be observed for all studied films independently of thermal treatments. The effect of yttrium additive is clearly seen: the shape and the intensity of the main absorption bands are considerably different than those of undoped ZnO. The main band of ZnO film is broad and strong centred at 372 cm^{-1} (300°C) and shifting to 395 cm^{-1} for higher annealing temperatures. The doped films manifest broader bands with doublets or triplets peaks depending on annealing above 300°C. In this spectral range, there is obviously an overlapping of IR lines, contributions both from ZnO and Y_2O_3 phases. Other authors [11, 12] reported that cubic yttrium oxide shows peaks at 375, 383, 395, 420, 435 and 468 cm^{-1} (for Y_2O_3 films). Some of these lines are seen in figure 2. This result confirms that there is yttrium oxide presence even at the lowest annealing temperature although the XRD analysis do not show any traces of this phase. It must be noted that the spectra of ZnO:Y 0.1 (for all thermal treatments (FTIR spectra of annealed films at 400, 500 and 700°C were measured and analyzed but not given here) matches ZnO spectrum without a sign for yttrium oxide with exception of the highest annealing at 800°C, where its main absorption band clearly indicates some contribution of Y-O phase.

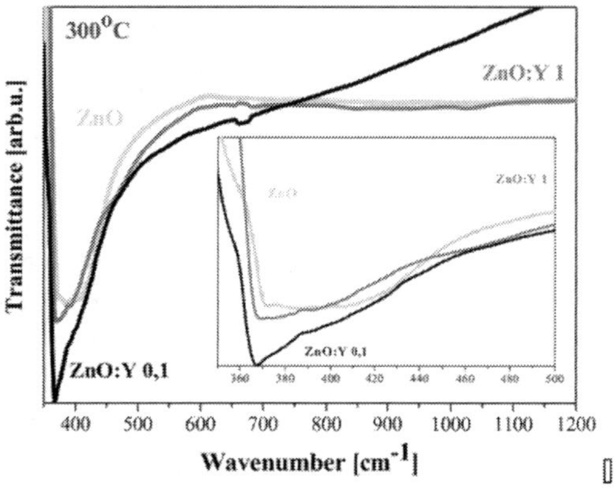

a)

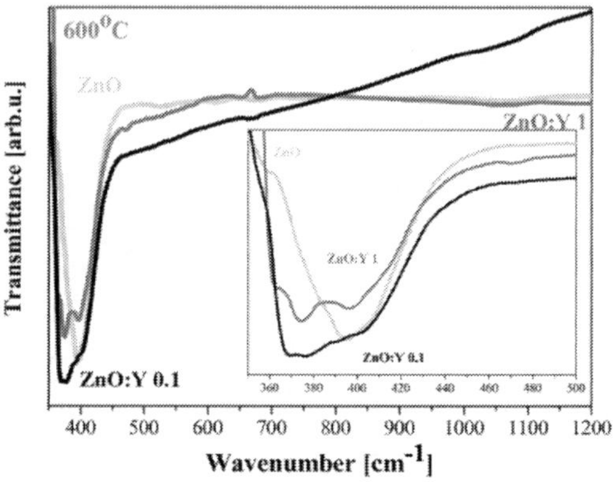

b)

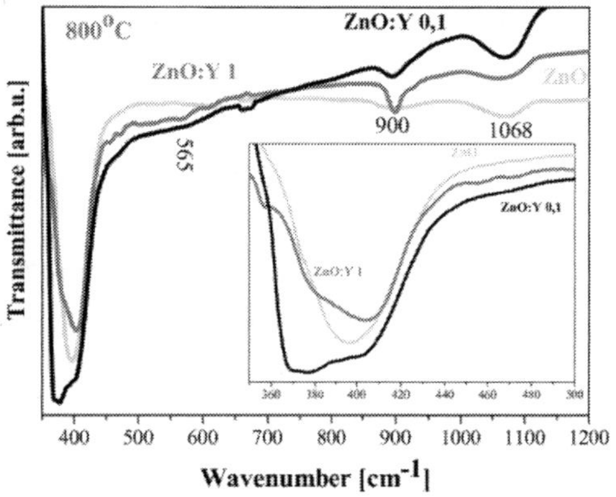

c)

Figure 2. FTIR spectra of ZnO and ZnO:Y films annealed at 300 (a), 600 (b) and 800°C (c). The inset figures show enlarged main absorption band in the spectral range 350-500 cm^{-1}.

The FTIR spectra of ZnO based films, annealed at 800oC shows some additional features. The clear band at 1068 cm^{-1} can be assigned to Si-O bonds and it is due to the Si substrate. New weak bands appeared at 471, 465 cm^{-1} and a clear line at 565 cm^{-1} for ZnO:Y 1 sample. The absorption band at 565 cm^{-1} is assigned to Y-O lattice vibration or Y-O stretching bond [13]. FTIR study shows that a fraction of yttrium oxide can exist in ZnO:Y 1 independently of the annealing temperatures.

The optical transmittance and reflectance spectra of sol-gel ZnO:Y films are shown in figure 3 (the films are deposited on quartz substrates). It can be observed that the thermal annealing influences the optical behavior of the films considerably. The transparency in the visible spectral range is decreasing with raising the annealing temperatures. It can be seen that the transmittance diminishes from 94 % (for 500°C annealed ZnO:Y 0.1 films) to 74 % (800°C) at the wavelength of 550 nm. The ZnO:Y 1 films show lower transmittance in comparison to ZnO:Y 0,1 films, but the optical transparency also shows strong reduction after high temperature treatment. This had been observed for sol-gel ZnO films, as their transmittance drops from 78% at 400°C annealing down to 65 % at 800°C.

The specific bands which have been observed in the optical spectra of ZnO:Y 0.1 (annealing temperatures 500°C and above) below the absorption edge are due to the excitonic absorption of ZnO. The excitonic feature of the bulk ZnO is located near to 373 nm. ZnO films reveal excitonic absorption features even after 300°C treatment. The appearance of excitonic peak is reported to be a sign of high optical quality and good crystallinity of ZnO [14]. The excitonic absorption peaks prove that these sol-gel ZnO and ZnO:Y 0,1 thin films have good optical properties. On the other hand, no excitonic band can be observed for ZnO:Y 1 even at the highest annealing temperatures.

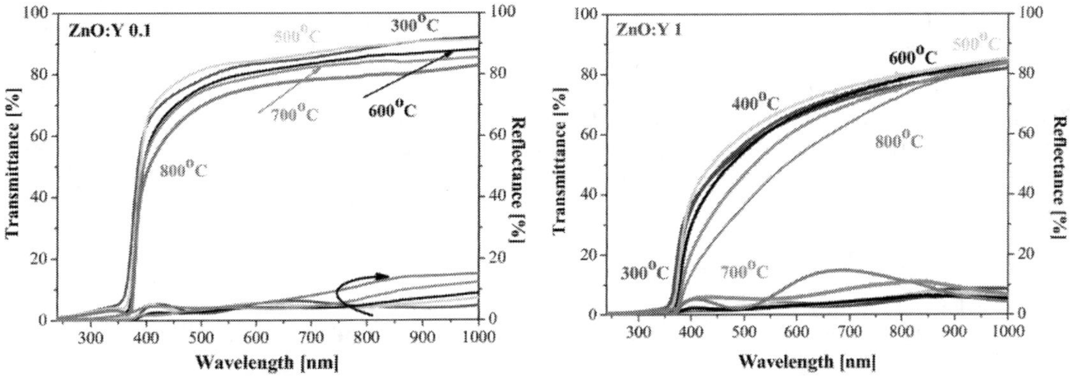

Figure 3. UV-VIS spectra of ZnO:Y 0,1 and ZnO:Y 1 films, annealed at different temperatures.

Figure 3 shows comparison of ZnO and ZnO:Y film transmittance for two annealing temperatures. The lower yttrium concentration leads to slight improvement of the optical transparency in respect of undoped ZnO. The higher yttrium doping results in lower transmittance of ZnO:Y 1 film which might be due to higher scattering.

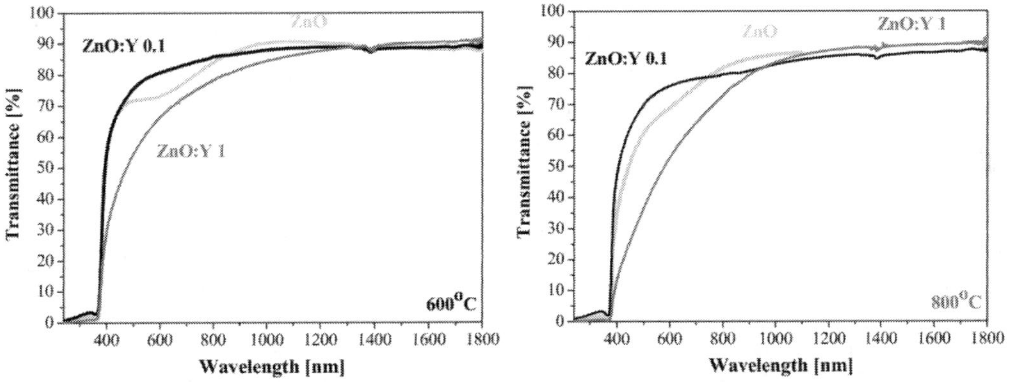

Figure 4. Transmittance spectra of ZnO and ZnO:Y films, annealed at 600 and 800°C.

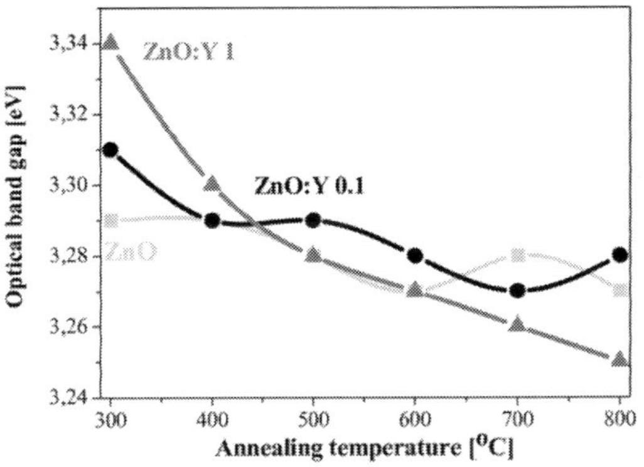

Figure 5. The dependence of the optical band gap with the annealing temperatures.

The optical band gaps (E_g) versus the annealing temperatures are presented in figure 5. For ZnO films, the obtained E_g values are in the range of 3.27 to 3.29 eV, which is below to the reported value

(3.37 eV) of bulk ZnO [15] but close to those given in the literature for ZnO thin films [16]. The bandgap difference between the thin film and crystal can be related to grain boundaries and imperfections of the polycrystalline film structure. The optical band gap values of ZnO:Y 0.1 and ZnO:Y 1 films decrease with increasing the annealing temperatures. This can be attributed to increase of the crystallites size (as it is proved by the XRD analysis); it is known that optical band gap increased with decreasing particle size [17]. Other authors claim that yttrium contribution in sol-gel ZnO films can lead either to higher or smaller optical band gap [18]. In this study, it can be seen a strong correlation with the annealing temperatures and yttrium concentration. The optical band gap of Y_2O_3 is reported 5.5 eV [11] so it seems logically to expect that that ZnO:Y films will show higher Eg values than undoped ZnO. A more detail optical study will be further performed for sol-gel ZnO:Y films with higher yttrium doping.

4. Conclusions
Sol-gel technology is found to be very suitable approach for obtaining nanocrystalline thin films and for deposition of undoped and doped ZnO films. The prepared sols remain stable and were usable for spin-coating within 3 months. ZnO and ZnO:Y films are polycrystalline and have wurtzite structure with a small trace of yttrium oxide phase for ZnO:Y 1 annealed at 800°C. This conclusion is confirmed by FTIR investigation. The optical transparency is improved by a low Y doping and then strongly decreases for higher yttrium concentration. The optical band gaps in ZnO and ZnO:Y films vary in the range of 3.25 eV to 3.34 eV. The obtained properties of ZnO and ZnO:Y films on Si and quartz substrates are promising for applications in optoelectronic devices or solar cells.

5. Acknowledgements
This paper was presented at INERA Conference "Light in Nanoscience and Nanotechnology 2015", 20-22 October 2015, Hissar, Bulgaria. The Conference is part of the Program of INERA REGPOT Project of Institute of Solid State Physics, Bulgarian Academy of Sciences.

6. References
[1] You Q, Cai H, Hu Z, Liang P, Prucnal S, Zhou S, Sun J, Xu N and Wu J 2015 *J. Alloys Compounds* **644** 528–533
[2] Choi H W, Lee K, Theodore N and Alford T 2013 *Sol. Energy Mater. Sol. Cells* **117**. 273 -278
[3] Wang Y, Yang J, Kong J, Jia H and Yu M 2015 *Superlatt. Microstr.* **86** 228–235
[4] Kaur R, Singh A V and Mehra R M 2006 *J. Non-Cryst. Solids* **352** 2335–2338
[5] Zheng J H, Song J L, Jiang Q and. Lian J S 2012 *App. Surf. Sci.* **258** 6735–6738
[6] Heo S, Sharma S, Lee S, Kim Ch, Lee B, Lee H and Kim D Y 2014 *Thin Solid Films* **558** 27
[7] Anandan S and Muthukumaran S 2013 *Optical Mater.* **35** 2241–2249
[8] Ivanova T, Harizanova A, Koutzarova T and Vertruyen B 2010 *Mater. Lett.* **64** 1147-1149.
[9] Khachatourian A M, Golestani-Fard F, Sarpoolaky H, Vogt C and Toprak M S 2015 *Ceramics Inter.* **41** 2006-2014
[10] Ergin B, Ketenci E and Atay F 2009 *Int. J. Hydrogen Energy* **34** 5249
[11] Chopade S, Barve S, Raman K H T, Chand N, Deo M N, Biswas A, Rai S, Lodha G, Rao G and Patil D S 2013 *Appl. Surf. Sci.* **285** 524
[12] Barve S, Jagannath N, Mithal M, Chand N, Bhanage B M, Gantayet L and Patil D S 2010 *Surf. Coatings Techn.* **204** 3167
[13] Gupta A, Brahme N and Bisen D 2014 *J. Electroluminescence* **155** 112
[14] Xu L, Zheng G, Miao J and Xian F 2012 *Appl. Surf. Sci.***258** 7760
[15] Tsay Ch-Y and Wang M C 2013 *Ceramics Inter.* **39** 469
[16] Zhang Y, Ji H, Li P, Yang F and Zheng Z 2011 *Optics Commun.* **284** 236
[17] Thongsuriwong K, Amornpitoksuk P and Suwanboon S 2013 *Adv. Powder Techn.* **24** 275
[18] Turgut G, Duman S and Keskenler E F *2015 Superlatt. Microstr.* **86** 363

Orientation of the nanocrystallites in AlN thin film determined by FTIR spectroscopy

K Antonova[1], A Szekeres[1], L Duta[2], GE Stan[3], N Mihailescu[2], IN Mihailescu[23]
[1] Institute of Solid State Physics, Bulgarian Academy of Sciences, Tzarigradsko Chaussee 72, Sofia 1784, Bulgaria
[2] National Institute for Lasers, Plasma, and Radiation Physics, 409 Atomistilor Street, 077125 Magurele, Romania
[3] National Institute of Materials Physics, 105 bis Atomistilor Street, 077125 Magurele, Romania

E-mail: szekeres@issp.bas.bg

Abstract. Aluminum Nitride (AlN) films were deposited at 450°C in nitrogen ambient at a pressure of 0.1 Pa and at a laser incident fluence of ~3 J/cm^2 and pulse repetition rate of 40 Hz. Grazing Incidence X-ray Diffraction patterns evidenced the presence of nanocrystallites in the amorphous AlN matrix. In the FTIR spectra the characteristic Reststrahlen band of AlN crystal with a hexagonal lattice is observed but it is quite broadened (950-550 cm^{-1}). The angular dependence of the reflectance spectra in ***p***-polarised incidence radiation demonstrates the sensitivity of the A_1LO phonon mode of the AlN nanocrystallites to their orientation toward the normal to the substrate surface. With decrease of the incidence beam angle the intensity of the A_1LO phonon mode diminishes and softening of the resonance frequency occurs.

1. Introduction

Aluminum nitride (AlN) films have great potential for various applications due to their excellent electronic, optical, mechanical and thermal properties, which can be tailored in a wide range. The successful fabrication of devices based on AlN thin films requires better understanding of the film properties in relation to growth condition. Various techniques are exploited for the deposition of thin AlN films, such as metal organic chemical vapor deposition (MOCVD) [1], reactive magnetron sputtering [2] and pulsed laser deposition (PLD) [3-5]. Compared to MOCVD technique, the main advantage of PLD is the ability to facilitate the growth of thin films with good crystallinity and stoichiometry at relatively low temperatures.

In our recent studies PLD of AlN films on silicon substrates were optimized by using different nitrogen pressures at a temperature of 800°C [5]. The crystallinity of the AlN structure was found to be promoted by using higher repetition rates and lower nitrogen pressures for ablation. Further our aim was to obtain AlN films with considerably good crystallinity at lower deposition temperatures starting at 450°C.

Journal of Physics: Conference Series **682** (2016) 012024 doi:10.1088/1742-6596/682/1/012024

In this work we present results on the study of the structure of PLD AlN films deposited at 450°C again using high repletion rate and low nitrogen pressure. The obtained films structure was analysed by Fourier Transform Infrared (FTIR) spectroscopy. Measurements with polarised radiation give more effective results for investigations of the electromagnetic wave interactions with the phonons in the nanocrystallites, thus delivering information about the stoichiometric state of the layer. IR reflectance spectra are very informative in the spectroscopy of crystals, especially in the spectral range of the Reststrahlen band arising as a result of the TO and LO crystal phonon modes. By measuring the angular dependence of the Reststrahlen band spectra in ***p***-polarised IR radiation, the orientation of nanocrystallites in polycrystalline AlN thin film was qualitatively determined. The results were supported by analysing the Grazing Incidence X-ray Diffraction (GIXRD) patterns.

2. Experimental details

2.1. *Sample preparation*

AlN films were synthesized on Si(100) substrates by laser ablation of a polycrystalline AlN target with a KrF* excimer laser model COMPexPro 205 (λ = 248 nm, $\tau_{FWHM} \leq 25$ ns), running at a pulse repetition rate of 40 Hz. The incident laser fluence was set at ~3 J/cm^2. The laser beam was oriented with an inclination angle of 45° with respect to the target surface. The substrate-target distance was 5 cm. Before deposition, the target was cleaned by short time ablation with 1000 laser pulses. During this procedure, a shutter was placed between the target and the substrate. The target was continuously rotated at 0.4 Hz and translated along two orthogonal axes to avoid the formation of deep craters in order to ensure a uniform deposition during the pulsed laser irradiation. 15 000 subsequent laser pulses were applied resulting in a 370 nm AlN film as determined from the ellipsometric measurements.

Prior to loading the samples into the deposition chamber, the Si substrates were cleaned in diluted HF (5%) solution in order to eliminate the native oxide layer. During deposition, the substrates were heated at 450 °C using a PID-EXCEL temperature controller.

2.2. *GIXRD patterns*

The identification of crystalline phases was accomplished by GIXRD using a Bruker D8 Advance diffractometer, in parallel beam setting, equipped with Cu target X-ray tube. The incidence angle was set at 2°, and the scattered intensity was scanned in the range of 25–65° (2θ), with a step size of 0.04°, and 10 seconds per step.

2.3. *FTIR spectroscopy*

FTIR reflectance spectra were measured using ***p***-polarised incident IR radiation (electric vector ***E*** in the plane of incidence) at different angles: 12, 30, 45, 60 and 70 degrees. In this sense, more correct is to denote the orientation of ***E*** toward the normal to the film surface *z*, instead to the optical crystal axis ***c*** which, for every nanocrystallite, could be oriented to a random direction. The spectra were measured in the spectral region of 1200 – 400 cm^{-1} with a resolution of 2 cm^{-1} and number of scans 64. A Bruker Vertex 70 instrument (10 000 – 30 cm^{-1}) and a reflectance accessory Bruker A513/Q were used.

3. Results and discussion

The GIXRD pattern of the AlN film deposited at 450°C is given in figure 1. The presence of well-defined but broad diffraction peaks attributed to hexagonal (wurtzite) AlN phase (ref. file ICDD: 01-070-2543, inserted at the bottom of figure 1) testified for the formation of randomly oriented AlN hexagonal crystallites in the amorphous film matrix. The amorphous AlN phase is evidenced by the broad background in the GIXRD pattern on which the 100, 002 and 101 diffraction maxima are superimposed on.

From the GIXRD results it is clearly shown that at the given deposition conditions nanocrystallites could be formed in the amorphous AlN matrix at such a low temperature as 450°C. However, the broad diffraction maxima indicate that the film structure has low crystallinity.

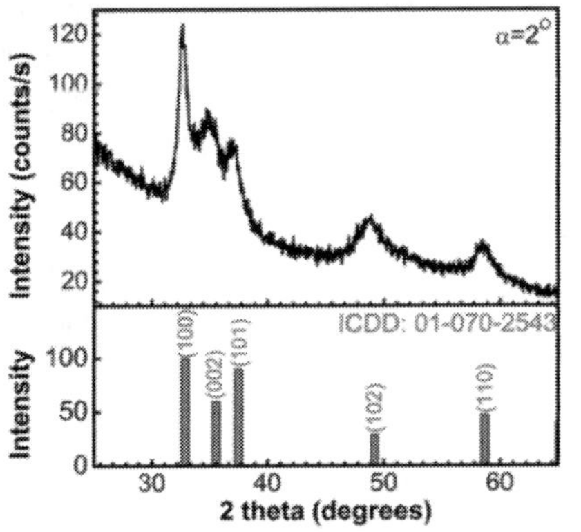

Figure 1. GIXRD pattern of AlN film deposited in nitrogen ambient at a pressure of 0.1 Pa and at a substrate temperature of 450°C.

FTIR reflectance spectra at different angles of incidence radiation: 12°, 30°, 45°, 60° and 70° are presented in figure 2. Despite its polycrystalline structure, the FTIR spectra of AlN thin film still exhibit the characteristic Reststrahlen band of AlN crystal with a hexagonal lattice. For hexagonal AlN, the IR active phonon vibrational modes have been determined [6-8] at ~611, ~670, ~890 and ~912 cm^{-1}, arising from A_1(TO), E_1(TO), A_1(LO) and E_1(LO) respectively. In our measurements, the band is broadened (950-550 cm^{-1}) with a complicated structure. For the analysis of the spectra, we used the Levenberg-Marquardt deconvolution method with four peaks. The accuracy after the procedure of the peak finding was in the order of 10^{-3}. It turned out that the only band which changed its position with the incidence angle variation was that at 860 cm^{-1}.

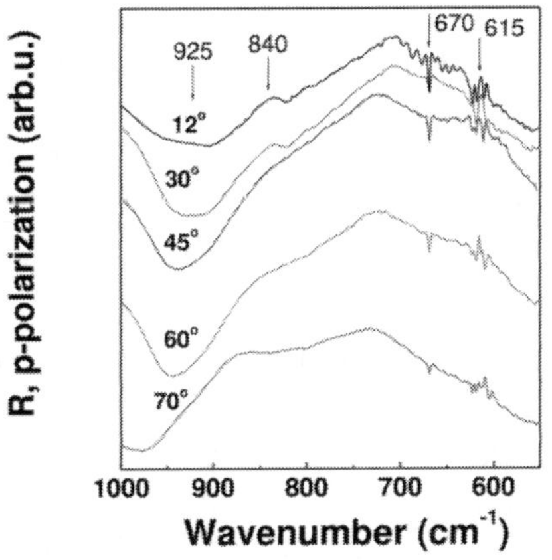

Figure 2. Angular dependence of the FTIR reflectance, R, spectra in ***p***-polarised incidence radiation.

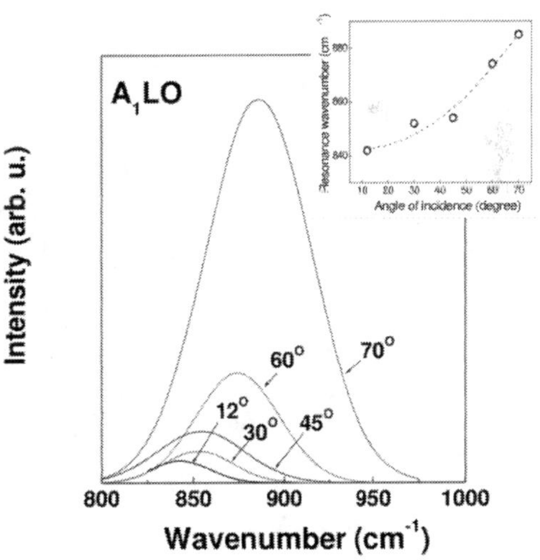

Figure 3. Angular dependence of the intensity of A_1LO phonon mode. Insert: softening of A_1LO frequency.

We assign the band around 860 cm^{-1} to the A_1LO phonon mode of the AlN nanocrystallites. The polarisation of this mode is parallel to the crystal axis ***c*** [9]. If there is a considerable quantity of nanocrystallites with ***c*** oriented close to the Si surface normal, the band intensity will show most pronounced change in ***p***-polarisation, i.e. its intensity will increase with the increase of the angle of the incidence radiation: first, because the ***E//z*** component of the radiation in this geometry increases and second, the beam optical path in the layer at large angles is maximal. The presence of crystalline AlN phase of such orientation is supported by the GIXRD analysis (see figure 1). It is seen that the corresponding peak (002) to an orientation ***c//z*** [10] is relatively intensive. The intensity increase of the band around 860 cm^{-1} with increasing the angle is presented in figure 3 demonstrating qualitatively the sensitivity of the A_1LO phonon mode to the nanocrystallite lattice growth in z – direction. Moreover, the polariton resonance arising at the interaction of the $A_1(LO)$ mode with the electric vector ***E*** of the IR radiation decreases in frequency with the incidence beam angle decrease (insert in figure 3). One possible explanation of this result could be that small incidence angles make the

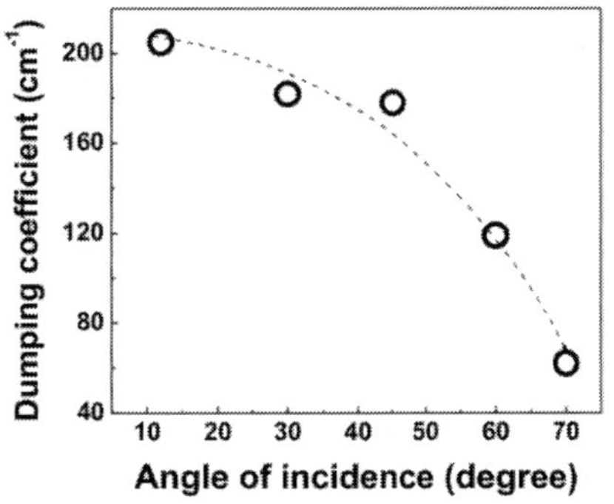

Figure 4. Angular dependence of the resonance dumping coefficient, γ.

p-polarized IR radiation sensitive also to phonon vibrations of perpendicular polarization in the x,y substrate plane which includes the nanocrystallites with oblique ***c***-orientation close to this plane. This is equal to a force dissipation of ***E*** leading to a dump of the polariton resonance oscillations in z–direction. In this case, the resonance wavenumber ν becomes dependent on the dumping coefficient γ [11]: $\nu = (\omega_o^2 - 2\gamma^2)^{1/2}$, at $\gamma << \omega_o$; ω_o is the resonance polariton wavenumber of 860 cm^{-1} at 70°. The ν-softening with the decrease of incidence beam angle is shown in the insert of figure 3. The obtained angular dependence of the dumping coefficient γ is presented in figure 4. This dependence characterizes qualitatively the degree of disorientation of crystallites in the AlN film.

4. Conclusions

The efficiency of FTIR reflectance spectroscopy in polarised radiation at different incidence angles for the characterisation of the PLD AlN film growth quality is demonstrated. Despite its poor crystalline structure, revealed by GIXRD, the FTIR spectra of AlN thin film still exhibit the characteristic Reststrahlen band of AlN crystal with a hexagonal lattice which, in our measurements is broadened. The sensitivity of the A_1LO phonon mode of the AlN nanocrystallites to their orientation toward the surface normal is illustrated: its intensity increases with the incidence angle increasing; softening of the resonance at the interaction of the mode with the electric vector ***E*** of the IR radiation occurs by decreasing the incidence angle.

Acknowledgements

All authors acknowledge with thanks the support of this work under the 2012–2014 Collaboration Agreement between the Bulgarian and Romanian Academies of Science. The Bulgarian co-authors thank the European Regional Development Fund, Ministry of Economy of Bulgaria, Operational Programme "Development of the Competitiveness of the Bulgarian economy" 2007 – 2013 for purchasing the FTIR instrument, as well as the partial support from the Project INERA under the contract REGPOT 316309. LD acknowledges with thanks the support of this work from the Romanian National Authority for Scientific Research, UEFISCDI, under the Contract no. PN-II-RU-TE-2014-4-1570. NM and INM acknowledge the financial support from the UEFISCDI Project No. 7-083/2014.

References

[1] Kung P, Saxler A, Zhang X, Walker D, Wang T C, Furguson I and Razeghi M 1995 *Appl. Phys. Lett.* **66** 2958

[2] Auger M A, Vázquez L, Jergel M, Sánchez O and Albella J M 2004 *Surf. Coat. Technol.* **180 – 181** 140

[3] Ristoscu C, Ducu C, Socol G, Craciunoiu F and Mihailescu I N 2005 *Appl. Surf. Sci.* **248** 411

[4] Ren Z M, Lu Y F, Ni H Q, Liew T Y F, Cheong B A, Chow S K, Ng M L and Wang J P 2000 *J. Appl. Phys.* **88** 7346

[5] Bakalova S, Szekeres A, Fogarassy Zs, Georgiev S, Ivanov T, Socol G, Ristoscu C and Mihailescu I N 2014 *Micro and Nanosystems* **6** (1) 1

[6] Davydov V Yu, Kitaev Yu E, Goncharuk I N, Smirnov A N, J. Graul, Semchinova O, Uffmann D, Smirnov M B Mirgorodsky A P and Evarestov R A, 1998 *Phys. Rev.* B **58** 12 899

[7] Prokofyeva T, Seon M, Vanbuskirk J and Holtz M, 2001 *Phys. Rev.* B **63** 125313

[8] Lu Y F, Ren Z M, Chong T C, Cheong B A, Chow S K and Wang J P 2000 *J. Appl. Phys.* **87** 1540

[9] Born M and Kun H, Ser. Oxford Classic Texts *in the Physical Sciences, Dynamical Theory of Crystal Lattices*, Clarendon Press, 1998

[10] Szekeres A, Fogarassy Zs, Petrik P, Vlaikova E, Cziraki A, Socol G, Ristoscu C, Grigorescu S and Mihailescu I N 2011 *Applied Surface Science* **257** 5370

[11] Landau L D and Lifshitz E M, Mechanics (Volume 1 of A Course of Theoretical Physics) Pergamon Press 1969

Optical characterization of composite layers prepared by plasma polymerization

E Radeva[1], T Hikov[1], D Mitev[1], H Stroescu[2] , M Nicolescu[2], M Gartner[2], R Presker[3] and L Pramatarova[1]

[1]Institute of Solid State Physics, Bulgarian Academy of Sciences, 72 Tzarigradsko Chaussee Blvd., 1784 Sofia, Bulgaria
[2]Institute of Physical Chemistry of the Romanian Academy, Spl. Independentei 202, Bucharest 060021, Romania
[3]Faculty of Electrical Engineering, University of Ljubljana, Ljubljana, Slovenia

E-mail: eradeva@issp.bas.bg

Abstract. Thin composite layers from polymer/nanoparticles (Ag-nanoparticles and detonation nanodiamonds) were prepared by plasma polymerization process on the base of hexamethyldisiloxane. The variation of the layer composition was achieved by changing the type of nanoparticles. The optical measurement techniques used were UV-VIS-NIR ellipsometry (SE), Fourier-transformed infrared spectroscopy (FTIR) and Raman spectroscopy. The values of the refractive index determined are in the range 1.30 to 1.42. All samples are transparent with transmission between 85-95% and very smooth. The change in Raman and FTIR spectra of the composites verify the expected bonding between polymer and diamond nanoparticles due to the penetration of the fillers in the polymer matrix. The comparison of the spectra of the corresponding NH_3 plasma treated composites revealed that the composite surface becomes more hydrophilic. The obtained results indicate that preparation of layers with desired compositions is possible at a precise control of the detonation nanodiamond materials.

1. Introduction

The producing of biomaterials that support appropriate cellular behavior and functions is critical for tissue engineering and cell therapy. There is a growing interest in functional organic/inorganic composites where a small amount of nanometer-sized material yields better physicochemical properties for cells [1, 2, 3]. Plasma polymers are easy to prepare by a well established technology of plasma polymerization. It has a long history of exploitation in a variety of applications, because it is non-toxic, transparent, with a very low surface tension, very good adhesion to any substrate, flexible, and it neither dissolves nor swells in a cell culture medium, insoluble in cold and hot solvents [4, 5, 6]. From the other hand nanodiamonds are of interest due to the combination of unique properties inherent to diamond and the specific surface structure of particles facilitating its functionalization [1].

In the present study the polymer/nanodiamond composite layers based on hexamethyldisiloxane and detonation-generated nanodiamond particles embedded into a polymer matrix, were prepared and characterized by UV-VIS-NIR ellipsometry, FTIR and Raman spectroscopy.

2. Materials

The substrates used were cover glass (CG) discs with a diameter of 15 mm.

Journal of Physics: Conference Series **682** (2016) 012025 doi:10.1088/1742-6596/682/1/012025

In order to prepare composites with different properties we have used different types of nanoparticles. The detonation nanodiamonds (DND), synthesized by a detonation of carbon-containing explosives, are produced with particles of mostly 4 nm in size and can be easily modified by chemical reactions. In the present study the following diamond nanoparticles were used:

6-DND powder: The initial nanodiamond's synthesis was carried out by use of carbon-containing explosive mixtures – conical cast charges consisting of TNT/RDX in ratio 70/30. The purification of the diamond powder was done as follows: boiling in a solution of stoichiometrically calculated amounts of potassium dichromate in sulfuric acid at 100-115°C, followed by several water washings and additional removal of metal impurities by HCl-treatment (80-100°C) and washing with water until pH=7.

Ag-DND powder: The ammonia complex of silver $[Ag(NH_3)_2]^+$was added to 6-DND-suspension under constant stirring at room temperature followed by adding of a dextrose solution. The temperature of the mixture was raised to 50°C until the Ag incorporation into the DND surface was finalized.

Si-DND powder: The silanization of 6-DND surface through an attachment of trimethylsilyl groups was done by mixing of a dehydrated 6-DND with ethyl acetate containing hexamethyldisilazane and trimethylchlorosilane[7].

Ag nanoparticles: "AgBion" (Ag-nano): prepared by bio-chemical synthesis in a form of colloidal solution of Ag-nano particles (3-16 nm in size), stable in various organic solvents (Dodecane and Octane) as well as in water and mixture of water-ethanol (stabilized by surface-active substances). "AgBion" have antibacterial, viral and fungus characteristics. Also it shows negative influence over mould and blue-green duckweed.

Plasma Poly-hexamethyldisiloxane (PPHMDS) films: prepared by plasma polymerization technology previously described [5]. The deposition of the composites was carried out in the same plasma polymerization equipment by the following procedure: the mixture of DND powder and HMDS was ultrasonically treated for 15 minutes before and during the plasma procedure.

In order to investigate the influence of chemistry of the composite surface on its properties the samples were subsequently treated by NH_3 plasma for 5 min in the same plasma reactor [7].

3. Characterization

The composite layers were characterized by UV-VIS-NIR ellipsometry (SE), FTIR and Raman spectroscopy.

3.1. UV-VIS-NIR ellipsometry (SE)

The measurements were performed using a variable angle spectroscopic ellipsometer (model VASE from the J. A. Wooman Co., Inc.) with 5 nm steps, in the spectral range of 190-1700 nm at the angles of incidence (AOI) for the composites of 56°, 57°, 58°. The model used for fitting the ellipsometric data is based on General Oscillator model and is composed of 2 Tauc - Lorentz and 6 Gauss oscillators.

3.2. Fourier-transformed infrared spectroscopy (FTIR)

FTIR spectra of the of PPHMDS and nanoparticles/PPHMDS composites were registered by Brucker FTIR spectrometer at ambient temperature in the range of 400 to 4000 cm^{-1}, using OPUS software, average of 64 scans and a resolution of 2 cm^{-1}. The assignment of the absorption bands was based on experience with organic compounds and the literature data.

3.3. Raman spectroscopy

Room temperature resonance Raman spectra of PPHMDS and DNDs/PPHMDS structures, grown on Si substrate were measured using a micro Raman spectrometer (Jobin-Ivon, HR 800) with a grating 600 gr/mm.

4. Results and Discussion

4.1. Optical transmission and refractive index

The optical transmission spectra of the composite films were measured with Woollam VASE ellipsometer in the wavelength range from 250 to 850 nm and the values are in the range of 85-95% as it is shown in figure 1. This is important for practical use when the coating needs to be seen transparent.

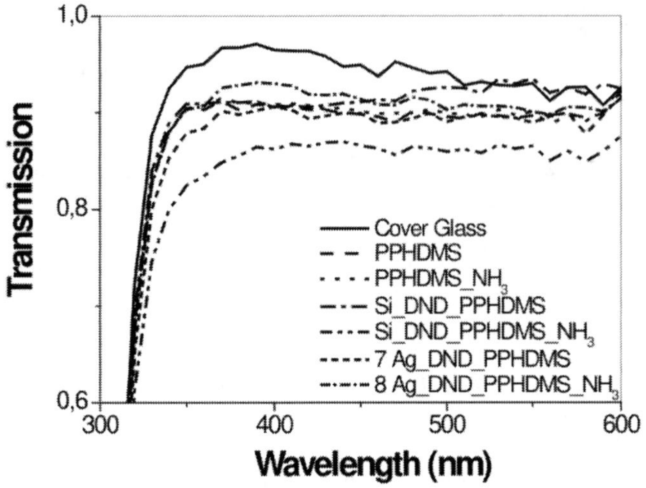

Figure 1. Measured optical transmission spectra of the composite thin films as a function of wavelength.

An important optical characteristic of this type of materials is their refractive index. The highest refractive index belongs to the composite containing Ag nanoparticles as filler (figure 2, approximately 1.43 at 630 nm).

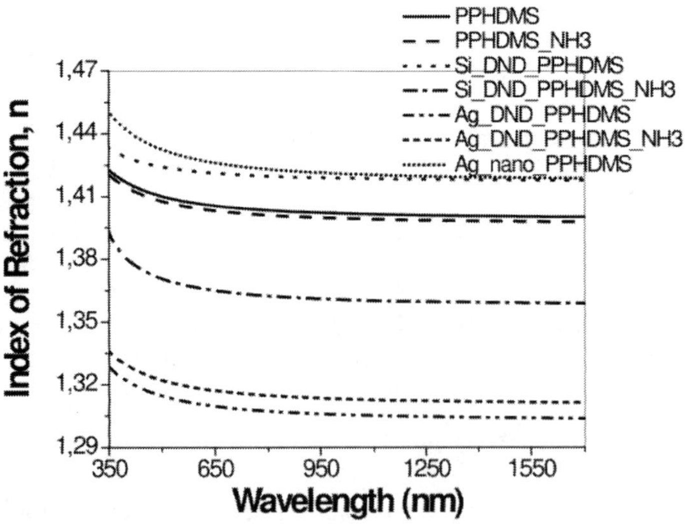

Figure 2. The refractive index dispersion

The lowest refractive index is obtained for the samples where Ag modified nanodiamonds is (1.30 at 630 nm). It is shown that this physical property can be varied by changing the type of the particles used to fill the polymer network.

In table 1 summarized data of the experiments above as well as measurements for the thickness and roughness of the layers were shown. It is found that modification of the composite with ammonia plasma is possible to change its surface chemistry. This method for polymerization yields very smooth layers as shown in the table.

Table 1. Summarized data of the ellipsometry measurements.

No.	Samples Nomenclature	d_{film} (nm)	d_{rough} (nm)	MSE	Transmission, %	Refractive index
1	CG			3.14	95	1.51
3	PPHMDS	33.53	0.0	2.04	90	1.41
4	PPHMDS_NH3	22.68	0.0	2.11	90	1.41
5	Si_DND_PPHMDS	287.27	2.2	6.56	90	1.43
6	Si_DND_PPHMDS_NH3	242.26	1.1	6.58	85	1.37
7	Ag_DND_PPHMDS	106.51	0.0	2.70	89	1.32
8	Ag_DND_PPHMDS_NH3	97.80	2.1	1.25	90	1.32
9	Ag_nano_PPHMDS_Si	963.59	20.5	4.11		1.43

4.2. FTIR spectra of the plasma polymer and composites.

For DND/PPHMDS films the characteristic peaks for sp^3 and sp^2 bonded C, new (C-Si-C) bonds and different CH_3 bonds proved the existence of both hydrophobic and hydrophilic centers. For Si-DND/PPHMDS composite definite peaks at around 500, 1200-1300, 1470 and 1600 cm^{-1} pointed for an increased content of amorphous C. The appearance of an intensive peak at 550cm^{-1} in the spectrum of Ag-DND/PPHMDS, the splitting of the peak at1550 cm^{-1} in to two peaks and the appearance of a broad band at 1800 cm^{-1} confirmed the location of Ag ions at the grain boundaries of DND (figure 3a).

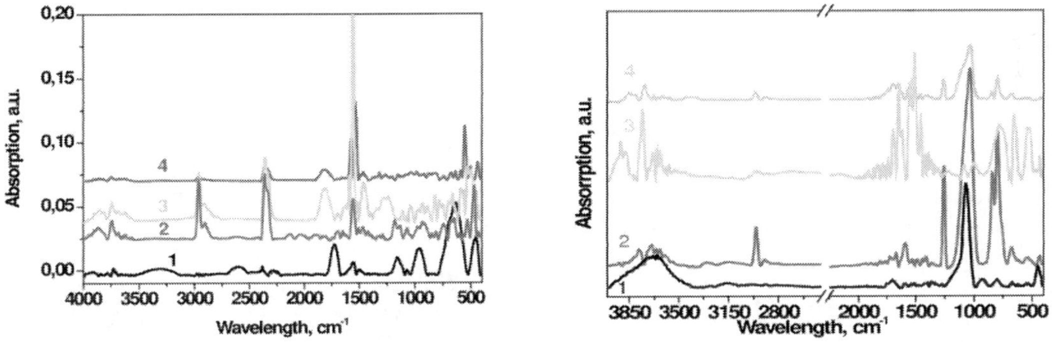

a) b)

Figure 3. FTIR spectra of a) (1)PPHMDS; (2)6-DND/PPHMDS; (3) Si-ND/PPHMDS; (4) Ag-DND/PPHMDS;
b) (1) PPHMDS/NH_3;(2) 6-DND/PPHMDS/NH_3; (3) Si-DND/PPHMDS/NH_3;(4) Ag-DND/PPHMDS/NH_3.

A change matrix of polymer characteristic bands is explained by the penetration of DND particles in the polymer. The comparison of the spectra of the corresponding NH_3 plasma treated composites revealed the decrease in intensity of the bands for methyl groups and the composite surface becomes more hydrophilic (figure 3b).

4.3. Raman spectra of the plasma polymer and composites.

For detailed analysis of characteristic bands, associated with bonding between the DNDs and PPHMDS, Raman microspectroscopy is used. In the figure 4a,b,c Raman spectra of DNDs/PPHMDS composites are presented.

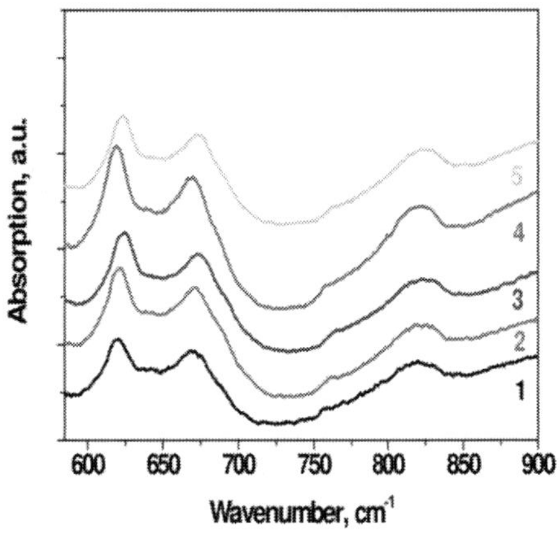

a)

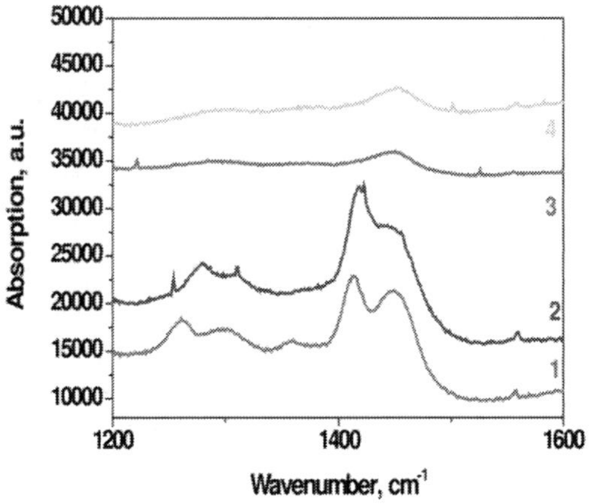

b)

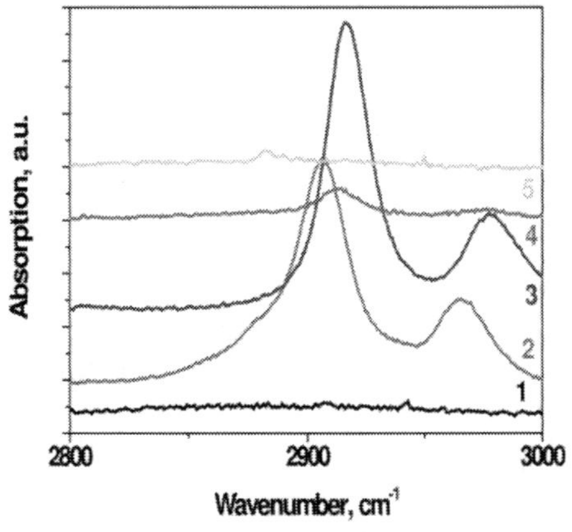

c)

Figure 4. Raman spectra:
a) (600 – 900 cm^{-1}): (1) PPHMDS;(2) 6-DND/PPHMDS;(3) 7-DND/PPHMDS; (4) Ag-DND/PPHMDS;(5) Ag-DND/PPHMDS/NH_3;
b)(1200 – 1600 cm^{-1}): (1) 6-DND/PPHMDS; (2) 7-DND_PPHMDS(2);(3) Ag-DND/PPHMDS;(4) Ag-DND/PPHMDS/NH_3;
c) (2800– 3000 cm^{-1}): (1) PPHMDS;(2) 6-DND/PPHMDS;(3) 7-DND/PPHMDS(2); (4) Ag-DND/PPHMDS;(5) Ag-DND/PPHMDS/NH_3.

The assignment of characteristic vibrational bands is based on a comparison with the results in [3, 8]. The repeating unit of PPHMDS is -($Si(CH_3)_2$-O)- with an Si-O-Si backbone and methyl side group attached to Si atoms. The Raman shift related to CH_3 vibration bands are as follows: 2967 and 2906 cm^{-1} (stretching) (figure 4c), 1413 and 1260 cm^{-1} (deformation) (figure 4b) and 865,756, and 688 cm^{-1} (rocking) (figure 4a). Raman spectra related to the C-Si-C groups are 845 and 790 cm^{-1} (asymmetric stretching) and 706 cm^{-1} (symmetric stretching). All characteristic Raman vibration bands of PPHMDS are also present in DNDs/PPHMDS composites. A new relatively intensive band at 1450 cm^{-1} is observed in all composites containing DND (figure 4b). To appreciate the possibilities of DNDs to modify the structure and the nature of chemical bonding of the polymer matrix (PPHMDS) further analysis of relative intensities of normal modes was performed following procedure described in [3]. The ratio of relative intensities I_{2967}/I_{2905} of CH_3 was lower in the composites (DNDs/PPHMDS) in comparison with PPHMDS. It was pronounced in the composites containing DND particles without any modification (6-DND and 7-DND). Such a tendency of decreases was also present in the vibrations of side chains (C-Si-C). The observed changes in Raman and FTIR spectra of DNDs/PPHMDS verify the expected bonding between PPHMDS and DNDs nanoparticles.

5. Conclusion

Original technology for producing composites on the base of hexamethyldisiloxane and detonation nanodiamond particles by plasma polymerization was developed. The characterization of the obtained new materials proved that the nanoparticles were embedded in the polymer matrix. Moreover by varying the type of diamond or silver nanoparticles in the polymer matrix and using ammonia plasma modification, thin layers with desired properties could be obtained. The results are very promising for applications of composites in biocompatible materials.

Acknowledgements
This work was supported by REGPOT-2012-2013-1 NMP INERA "Research and Innovation Capacity Strengthening of ISSP-BAS in Multifunctional Nanostructures" Grant agreement no: 316309 and Bulgarian National Scientific Research Fund project №: DFNI_B01_18/2012.

References

[1] Shenderova O, Zhirnov V and Brenner D 2002 Carbon nanostructures *Solid State Mater. Sci.* **27** (3/4) 227-356
[2] Dolmatov V 2007 Composite materials based on elastomer and polymer matrix, filled with detonation nanodiamonds *Russian Chemical Reviews* **70** (7) 607-626
[3] Borjanovic V, Bistricic L, Vlasov I, Furic K, Zamboni I, Jaksic M and Shenderova O 2009 Influence of proton irradiation on the structure and stability of poly(dimethylsiloxane) and poly(dimethylsiloxane)-nanodiamond composite *J. Vac. Sci. Technol. B.* **27** (6) 2396-2430
[4] Vasilev K, Sah V, Anselm K, Ndi C, Mateescu M, Dollmann B, Martinek P, Ys, Ploux L and Griesser H J 2010 Tunable Antibacterial Coatings That Support Mammalian Cell growth *Nano Letter* **10** 202-207
[5] Radeva E, Tsankov D, Bobev K and Spassov L 1993 Fourier Transform Infrared Analysis of Hexamethyldisiloxane Layers Obtained in Low-frequency Glow Discharge. *J. Appl. Polym. Sci.* **50** 165-171
[6] Min-Hsien 2009 Simple poly(dimethylsiloxane) surface modification to control cell adhesion *Surf. Interface Anal.* **41** 11–16
[7] Pramatarova L, Radeva E, Pecheva E, Hikov T, Krasteva N, Dimitrova R, Mitev D, Montgomery P, Sammons R and Altankov G 2011 The advantages of polymer composites with detonational nanodiamond particles for medical applications*on Biomimetics* ed L Pramatarova (InTech) chapter 14 p 297-320
[8] Pihan S, Tsukruk T and Forch R 2009 Plazma polymerized hexamethyl disiloxane in adhesion applications *Surface and Coatings Technology* **203**(13) 1856-1862

Investigation of multilayered polyelectrolyte thin films by means of refractive index measurements, FT-IR spectroscopy and SEM

I Bodurov[1], I Vlaeva[2], G Exner[1], Y Uzunova[3], S Russev[4], B Pilicheva[3], A Viraneva[1], T Yovcheva[1], Ts Grancharova[1], S Sotirov[1] and M Marudova[1]

[1]Plovdiv University "Paisii Hilendarski", 24 Tsar Assen str., 4000 Plovdiv, Bulgaria
[2]University of Food Technologies, 26 Maritsa blvd., 4002 Plovdiv, Bulgaria
[3]Medical University Plovdiv, 15A Vassil Aprilov blvd., 4002 Plovdiv, Bulgaria
[4]Sofia University "St. Kliment Ohridski", 5 James Bourcher blvd., 1164 Sofia, Bulgaria

E-mail: bodurov@uni-plovdiv.net

Abstract. Multilayered polyelectrolyte films are promising structures in the biomedical field. In order to meet the demands for biomedical applications, the structures have to be built from biocompatible and/or biodegradable, nontoxic starting materials, possessing some specific functional properties, depending on the particular application. In the present study, the multilayered polyelectrolyte films with potential use as buccal bioadhesive drug delivery systems were investigated. They were prepared via layer-by-layer deposition of successive nanolayers onto substrate. Three different biopolymers were used. The substrate, from poly(lactic acid), was solvent casted. After that, it was subjected to corona treatment, which ensures surface charge excess for the multilayer deposition. The nanolayers were prepared either from 0.01 g/L solutions of chitosan or 0.05 g/L xanthan. Acetate buffer (pH 4.5 and ionic strength 0.1 M) was used as a solvent. The substrate was dipped successively into one of the solutions, allowing formation of polyelectrolyte complexes of chitosan (polycation) and xanthan (polyanion). The substrates was treated in negative corona. The multilayered structures consisted of 8, 9, 14, 15 or 20 nanolayers. Number of techniques, such refractive index measurements, FT-IR spectroscopy and SEM morphology were employed in order to monitor the properties of the so prepared multilayered polyelectrolyte films.

1. Introduction

In the last decades, polyelectrolyte multilayered thin films (MPTF) prepared by means of layer-by-layer (LbL) self-assembling have been widely used because of their fabrication easiness and great potential for application in biomedicine [1], drug delivery [2, 3], optics [4], food science [5], as biomembranes [6], etc. The technique is based on the consecutive deposition onto solid substrate via electrostatic interaction of polyions [7, 8]. LbL self-assembling process involves only molecular adsorption from solution onto charged surfaces, and it is applicable to a vast class of molecules, polymers and others. During the self-assembling of polyelectrolyte from solution, electrostatically bounded complexes are formed between the oppositely charged polyelectrolyte functional groups. Due to overcompensation during this process, an excess opposite to that of the substrate appear. As a result, irreversibly adsorbed polyelectrolyte multilayers could be deposited one over another. As far as the deposition occurs by

Journal of Physics: Conference Series **682** (2016) 012026 doi:10.1088/1742-6596/682/1/012026

natural adsorption, there are no restriction with respect to substrate material type, size and topology in the application of the technique [7, 8]. Varying the type of adsorbing material, processing conditions (such as polyelectrolyte charge, ionic strength, pH, polymer concentration, and adsorption time) and the method of assembling one can manipulate the molecular interactions and achieve controlled nanostructures with well predictable thicknesses and interpenetration [9].

Self-assembling could be attained by several techniques immersion, spinning, spraying, electromagnetic and fluidic assembling – each of which offers different material and processing advantages [10]. Commonly used dipping-based LbL self-assembling consists of three steps: (1) polyelectrolyte diffusion of from the solution volume to the substrate surface, (2) adsorption of the polyelectrolyte onto the surface via molecular interactions, (3) molecular rearrangement of polymeric chains along the surface and equilibrium phase formation [7].

Although the feasibility of the dipping-based process has been confirmed, it is recognized as time-consuming and to accomplish the process of adsorption equilibrium hours or even up to days could be needed [7]. LbL assembling via spin coating technique ("spinning-assembling", SA) makes use of spinning a substrate to facilitate the deposition process [11]. The SA is performed by either casting the solution onto a spinning substrate [12] or casting onto stationary substrate, which is then after spun [13]. Spinning quickens the assembling process considerably, due to the various forces governing the process, allowing for layers to be deposited in time limit of about 30 s. SA typically results in more homogenous films in comparison with immersion assembling, thanks to the electrostatic interactions, providing the adsorption and rearrangement of polymers, and the centrifugal forces, air shear, and viscous forces, leading to desorption of weakly bound polymers and dehydration of the films [13].

In the present study, the self-assembling process and its effect on the thickness and the morphology of chitosan/xanthan polyelectrolyte films deposited onto biodegradable poly(lactic) acid substrate is investigated by means of refractive index measurements, FT-IR spectroscopy and SEM.

2. Experimental

2.1. Poly(lactide) substrates formation

Poly(DL-lactide) (PDLA) (inherent viscosity 0.55-0.75 dL/g), purchased from Lactel Absorbable Polymers (USA), was used for the preparation of the biodegradable substrates. The substrates were casted from 2% w/v PDLA chloroform solution and then dried at 35°C for 48 hours. The substrates were then kept in exicator, at room temperature at relative humidity (RH) 54%. Immediately prior to the deposition process, the substrates were treated in corona discharge system, in order to achieve negative surface charge excess. The corona discharge system consisted of corona electrode (needle), grounded plate, and a metal grid placed between them. Substrates were placed onto the grounded plate electrode. A voltage of –5 kV was applied to the corona electrode and –1 kV to the grid. The samples were charged under standard room conditions (T = 21÷23 °C and RH = 40-60%) for 1 min.

2.2. Layer-by-layer deposition of chitosan/xanthan multilayers

Chitosan (low molecular mass, degree of deacetylation > 75%) and xanthan gum were purchased from Sigma-Aldrich and were used as received. For LbL assembling process 0.1% w/v chitosan and 0.05% w/v xanthan solutions of acetate buffer (pH 4.5 and ionic strength 0.1 M) as solvent were prepared. Either dip-coating or spin-coating was employed for the deposition process. In both cases, the first built-up layer always possesses opposite to the substrate electric charge.

For dip-coating assembling a programmable slide stainer (Poly Stainer IUL, Spain) was used with deposition program: (1) 15 min solution dipping - adsorption of polyelectrolyte molecules; followed by 5 min washing in the acetate buffer (pH 4.5 and ionic strength 0.1 M); (3) 15 min solution dipping - adsorption of polyelectrolyte molecules of opposite charge; (4) 5 min washing in the same acetate buffer. The procedure was repeated and 8, 9, 14, 15 or 20 chitosan/xanthan multilayers were built up. After the MPTF were dried in hot air and then stored in exicator, at RH 54%.

Spin-coating assembling was performed on DDR Labortechnik T54 spin-coater. The polyelectrolyte solutions of chitosan and xanthan were spin coated one over the other (15 s at 1000 rpm) onto the preliminary charged PDLA substrates with an intermediate washing step (acetate buffer) between the deposition of the next layer.

2.3. FTIR spectroscopic measurements
In order to prove the successful self-assembling of the MPTF, FTIR-ATR spectra were collected. Thermo Scientific™ Nicolet™ iS™ 10 FT-IR spectrometer, equipped with a diamond ATR accessory in the range from 650-4000 cm^{-1}, giving a resolution 4 cm^{-1} and 64 scans was used. The integral intensity was measured using OMNIC software package.

2.4. Surface refractive index measurements
As a crosscheck prove, as well as to estimate the thicknesses of the nanolayers (see below), the refractive indices of MPTF were measured. For all samples, the surface refractive indices (sRI) were measured by mean of disappearing diffraction pattern method [14], using a laser refractometer at two wavelengths (532 nm and 635 nm, each with a power 5 mW). As shown earlier [15], the penetration depth of the evanescent field in this case is less than 1 μm. Hence, the sRI measurements correspond to a surface layer of about 1 μm thickness. When the laser beam reaches the sample at an angle smaller than the critical angle, α_{cr}, the beam may pass through, reaching a reflection grating, placed behind giving rise to a diffraction pattern. When the incidence angle equals or exceeds α_{cr} a single spot of the totally internal reflected beam is observed. Therefore, α_{cr} can be measured as the angle at which the diffraction pattern disappears and one may calculated sRI by the following formula:

$$sRI = N\sin\left(A \pm \arcsin\frac{\sin\alpha_{cr}}{N}\right), \quad (1)$$

where A and N are the refracting angle and refractive index of the prism, used in the laser refractometer, respectively. In the present case, $A = 64.7°$, N(532 nm) = 1.7480, N(635 nm) = 1.7347.

2.5. *Scanning electron microscopy (SEM)*
The general morphology of the obtained MPTF was revealed by means of SEM. A scanning electron microscope Lyra 3 XMU (Tescan) was employed. The working voltage was 30 kV. Prior to the measurements, the samples were covered with a thin film of gold (about 30 nm).

3. Results and discussion
In figure 1, ATR-FTIR spectra of MPTF composed of negatively (-) charged PDLA – substrates with dipping-assembled multilayers are presented.

PDLA substrates spectrum show two strong bands (at 2946 and 2881 cm^{-1}) belonging to the CH stretching region. The same bands are preserved in the MPTF spectra as well but they are broader and have lower intensity. The band at 2996 cm^{-1} in PDLA substrate appears as a law intensity shoulder, whereas its intensity increases for the dipping-assembled multilayered films. The broad band, positioned between 3000 and 3700 cm^{-1}, corresponds to the superposition of two overlapping vibrations: -OH and amide groups in the carbohydrates structures 1594 cm^{-1} band appears only in MPTF spectra and should be attributed to the NH_2 group [16]. Its intensity rises with increasing number of layers. Peaks that correspond to asymmetric and symmetric vibrations of $NH3^+$ (at 1632 cm^{-1} and 1522 cm^{-1}) of chitosan and ionized carboxyl group (COO–) of xanthan at 1620 cm^{-1} were not observed suggesting the formation of complex between both polymers, caused by ionic interactions between afore-mentioned groups [17].

The intensities of all other observed characteristic bands increase with increasing the number of layers.

Figure 2 represent the ATR-FTIR spectra of MPTF obtained by spin-coating assembling. Although, the bands intensities are not as high as in the case of dipping-assembling the general tendency of intensities increase is preserved in these spectra as well.

Journal of Physics: Conference Series **682** (2016) 012026 doi:10.1088/1742-6596/682/1/012026

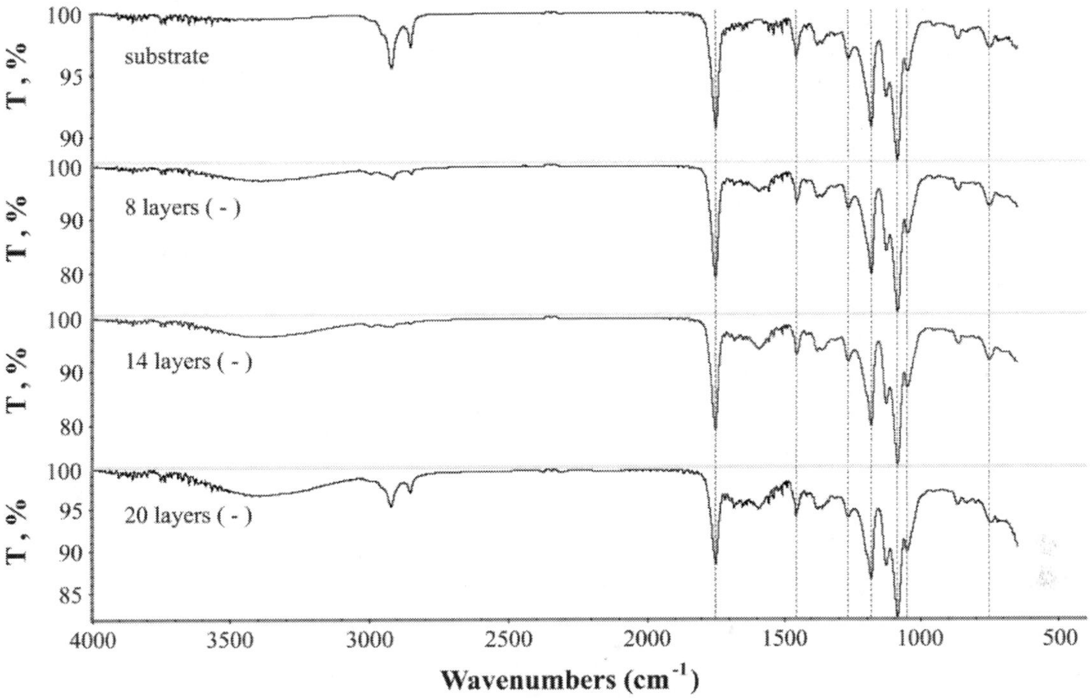

Figure 1. ATR-FTIR spectra of MPTF - negatively charged PDLA with dipping-assembled multilayers (first layer chitosan).

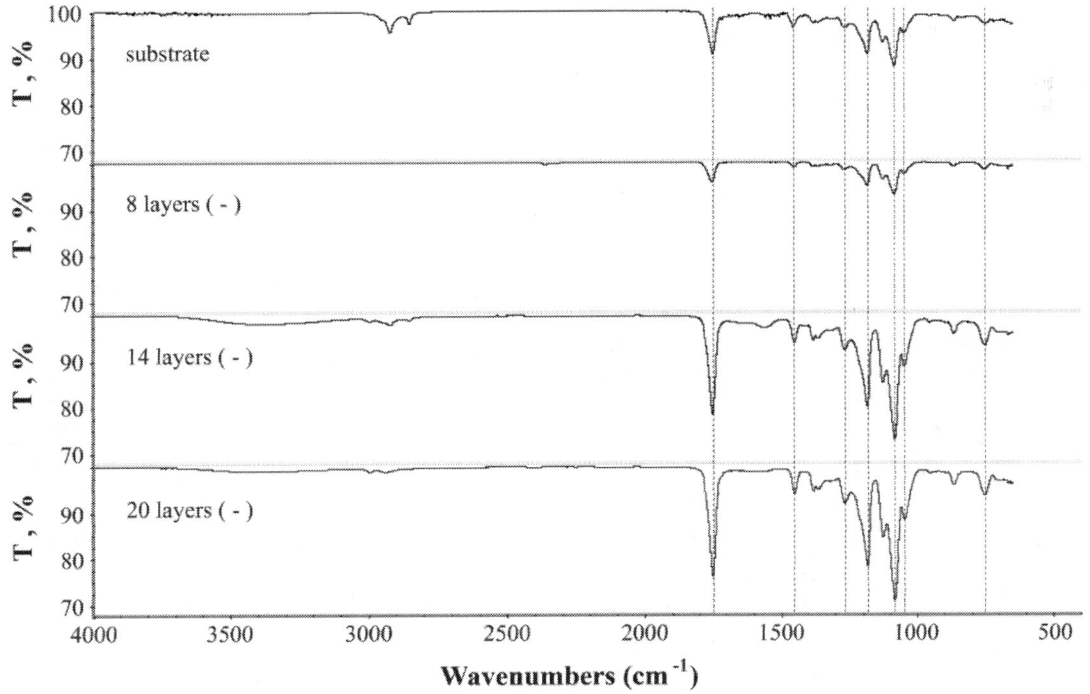

Figure 2. ATR-FTIR spectra of MPTF - negatively charged PDLA with spin-assembled multilayers (first layer chitosan).

Interestingly, the total integral FTIR intensity of MPTF with 20 layers is lower than these of MPTF with 8 and 14 layers. It is plausible that specific physical interactions and/or weak ionic interactions between the functional groups of the carbohydrates could explain the observation [18].

In summary, ATM-FTIR results suggest that:

- Chitosan and xanthan were successfully deposited onto the charged PDLA substrates and the adsorption process is more or less irreversible;
- The increase of number of layers leads to an increase of the amount of deposited hydrocolloids;
- Chitosan and xanthan molecules interact predominantly by ionic interactions, forming polyelectrolyte complexes.

Thicknesses of the MPTF layers were estimated based on measurements of surface refractive index. Figure 3a present sRI of negatively charged PDLA substrates with dipping-assembled multilayers (chitosan-xanthan) depending on the number of layers. Figure 3b represent the same dependences but in the case of spin-assembled MPTF.

Refractive index changes display similar behavior for all samples. The sRIs shift to higher values with increasing number of deposited layers. Hence, the refractive indices confirm once again the presence of multilayers, as it was shown in FTIR experiments. The average sRI over the penetration depth d_p of multilayer films, where Lorentz-Lorenz equation holds, reads:

$$d_p \frac{n_s^2-1}{n_s^2+2} = d\frac{n^2-1}{n^2+2} + Nd_1\frac{n_2^2-1}{n_2^2+2} \qquad d_p = d + N(d_1+d_2), \tag{2}$$

where d_p is the average evanescent wave penetration depth at total internal reflection, ($d_p = 0.82$ nm for 532 nm; $d_p = 0.98$ nm for 635 nm [15]); N is the number of one type of the layers; d_1, n_1 are the thickness and the refractive index of first type of layers (chitosan); d_2, n_2 are the thickness and the refractive index of the second type of layers (xanthan); d is the thickness of the PDLA substrate at evanescent wave penetration; n is the refractive index of the charged PDLA substrate. When the number of layers in MPTF increase, the PDLA substrate refractive index impact decreases. The corresponding values for 532 nm are $n_1 = 1.52$ [19], $n_2 = 1.50$ [20], and $n = 1.47$. Solving equation (2) with the particular values shows an increase in the refractive index of the MPTF with increasing number of the layers. By the assumption of equal thickness of all layers, the average layer thickness of single polyelectrolyte layer was estimated to be about (5 ± 0.5 nm) for the spin-assembled MPTF and about (10 ± 1.0 nm) for the dipping-assembled MPTF.

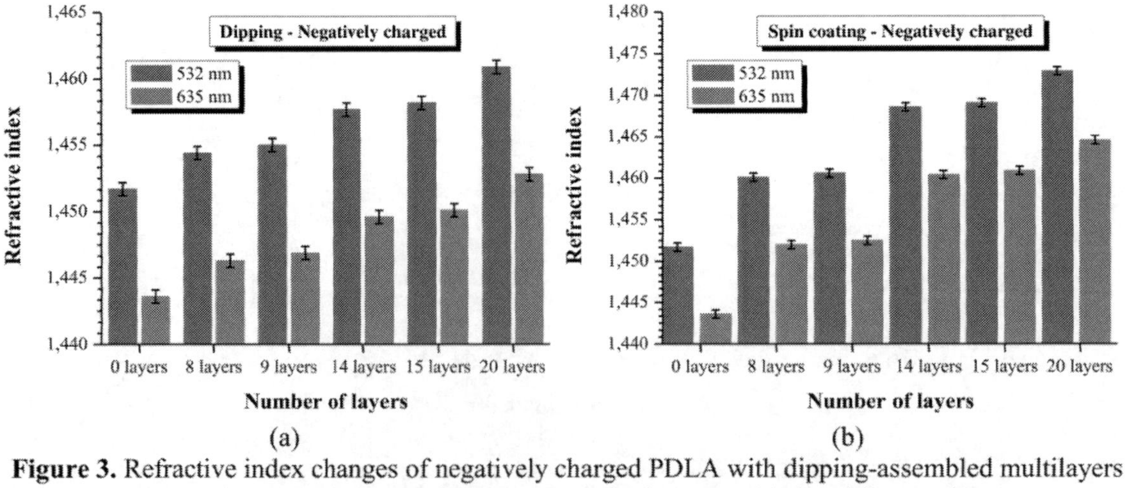

Figure 3. Refractive index changes of negatively charged PDLA with dipping-assembled multilayers (a) and negatively charged PDLA with spinning-assembled multilayers (b).

The results suggest faster thickness growth of the dipping-assembled films. This observation is in agreement with previous investigation on deposition process of weakly charged polyelectrolytes [21] and could be explained by specific interaction such as interdiffusion and interpenetration of the polyelectrolyte molecules. At pH 4.5 of deposition process the molecules of both chitosan and xanthan are partially charged (the pKa value of the chitosan is about 6.5 [22] and the pKa values of xanthan is 3.1 [23]) and part of their chains are free to move in the interior part of the polyelectrolyte complexes. Therefore, an excess amount of polyelectrolyte can be accumulated inside the layer volume and its thickness increases. In contrast, the high centrifugal force causes the film thickness to be lower because most of the unbounded polymer molecules are expelled.

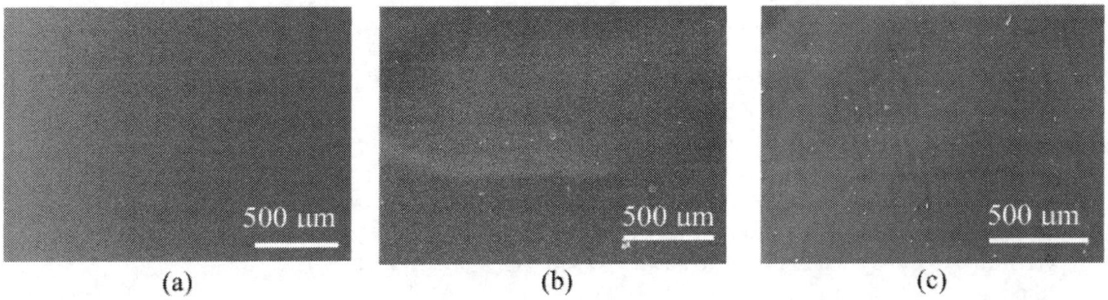

Figure 4. Scanning electron microscopy images of (a) PDLA substrate; (b) MPTF of negatively charged PDLA with 20 dipping-assembled multilayers (first layer chitosan); (c) MPTF of positively charged PDLA with 20 dipping-assembled multilayers (first layer xanthan).

The surface morphology of the bare substrates is quite smooth, consisting of some very fine grain-like structures (figure 4a). The surface morphology does not change with the layer deposition independent of the substrate charge or the deposition method. Representative SEM photograph for MPTF of negatively charged PDLA with 20 dipping-assembled multilayers is shown in figure 4b and photograph for MPTF of positively charged PDLA with 20 dipping-assembled multilayers is shown in figure 4b. The main difference are only the appearance of small impurities or defects onto the surface, which are very common for biopolymers. Their presence confirm once more the successful deposition of multilayers.

4. Conclusions

It is shown that dipping (dipping-assembling) and spin-coating (spinning-assembling) methods are successfully applied for deposition of chitosan and xanthan multilayers on PDLA corona charged substrates. In both cases the binding is irreversible and multilayer growth is predominantly due to ionic interactions and formation of polyelectrolyte complexes. The higher film thickness and surface roughness of dipping-assembled multilayers indicate the presence of inter-polymer diffusion and interpenetration.

Acknowledgements

This study was financial supported by Project No DFNI B-02/7 of Bulgarian National Scientific Fund.

REFERENCES

[1] Costa R R and Mano J F 2014 *Chem. Soc. Rev.* **43** 3453
[2] Gunjkar V N, Patwekar S L and Dhage S P 2015 *World Journal of Pharmacy and Pharmaceutical Science* **4** 216
[3] Tian F, Min J, Kanka J, Hammond PT and Du H 2015 *Proc. SPIE* **9480** 948005
[4] Bharadwaj R, Mukherji S and Mukherji S 2014 *OSA Technical Digest (online)* paper SeM3C.5

[5] Poverenov E, Danino S, Horev B, Granit R, Vinokur Y and Rodov V 2013 *Food Bioprocess. Technol.* **7** 1424
[6] Pasco E V, Shi H, Xagoraraki I, Hashsham S A, Parent K N, Bruening M L and Tarabara V V 2014 *J. Membrane Sci.* **469** 140
[7] Decher G, Hong J D and Schmitt J 1992 *Thin Solid Films* **210/211** 831
[8] Decher G 1997 *Science* **277** 1232
[9] Lee Y M, Park D K, Choe W S, Cho S M, Han G Y, Park J and Yoo P J 2009 *J. Nanosci. Nanotechnol.* **9** 7467
[10] Richardson J J, Björnmalm M and Caruso F 2015 *Science* **348** 411
[11] Thomas I M 1987 *Appl. Opt.* **26** 4688
[12] Chiarelli P A, Johal M S, Casson J L, Roberts J B, Robinson J M and Wang H L 2001 *Adv. Mater.* **13** 1167
[13] Cho J, Char K, Hong J D and Lee K B 2001 *Adv. Mater.* **13** 1076
[14] Sainov S and Dushkina N 1990 *Appl. Opt.* **29** 1406
[15] Yovcheva T, Sainov S and Meksihev G 2007 *J. Optoelectron. Adv. Mater.* **9** 2087
[16] Kulkarni N, Wakte P and Naik J 2015 *Journal of Pharmaceutical Investigation* **5** 73
[17] Fukuda M, Peppas N A and McGinity J W 2010 *International Journal of Pharmaceutics* **310** 90
[18] Dehghan M H G and Marzuka M 2014 *Iranian Journal of Pharmaceutical Research* **13** 769
[19] Azofeifa D E, Arguedas H J and Vargas W E 2012 *Optical Materials* **35** 175
[20] Panchev I, Nikolova K and Sainov S 2007 *Proc. SPIE* **5449** 35
[21] Lee Y M, Park D K, Choe W-S, Cho S M, Han G Y, Park J Y and Pil J 2009 *Journal of Nanoscience and Nanotechnology* **9** 7467
[22] Wu S-J, Don T-M, Lin C-W and Mi F-L 2014 *Mar. Drugs* **12** 5677
[23] Oprea A-M, Nistor M-T, Profire L, Popa M I, Lupusoru C E and Vasile C 2013 *Journal of Biomaterials and Nanobiotechnology* **4** 123

Flexo-dielectro-optical spectroscopy of PDLC films modified by nano-rubbed PTFE layers

Lidia T Popova
Georgi Nadjakov Institute of Solid State Physics, Bulgarian Academy of Sciences, 72 Tzarigradsko chaussee, Sofia 1784, Bulgaria

E-mail: tlidia@issp.bas.bg

Abstract. The electro-optical (EO) response of planar single layers of polymer-dispersed liquid crystal (PDLC) composites of relatively large nematic microdroplets modified by layers of teflon (PTFE), was studied. The PDLC films were prepared from liquid crystal E7 and photopolymer NOA-65 in cells assembled with parallel or orthogonal PTFE-covered glass plates. The influence of nanostructured PTFE polymer nanolayers on both the polarized and depolarized component of laser light transmitted through PDLC cells of both geometry of layer rubbing directions was determined. Flexo-dielectro-optical spectroscopy in the range of 10 Hz – 1 kHz was applied to examine the amplitude-frequency EO modulation by PTFE-modified PDLCs in dependence on the applied alternating-current electric field. Specific fall-downs in the frequency spectra of the first and second harmonic EO modulation by PTFE-modified PDLCs were observed, that could be tuned by the driving electric field.

1. Introduction

Due to their attractive electro-optical (EO) properties, the polymer-dispersed liquid crystal (PDLC) composite films [1] have found a number of advanced applications, such as "light valves" in integrated-optic circuits, smart EO materials in switchable glasses and intelligent glass facades, for displays and large area flexible displays, for devices for active control of light, etc. [2-5]. The EO response of single-layered PDLC films consisting of large microdroplets of liquid crystal (LC) disposed in an optically-transparent polymer matrix have also been investigated in view of their applications in switchable diffractive, adaptive and micro optics, for beam steering devices, as well devices for splitting monochromatic beams and beam multiplexing [6-8].

In this work, PDLC composites of relatively large micrometer-sized droplets of the LC commercially known as E7 (a room-temperature nematic; nematic phase from -10° C to 60° C) dispersed in a transparent matrix of the photocurable polymer NOA-65, were studied. In particular, single layers of LC droplets confined in thin films (6 μm thickness) between transparent conducting substrates coated with teflon (polytetrafluoroethylene, PTFE) nanolayers, were produced. Recently, a band-pass selective amplitude-frequency electro-optical modulation by such PDLC films aligned by PTFE nanolayers, was reported [9]. Here, the polarization properties and the amplitude-frequency EO modulation behaviours of E7/NOA65 single-layered PDLC films aligned in this manner, are reported. To obtain the frequency characteristics of the PDLC films, flexo-dielectro-optical spectroscopy was applied. Previously, this experimental technique has been successfully used for investigation of flexoelectric and dielectric properties of PDLCs [10-12].

2. Experimental

The preparation of E7/NOA65 PDLC single layered films modified by nano-rubbed layers of PTFE was described in details elsewhere [13]. The microscale LC-polymer composite films with a size of 1 cm × 1 cm and a thickness of 6 μm were formed in cells assembled from two glass plates, to serve as electrodes, each coated inside by transparent conductive layers of indium tin oxide (ITO). The ITO-glass substrates were beforehand treated by PTFE – over the ITO-glass surfaces, a nanolayer of PTFE was deposited by hot 'sliding on' method [14,15]. The PTFE overcoating had a unidirectional rubbing. Two kinds of LC cells were assembled: with parallel (P-PDLC) and orthogonal (T-PDLC) orientation directions of the rubbing of both substrates.

The experimental set-up for EO measurements has been given in [16]. The EO response of E7/NOA65 PDLC films to an alternating-current (AC) electric field was investigated by use of He-Ne laser (λ = 632.8 nm, 1 mW) and computer-controlled SR830 lock-in amplifier for excitation of the PDLC films (by a sinusoidal AC voltage) and registration of the intensity of He-Ne laser beam transmitted through the films (measured with a photodiode). The polarization ratio of the PDLC transmittance was calculated as $\gamma = I_{A\|P} / I_{A\perp P}$, where $I_{A\|P}$ and $I_{A\perp P}$ are the transmitted intensity measured in the saturation region of the voltage-dependent transmittance of the PDLC films, at parallel or orthogonal analyzer and polarizer, respectively. The interval between the data acquirement was equal to 8 s (the same applies to the recording of frequency spectra of the amplitude of the first or the second harmonic of the modulated transmitted light by use of variable (sweep) frequency in the range of 10 Hz – 1 kHz, provided by the lock-in amplifier.

3. Results and Discussion

The conclusions from the optical microscope observations of the morphology of PDLCs in the two cell types studied here, are, in principle, the same as discussed in [13]. For example, figure 1 shows micrographs of P-PDLC and T-PDLC taken with no polarizers and under crossed polarizers. As seen, the single-layered PDLC composites contain almost uniformly distributed relatively large oval-shaped LC droplets whose size exceeds the layer thickness. The average droplet size was about 10 μm and 20 μm for P-PDLCs and T-PDLCs, respectively. The polarizing microscope inspection of PDLCs reveals a bipolar configuration of the droplets (planar anchoring). The PTFE nanolayers determine the surface anchoring and the orientation of the LC in the droplets. The use of rubbed PTFE nanolayers as orienting surfaces for LC orientation is well known [17]. In our case, the modification (both morphology change and LC alignment) of the PDLC system occurs by the unidirectional grooves of the rubbed PTFE with an adjacent spacing in the submicrometer range, an average thickness of a few nm, and a width of about 100 nm [15]. Thus, owing to the two parallel rubbed surfaces, the morphology of P-PDLC (figure 1 a, b) should be characterized by nearly homogeneous planar alignment [18]. In contrast, due to the orthogonal rubbing of both plates of the T-PDLC cells and orthogonal anchoring, the T-PDLC single-layered films exhibit a predominantly twisted alignment of the LC in the droplets (figure 1 c, d).

As known, the PDLCs are switched between their translucent and transparent states by applying an external AC electric field [1]. Due to the positive dielectric anisotropy of E7 ($\Delta\varepsilon_{\|}$ = 19 and $\Delta\varepsilon_{\perp}$ = 5.2 at 20° C and 1 kHz frequency [19]), the electric field forces the nematic director reorientation towards the field direction (and the direction of the light propagation, as well). As a result, a refractive index mismatch between the droplets and the polymer matrix (translucent state) is removed (transparent state). The ordinary and extraordinary refractive indices of the nematic E7 at 20° C and λ = 633 nm are: n_0 = 1.5185 and n_e = 1.737, respectively [20]. If the effective (average) refractive index (n_{eff}) of the aligned LC matches the refractive index of the transparent polymer matrix (n_p), then the bulk PDLC material appears transparent (ON). Removal of the applied electric field makes the film revert to its translucent OFF-state. Clearly, in order to achieve high ON-state transmittance, one has to ensure that $n_0 \approx n_p$. The index-matching problem can be resolved by a proper choice of both the LC and the polymer matrix materials. In our case, n_p = 1.524 (λ = 633 nm and at 20° C).

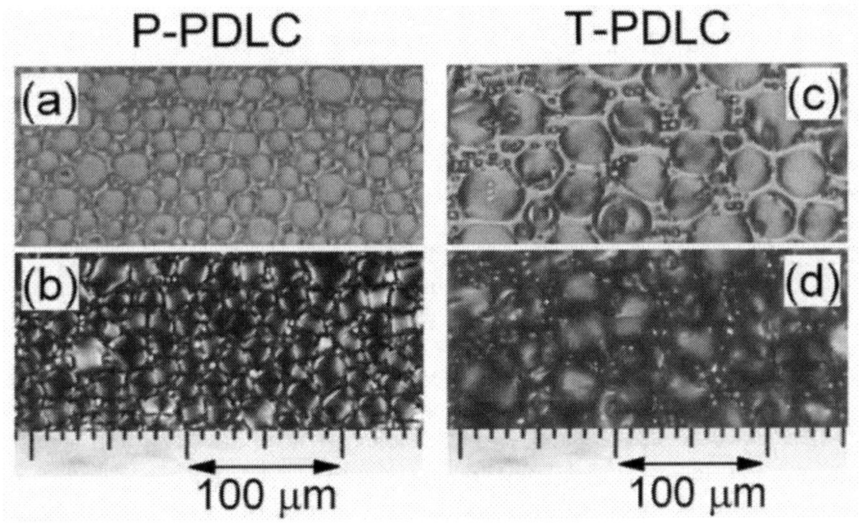

Figure 1. Optical microscopy images of PDLC single layers studied here: (a) P-PDLC, no polarizers; (b) P-PDLC, crossed polarizers; (c) T-PDLC, no polarizers; (d) T-PDLC, crossed polarizers. The temperature of the films was 25 °C.

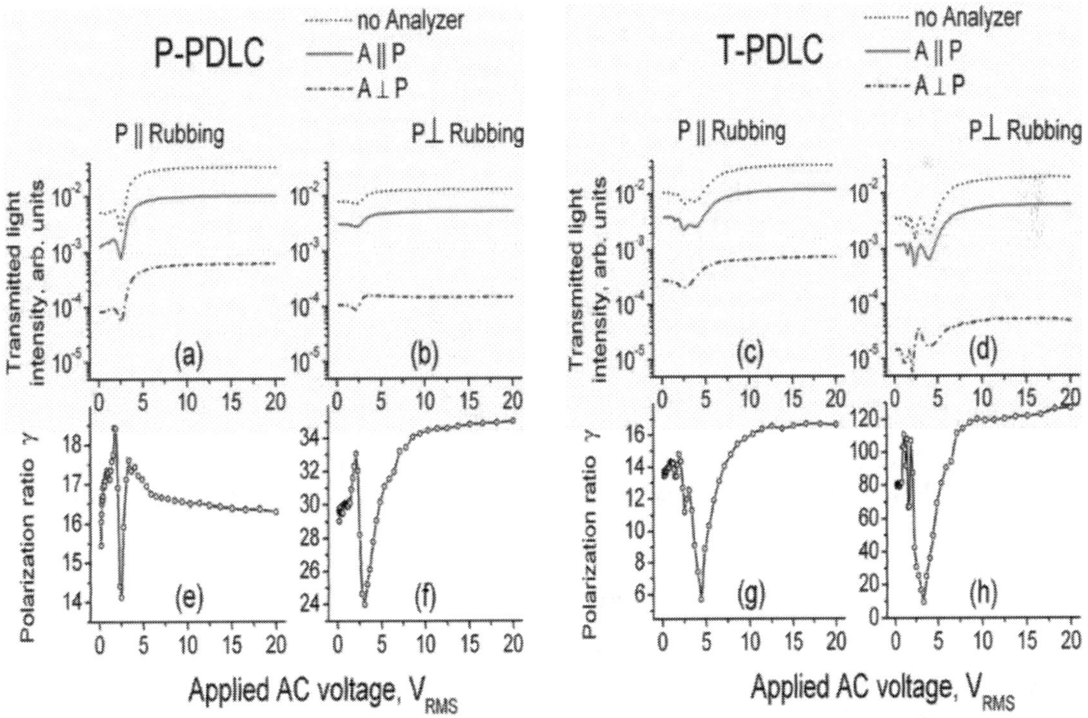

Figure 2. Up: the voltage-dependent light transmittance curves for P-PDLC film (a, b) and T-PDLC film (c, d) measured at two polarization configurations – by polarization (P) of the incoming He-Ne laser beam parallel (a, c) or orthogonal (b, d) to the rubbing of the front plate of the PDLC cell. Data are obtained by use of analyzer parallel or orthogonal to P. The other experimental conditions were identical, the temperature was 25° C. Down (e-h): corresponding curves for the polarization ratio γ.

Figure 2 reports the voltage-dependent light transmittance curves together with the corresponding curves for the polarization ratio *P* for P-PDLC and T-PDLC cells measured when the polarization of the incoming He-Ne laser beam was parallel (i) or orthogonal (ii) to the rubbing of the plate of the PDLC cell on which the light impinges first (hereafter called "front plate"), keeping identical the other experimental conditions. In both cases, a relatively high level of the light scattering in the voltage-off state was observed, as well as oscillations and deep interference minima at low voltages. These features are well known characteristics of the microscale PDLC single layers [21-24]. As seen from figure 2 (e,g), in the case (i) both P-PDLC and T-PDLC cells exhibited the same polarization ratio (γ ~ 16-17), i.e. no sense of use of T-PDLC. In contrast, in case (ii) for T-PDLC was measured γ up to 100-120 (figure 2 h), significantly higher than γ ~ 34 for P-PDLC (figure 2 f). Thus, this polarization configuration utilizes the polarizing properties of T-PDLC films and they can be used as efficient polarizing optical elements. Further, this property is electrically-controllable (the effect in our case can exceeds 10 times (a change from γ ~ 10 at 3.5 V to γ = 120 at 10 V, as shown in figure 2 h).

The EO amplitude-frequency modulation by considered single-layered PDLC films modified with nano-rubbed teflon layers, was also examined. Figure 3 presents the first- and the second-harmonic EO modulation of transmitted laser light (the modulation amplitude as a function of the frequency of the applied AC electric field, when the magnitude of the latter was kept constant) measured for T-PDLC film. Whereas the flexoelectro-optical response of the PDLCs can be registered as linear electro-optic modulation (the first harmonic) [25,26], the second-harmonic EO modulation response of PDLC is relevant mainly to the dielectric oscillations of the nematic director with respect to the frequency of the applied electric field.

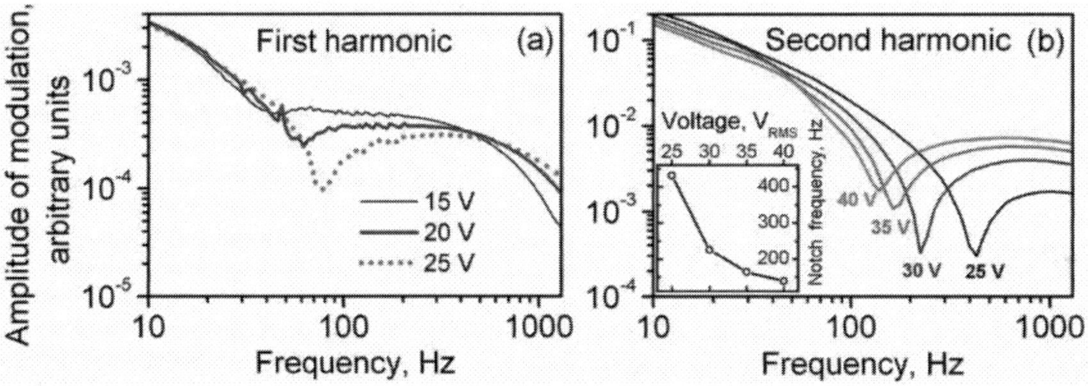

Figure 3. Flexo-dielectro-optical spectra measured for various values of the applied voltage (denoted in the graphs) under identical other experimental conditions: (a) first harmonic; (b) second harmonic (amplitude of EO modulation of circularly-polarized He-Ne laser beam transmitted through single-layered T-PDLC film, versus the electric-field frequency). In all cases, the temperature was 30° C. Insert in (b): notch-filter frequency vs applied voltage.

The flexo-dielectro-optical spectra of T-PDLC films studied here exhibit narrow minima at a given frequency, i.e. a notch filtering of modulation frequencies can be achieved. The minima are characteristics of the confined nematic system [1] depending on LC droplet size, temperature and applied voltage. The notch-filter-like spectral shapes are consistent with the features previously reported for other single-layered PDLCs [25]. For explanation of the minima, the model of selective diffraction filtering of propagated light due to formation of an internal refractive index contrast in the LC droplets of the LC-polymer dispersions, was proposed [25]. This process results from spatial deflection of modulated light components in the forward scattering direction. Macroscopic optical interference effects in such confined systems of single layers of PDLC have to be also taken into account [22,24]. It should be noted that the observed modulation characteristics are different from the specific band-pass-like selective EO modulation by T-PDLC films simultaneously observed at lower frequencies (in our case below 100 Hz [9]).

As seen from figure 3, the frequency of the sharp minima can be tuned in some frequency range by varying the strength of the driving AC electric field (e.g., from 150 to 400 Hz by lowering the applied voltage from 40 V to 25 V, see the insert in figure 3 b). The effect was more pronounced for the second-harmonic EO modulation (figure 3 b). The EO notch-like frequency filter performance of single-layered PDLC is of interest for specific applications.

4. Conclusion

The objects of this study were planar films (6 μm-thick) of E7/NOA65 PDLC where single microdroplets of nematic LC with diameters as larger as 10 μm or 20 μm were confined. The nematic director field in this microscale composite was efficiently modified by nano-rubbed PTFE nanolayers deposited on the conductive ITO glasses of the PDLC cells. Owing to the surface alignment by PTFE, the single-layered PDLC system exhibits an efficient electrically-controllable polarization of the transmitted light when the PDLC cell has orthogonally rubbed PTFE nanolayers.

Also, the flexo-dielectric response of the examined PDLC films was studied as depending on the applied voltage. Specific notch-filter characteristics of the amplitude-frequency modulation of light can be achieved by the PTFE-modified single-layered PDLC films. The effect is attributed to the selective diffraction by single-layered PDLCs. Within a certain frequency range, the sharp minima in the notch-filter behavious can be tuned by AC voltage applied on the PDLC films. This property can be useful for applications based on various schemes by exploiting the EO modulation by PDLCs.

Acknowledgments

The author acknowledges the EU FP7 project INERA REGPOT-2012-2013-1 "Research and Innovation Capacity Strengthening of ISSP-BAS in Multifunctional Nanostructures", for the support of her participation in LNN2015 conference. This study was carried out as a part of Indo-Bulgarian joint research project DNTS/In-01/4/2013 supported by the Department of Science and Technology, New Delhi and the Ministry of Education and Science, National Science Fund of Bulgaria.

References

[1] Drzaic P S 1995 *Liquid Crystal Dispersions* (Singapore: World Scientific Publishing)
[2] Wu S T and Yang D K 2001 *Reflective Liquid Crystal Displays* (New York: Wiley)
[3] Bloisi F and Vicari L 2003 Polymer-dispersed liquid crystals *Optical Applications of Liquid Crystals*, Series in Optics and Optoelectronics, ed L Vicari (Bristol: Institute of Physics Publishing) chapter 4 pp 148 – 200
[4] Jeon S, Kim B, Ye H and Hong S K 2014 *Mol. Cryst. Liq. Cryst.* **597** 159
[5] Song P, Gao Y, Wang F, Zhang L, Xie H, Yang Z and Yang H 2015 *Liq. Cryst.* **42** 390
[6] Marinov Y, Hadjichristov G B and Petrov A G 2007 *J. Optoelectr. Adv. Mater.* **9** 417
[7] Hadjichristov G B, Marinov Y G and Petrov A G 2009 *Opt. Mater.* **31** 1578
[8] Marinov Y G, Hadjichristov G B and Petrov A G 2009 *J. Optoelectr. Adv. Mater.* **11** 1186
[9] Marinov Y G, Hadjichristov G B, Petrov A G, Marino S, Versace C and Scaramuzza N 2013 *C. R. Acad. Bulg. Sci.* **66** 819
[10] Hadjichristov G B, Pavlov L, Marinov Y and Petrov A G 2008 *Proc. SPIE* **7027** 70271M
[11] Marinov Y G, Hadjichristov G B and Petrov A G 2010 *Opt. Las. Eng.* **48** 1161
[12] Marinov Y G, Hadjichristov G B, Marino S, Todorova L, D'Elia S, Versace C, Scaramuzza N and Petrov A G 2012 *Bulg. J. Phys.* **39** 92

[13] Marinov Y G, Hadjichristov G B, Petrov A G, Marino S, Versace C and Scaramuzza N 2013 *J. Appl. Phys.* **113** 064301
[14] Wittmann J C and Smith P 1991 *Nature* **352** 414
[15] Marinov Y, D'Elia S, Todorova L, Petrov A G, Versace C and Scaramuzza N 2006 *Liq. Cryst.* **33** 1219
[16] Hadjichristov G B, Marinov Y and Petrov A G 2010 *Mol. Cryst. Liq. Cryst.* **525** 128
[17] Takatoh K, Sakamoto M, Hasegawa R, Koden M, Itoh N and Hasegawa M 2005 *Alignment Technology and Applications of Liquid Crystal Devices* (London: CRC Press Taylor & Francis group)
[18] D'Elia S, Versace C, Scaramuzza N, Marinov Y and Petrov A G 2007 *Mol. Cryst. and Liq. Cryst.* **465** 301
[19] Drzaic P S and Muller A 1989 *Liq. Cryst.* **5** 1467
[20] Coates D, Greenfield S, Sage I C and Smith G 1990 *Proc. SPIE* **1257** 37
[21] Marinov Y G, Hadjichristov G B and Petrov A G 2009 *Cryst. Res. Technol.* **44** 870
[22] Hadjichristov G B, Marinov Y G and Petrov A G 2009 *J. Optoelectr. Adv. Mater.* **11** 1190
[23] Hadjichristov G B, Marinov Y G and Petrov A G 2010 *AIP Conf. Proc.* **CP1203** 182
[24] Hadjichristov G B, Marinov Y G and Petrov A G 2011 *Appl. Opt.* **50** 326
[25] Petrov A G, Marinov Y, D'Elia S, Marino S, Versace C and Scaramuzza N 2007 *J. Opt. Adv. Mat.* **9** 420
[26] Meyer R B 1969 *Phys. Rev. Lett.* **22** 918

Journal of Physics: Conference Series **682** (2016) 012028 doi:10.1088/1742-6596/682/1/012028

Raman Spectroscopy Investigation of the Polar Vibrational Modes in CuB2O4

V Tomov, P M Rafailov and L Yankova
Institute of Solid State Physics, Bulgarian Academy of Sciences, 72 blvd. Tzarigradsko shaussee, 1784 Sofia, Bulgaria

E-mail: vixren@issp.bas.bg

Abstract. We report results of polarized Raman spectroscopy of piezoelectric copper metaborate CuB2O4 (space group $I\bar{4}2d$, Z = 12) in the range from 80 to 1300 cm−1. Due to the dense population of the phonon spectra we carried out angular-dependent measurements by rotating the sample in different crystallographic planes thus performing gradual transitions from one polarization configuration to another. The different angular dependence of the scattering intensities of vibrational modes of different symmetries allowed us to obtain Raman signal from some transversal optical (TO) modes of E-symmetry and to demonstrate the single-mode behavior of the B_2 modes by means of transition from TO character to longitudinal optical (LO) character of these phonons.

1. Introduction

Copper metaborate CuB2O4 is an intriguing system with interesting magnetic, acoustic, dielectric, and magnetooptical properties. The complex crystal structure with magnetic Cu^{2+} ions in symmetrically nonequivalent positions leads to a variety of magnetic phases at low temperatures starting with an antiferromagnetic ordering below T_N = 21 K [1]. Commensurate and incommensurate magnetic structures arise as a result of interactions within and between the different copper sublattices. In connection to its interesting magnetic behavior, this material also demonstrates a rich variety of linear and nonlinear optical properties. A couple of examples are magnetic-field-induced second harmonic generation [2, 3] recently reported magnetic-field induced chirality [4], and magneto-electric dichroic signals [5].

Below 1.35 eV, CuB_2O_4 is transparent. A complex absorption band is seen in the range of 1.35–2.5 eV related to electronic transitions between crystal-field-split levels within two different 4b and 8d copper sublattices [6]. Above 2.5 eV, CuB_2O_4 becomes again transparent up to 3.8 eV where the fundamental absorption due to charge-transfer (CT) transitions begins [6]. This window defines a dark blue color of CuB_2O_4.

The study of low-temperature optical absorption spectra and second harmonic generation spectra established six zero-phonon (ZP) lines with exceptionally rich structure of phonon sidebands [6] which seems exotic for a 3d compound as CuB_2O_4. Motivated by the challenge to rationalize this rich vibronic structure, several thorough studies of the lattice dynamics and the vibronic spectra of CuB2O4 appeared recently [7, 8]. Detailed infrared (IR) and Raman investigations identified almost all zone-center optical phonon modes [7, 8]. Especially, in the Raman measurements of polar phonons the emphasize was mainly put on longitudinal optical (LO) modes.

In this study we exploit the abilities of polarized Raman spectroscopy combined with rotation of the sample about various crystal axes. This enabled us to obtain from Raman measurements the frequencies of several transversal optical (TO) phonons that were so far known only from IR studies. We also confirm the belonging of TO and LO components to each other for some polar modes with a larger LO/TO splitting.

2. Experimental

CuB_2O_4 single crystals were successfully grown by the high-temperature solution growth method. The starting materials CuO and B_2O_3 were taken in 1:20 molar ratio, B_2O_3 being used as a flux. The mixture was first homogenized for 48h at 980 °C and then cooled to 750 °C at a cooling rate of 1 °C/h. The crystals were grown in a platinum crucible and the process took place in a resistive heating oven in air atmosphere. The size of the crystals synthesized at the bottom and walls of the crucible was up to 7 × 3 × 3 mm^3.

The crystallographic characterization of the crystals was carried out by X-ray single-crystal diffractometry. A single crystal was mounted on a glass capillary and diffraction data were collected at room temperature by an ω -scan technique, on an Agilent Diffraction SuperNova dual four-circle diffractometer equipped with an Atlas CCD detector. A mirror-monochromatized Mo Kα radiation from a microfocus source was used ($\lambda = 0.7107$ Å). The determination of the cell parameters, data integration, scaling, and absorption correction were carried out using the program package CRYSALISPRO [9]. The structure was solved by direct methods (SHELXS -97) [10] and refined by full-matrix least-square procedures on F2. The single-crystal x-ray diffraction confirmed the $I\bar{4}2d$ crystal structure with lattice parameters a = 11 .4972 Å and c = 5.6290 Å and atomic site positions close to those of [11, 12].

For the spectroscopic measurements single crystals with elongated shapes along the *Z* axis with naturally grown {100} and {001} surfaces were selected which additionally displayed well shaped {110} surfaces. The Raman spectra were measured in the range of (80 - 1300) cm^{-1} on a HORIBA Jobin Yvon Labram HR visible spectrometer equipped with a Peltier-cooled CCD detector. The 1.95 eV line of a He-Ne laser was used for excitation, the absolute accuracy being 0.5 cm^{-1}. At this energy there is an absorption band due to crystal-field induced transitions. Therefore this excitation energy was chosen to minimize the path length of the laser light inside the crystal in order to avoid possible birefringence effects.

3. Results and Discussion

Copper metaborate CuB_2O_4 crystallizes in a noncentrosymmetric tetragonal structure with the point group symmetry $\bar{4}2m$, the space group $I\bar{4}2d$ (No. 122).2,3,5,30 [13]. The unit cell contains 12 formula units and the lattice parameters are a = b = 11.484 Å and c = 5.620 Å. The structure comprises a strongly coupled three-dimensional (3D) network of BO_4 tetrahedrons with cavities between them occupied by copper ions in the 4b and 8d positions. There are two types of boron atoms B1(16e) and B2(8d) positioned at the centers of tetrahedrons, and the bond lengths B-O vary between 1.444 and 1.487 Å. The tetrahedrons' vertices are linked by common oxygen ions. There are four nonequivalent oxygen positions in the unit cell, O1(16e), O2(8d), O3(8d), and O4(16e).

The primitive unit cell contains 42 atoms, which results in 126 normal modes.
After subtracting acoustic and silent modes one ultimately obtains the distribution of the Raman active modes as follows:

$$\Gamma_{Raman} = 13A_1 + 14B_1 + 17B_2(z) + 31E(x,y). \quad (1)$$

The modes B_2 and E are polar. B_2 modes vibrate along the z-axis and the vibrational patterns of the E modes lie in the (x,y) plane. The Raman tensors for the pertinent modes have only the following non-zero elements:

A_1: $\alpha_{xx} = \alpha_{yy} = a$, $\alpha_{zz} = b$;
B_1: $\alpha_{xx} = c$, $\alpha_{yy} = -c$;
$B_2(z)$: $\alpha_{xy} = \alpha_{yx} = d$;
$E(x)$: $\alpha_{yz} = \alpha_{zy} = e$;
$E(y)$: $\alpha_{xz} = \alpha_{zx} = e$.

For the polarized Raman measurements we use the notations *X* (100), *Y* (010) and *Z* (001) for the main crystal axes as well as *X'* (110) and *Y'* (-110). The spectra were measured by irradiating the crystal along the *Z* axis and along different directions in the in the *XY* plane as a function of the azimuthal angle in the *XY* plane and the pertinent rotational angle, respectively. We used the following scattering configurations in Porto notations:
(i) $Z(x_\varphi y_\varphi)Z$, where x_φ and y_φ are orthogonal variable directions in the *XY* plane, the azimuthal angle φ being the angle between *X* and x_φ;
(ii) $x_\varphi(Z\, y_\varphi)x_\varphi$, where x_φ and y_φ are orthogonal variable directions in the *XY* plane, the rotational angle φ being again the angle between *X* and x_φ.
(iii) $x'_\varphi(Y'Y')x'_\varphi$, where x'_φ is a variable direction in the *X'Z* plane, depending on the rotational angle φ between *X'* and x'_φ.

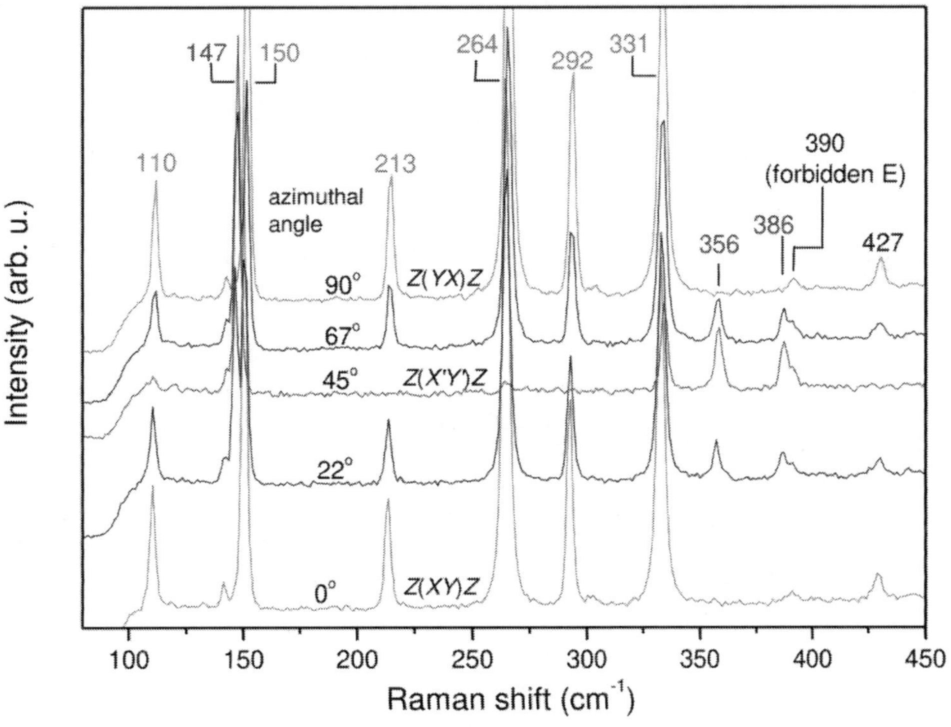

Figure 1. Azimuthal-angle dependence of the Raman spectra in $Z(x_\varphi y_\varphi)Z$ configuration. Frequencies of B_1 (B_2) modes are given in blue (red). Spectra in well defined scattering configurations are plotted in blue (B_1 modes allowed) and red (B_2 modes allowed).

The scattering intensity is $|\boldsymbol{e}_i.R.\boldsymbol{e}_s|^2$, where $\boldsymbol{e}_i$ and $\boldsymbol{e}_s$ are unit vectors in the directions of the incident and the scattered radiation, respectively, and R is the corresponding Raman tensor. The theoretically

Journal of Physics: Conference Series **682** (2016) 012028 doi:10.1088/1742-6596/682/1/012028

predicted Raman intensity dependence on the angle φ for the examined vibrational modes is given in table 1.

Table 1. Theoretical Raman intensities, variable configurations dependent on the rotational angle φ

	$Z(x_\varphi y_\varphi)Z$	$x_\varphi(Z\, y_\varphi)x_\varphi$	$X'(Y'Y')X'$	$Z(Y'Y')Z$
B_1	$\lvert c\rvert^2.\sin^2 2\varphi$	0	0	0
B_2	$\lvert d_{LO}\rvert^2.\cos^2 2\varphi$	0	$\lvert d_{TO}\rvert^2$	$\lvert d_{LO}\rvert^2$
E(TO)	0	$\lvert e_{TO}\rvert^2.\sin^2 2\varphi$	0	0
E(LO)	0	$\lvert e_{LO}\rvert^2.\cos^2 2\varphi$	0	0

With figure 1 we exemplify the advantage of the azimuthal-angle dependent polarized measurements in $Z(x_\varphi y_\varphi)Z$ configuration. The doublet at 386 – 390 cm^{-1} consists of a B_1 mode at 386 cm^{-1} [7, 8] and a symmetry forbidden E(TO) mode at 390 cm^{-1} with comparable intensity. Due to the relatively weak Raman intensity of this particular B_1 mode, its correct assignment in one fixed polarization configuration would be difficult. By means of the azimuthal-angle dependence of its intensity, however, it can be reliably assigned because the intensity of the forbidden mode remains constant.

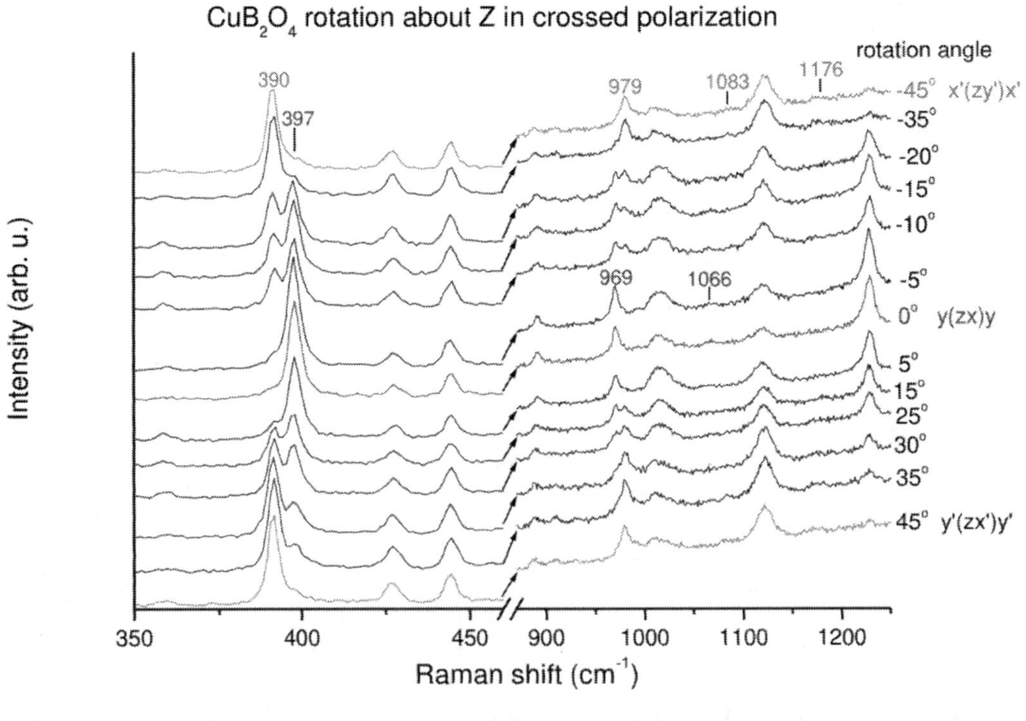

Figure 2. Raman spectra with intensity transfer from the TO to the LO component and vice versa (only E modes allowed by selection rules). Spectra in well defined scattering configurations are plotted in blue (LO allowed: φ = 0°) and red (TO allowed: φ = ±45°).

Journal of Physics: Conference Series **682** (2016) 012028 doi:10.1088/1742-6596/682/1/012028

Figure 2 displays Raman spectra in scattering configurations $x_\varphi(Z\ y_\varphi)x_\varphi$ selecting the E modes at different values of the rotational angle about the Z axis. This illustrates the gradual intensity transfer between the TO and the LO component of the E modes and facilitates identifying components with weak intensity as well as TO/LO pairs with larger splitting. In this way we provided a Raman verification of the TO/LO pair at 390 – 397 cm^{-1}, respectively, and of the TO components at 908, 979, 1083, 1122 and 1176 cm^{-1} which were so far only detected with IR spectroscopy [7].

The first-order Raman spectrum of CuB_2O_4 is densely populated with 75 modes distributed between 100 and 1250 cm^{-1}. Assignment of nondegenerate polar $B_2(z)$ modes with larger TO/LO splitting is therefore particularly difficult because they appear at different frequencies in different scattering configurations. We therefore measured a series of Raman spectra in configurations gradually passing from *X'(Y'Y')X'* , where B_2 modes are TO, to *Z(Y'Y')Z* , where B_2 modes are LO vibrations. This was accomplished by rotating the sample about the *Y'* direction. B_2 modes of tetragonal CuB_2O_4 as a function of the angle φ between the Z axis and incident laser direction K_i are depicted in figure 3. The spectra at different values for several elevated B_2 modes demonstrate a monotonic migration of the B_2 frequency from the TO to the LO value thus additionally confirming the belonging of TO and LO components to each other.

The spectra at different rotation-angle values for several elevated B_2 modes are depicted in figure 3 and demonstrate an approximately linear migration of the B_2 frequency from the TO to the LO value thus additionally confirming the belonging of TO and LO components to each other.

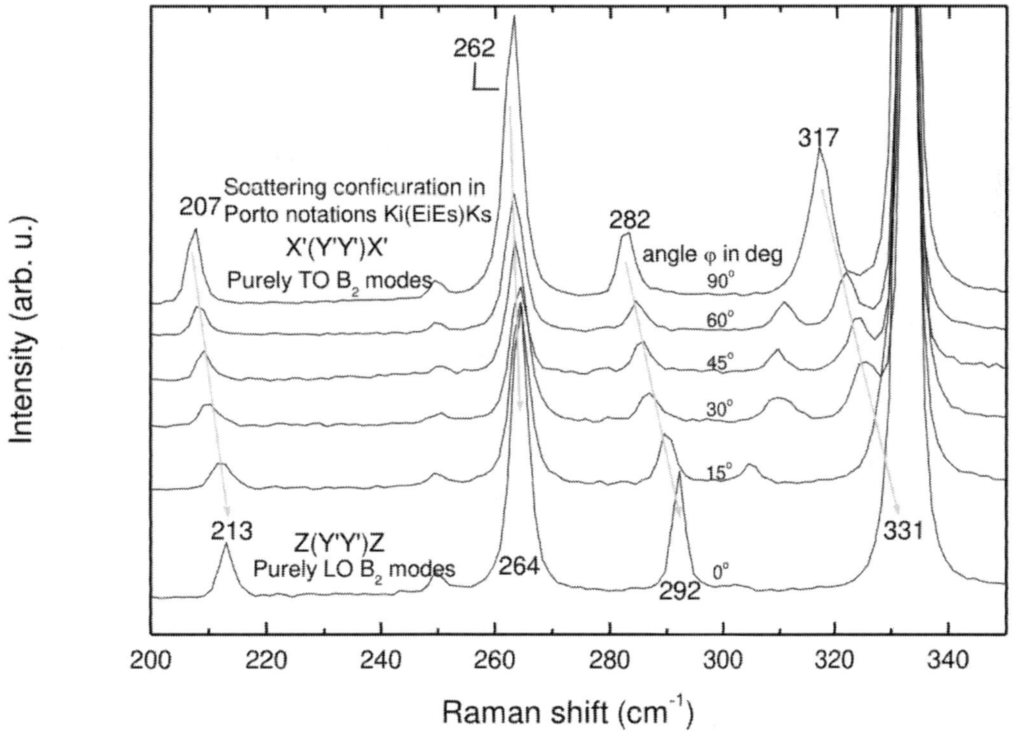

Figure 3. Raman spectra gradually passing from *X'(Y'Y')X'* to *Z(Y'Y')Z* scattering configuration (as shown by green arrows) by rotating the sample about the *Y'* direction. The rotation angle goes from 90 deg to zero from top to bottom

4. Conclusion

We obtained polarized Raman spectra of CuB_2O_4 single crystals in different scattering configurations dependent on rotational angles with respect to the main crystal axes. The different angular dependence of scattering intensity of vibrational modes with different symmetries enabled us to verify the belonging to each other of the TO/LO components of several polar modes and to confirm the frequency of some low-intensity TO components with Raman spectroscopy. These results demonstrate the power of Raman spectroscopy in characterization of crystal structures with lower symmetry.

5. Acknowledgments

Support from the Project INERA under the contract REGPOT 316309 is gratefully acknowledged.

6. References

[1] Petrakovskii G, Velikanov D, Vorotinov A M, Balev A D, Sablina K, Amato A, Roessli B, Schefer J and Staub U, 1999 *J. Magn. Magn. Mater.* **205** 105
[2] Pisarev R V, S¨anger I, Petrakovskii G A and Fiebig M 2004 *Phys.Rev. Lett.* **93** 037204
[3] Fiebig, V, Pavlov V and Pisarev R V 2005 *J. Opt. Soc. Amer.* **22** 96
[4] Saito M, Ishikawa K, Taniguchi K and Arima T 2008 *Phys. Rev. Lett.* **101** 117402
[5] Arima T 2008 *J. Phys.: Condens. Matter* **20** 434211
[6] Pisarev R V, Kalashnikova A M, Schöps O and Bezmaternykh L N 2011 *Phys. Rev.* B **84** 075160
[7] Ivanov V G, Abrashev M V, Todorov N D, Tomov V, Nikolova R P, Litvinchuk A P, and Iliev M 2013 *Phys. Rev.* B **88** 094301
[8] Pisarev R V, K. N. Boldyrev K N, Popova M N, Smirnov A N, Davydov V Yu, Bezmaternykh L N, Smirnov M B and Kazimirov V Yu 2013 *Phys. Rev.* B **88** 024301
[9] CRYSALISPRO 2010 *Agilent Technologies* UK Ltd.
[10] Sheldrick G M 2008 *Acta. Crystallogr., Sect.* A*: Found. Crystallogr.* **64** 112
[11] Martinez-Ripoll M, Martinez-Carrera S and Garcia-Blanko S 1971 *Acta Crystallogr., Sect.* B*: Struct. Crystallogr. Cryst. Chem.* **27** 677
[12] Nénert G, Bezmaternykh L. N, Vasiliev A. N and Palstra T. T. M, 2007 Phys. Rev. B**76**, 144401
[13] Hahn Th., 2006 International Tables for Crystallography, Springer, Vol A pp. 428-429,

Morphological study of lipid vesicles in presence of amphotericin B via modification of the microfluidic CellASIC platform and LED illumination microscopy

J Genova, M Decheva-Zarkova, and J I Pavlič
Institute of Solid State Physics, BAS, 72 Tzarigradsko Chaussee, 1784 Sofia, Bulgaria
E-mail: janez.pavlic@ijs.si

Abstract. Giant lipid vesicles (liposomes) are the simplest model of the biological cell and can be easily formed from natural or synthetic lipid species with controlled composition and properties. This is the reason why they are the preferred objects for various scientific investigations. Amphotericin B (AmB) is a membrane active drug, used for treatment of systemic fungal infections. In this work we studied the morphological behavior of giant SOPC vesicles in asymmetrical presence of amphotericin B antibiotic in the vicinity of the lipid membrane. The visualization of the vesicles was carried out via inverted phase contrast microscopy. The illumination source was modified in a way that tungsten light bulb was replaced by 10 W white LED chip. All the experiments were performed using CellASIC ONIX Microfluidic Platform. The setup has been modified thus opening new opportunities for a variety of experimental realizations. The performed morphological studies showed strong and irreversible effect on the vesicle shape at the presence of amphotericin B in concentration 10^{-5} g/l in the outer for the liposome's membrane solution. At concentration 10^{-3} g/l AmB the effect was less visible and in 15-20 minutes the vesicles regained its initial spherical shape.

1. Introduction

Investigations of effects of drugs on cells is mandatory part of creation and verification of medicines. Lipid vesicles are useful model systems, which membranes resemble real cell's membrane. To study basic interaction mechanisms that are responsible for the structure and function of biological membranes it is necessary to have simple composition of an artificial bilayer to facilitate a detailed examination of the membrane.

Amphotericin B (AmB) is an antifungal polyene antibiotic. It was originally extracted from Streptomyces nodosus, a filamentous bacterium, in 1955 at the Squibb Institute for Medical Research from cultures of an undescribed streptomycete isolated from the soil collected in the Orinoco River region of Venezuela. AmB is designated chemically as (1R, 3S, 5R, 6R, 9R, 11R, 15S, 16R, 17R, 18S, 19E, 21E, 23E, 25E, 27E, 29E, 31E, 33R, 35S, 36R, 37S) - 33 - [(3-amino-3, 6-dideoxy-β-D-mannopyranosyl)oxy] - 1, 3, 5, 6, 9, 11, 17, 37-octahydroxy-15, 16, 18-trimethyl-13-oxo-14, 39-dioxabicyclo [33.3.1] nonatriaconta- 19,21,23,25,27,29,31-heptaene- 36-carboxylic acid. The structural formula of AmB is shown on figure 1. The cell membrane is the site of action of this antibiotic. AmB is a representative of the group of chemotherapeutics which mechanism of action consists in change of membrane cell permeability. Destroying natural selective membrane

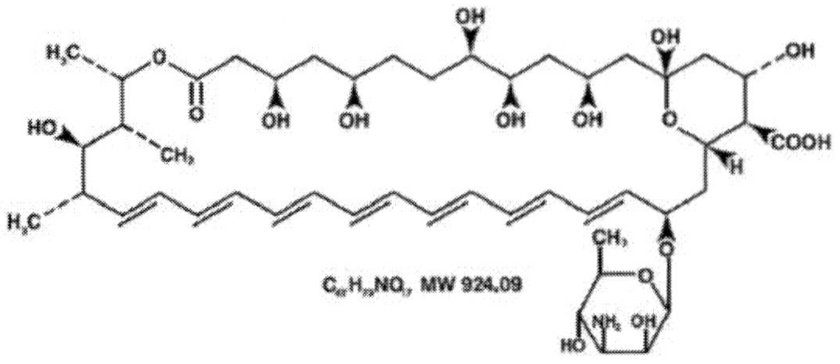

Figure 1. Molecule of amphotericin B (AmB).

permeability is an effect of incorporation of amphiphilic molecules of membrane-active antibiotic into the lipid phase and their interaction with lipid molecules [1]. Rod-shaped structure of AmB molecule, with the polar head mycosamine, the hydroxyl groups on the one side of the macrolide ring and the polyene fragment on the opposite side, makes it possible to interact both with the polar part of the lipid membrane and acyl chains. The consequence of such a structure of AmB is also a formation of molecular aggregates [2, 3]. Not only formation of porous molecular structures by AmB, but also modification of the physical properties of the lipid bilayers modulates membrane permeability to ions [4, 5]. Such mechanisms can be particularly effective at low concentrations of AmB, at which formation of aggregated molecular structures of the drug within the lipid phase cannot be expected [6].

2. Materials

Amphotericin B, sucrose (purity $\geq$ 99,5 GC), d-(+) glucose (purity $\geq$ 99,5 GC) were obtained from Sigma-Aldrich. SOPC (l-stearoyl-2-oleoyl-sn-glycero-3-phosphocholine (C18:0/C18:1)) was purchased from Avanti Polar Lipids Inc., USA).

All the chemicals were used without any further purification. The vesicles for the performed morphological studies were prepared via modified electroformation method [7]. The electroformation cell, used for the formation of vesicles contained two glasses, coated with transparent conductor, indium tin oxide (ITO; thickness of 100 $\pm$ 20 nm, resistively of 100 $\Omega/\Box$ acting as electrodes and a PDMS (polydimethylsiloxane) spacer.

The SOPC lipid was dissolved in chloroform 1 mg/ml. A number of small droplets of the lipid solution were laid on the glasses of the electroformation cell and put under vacuum. After the entire evaporation of the solvent the experimental cell was assembled and filled with 180 mM sucrose solution in double distilled water. A low frequency (10 Hz) sinusoidal alternative voltage was applied (1.5 V_{PP}) to the conductive glasses overnight. This procedure leaded to the formation of vesicles, appropriate for our experiment. After the formation the vesicles were taken out of the formation cell to a flask and 180 mM glucose solution in double distilled water was added to the flask. The flask was then carefully shaken and left for 1-2 hours so that the vesicles precipitate to the bottom of the flask. For the experiment the sample of the vesicles are taken from the bottom of the flask. We have chosen giant (diameter of the order of 20-40 μm) spherical vesicles for the study of the morphology due to the addition of antibiotic Amphotericin B. AmB was dissolved in 180 mM water solution of glucose in concentration 10^{-5} g/l and 10^{-3} g/l.

Journal of Physics: Conference Series **682** (2016) 012029 doi:10.1088/1742-6596/682/1/012029

3. Methods

3.1. LED illumination

In this work we present a modification of Nikon TMS inverted microscope's illumination. The recent advances and mass production of LEDs has brought prices of LED chips down to affordable level. LED power has increased drastically, thus their light intensity. LEDs are of compact size, consume less electrical energy, dissipate less heat and have very long lifetime compared to incandescent bulbs. In visual spectrum, LEDs do not emit in UV and IR spectrum, which reduces health risks, omits the need of filters and improves the light throughput of the system. LEDs output intensity can be easily and precisely controlled. When lowering the LED's drive current (dimming), there are no noticeable changes in emission spectrum and LED maintains practically the same colour temperature over a range of output intensity [8, 9, 10, 11]. All these advantages of the LED illumination motivate the implementation of LED in this experimental study.

In our case an incandescent light source as seen on figure 2 has been replaced with a white light emitting diode (LED) from SYNCHRO as seen on figure 3. The LED has a 10 W of electrical power with 850 lm luminous flux and has a colour temperature of 4000 K. To narrow down the emitted beam of light we used a collimating lens, which was put in front of the LED chip. Therefore we achieved tighter spot and brighter illumination (increase in light intensity).

The adaptation made to the inverted microscope is applied to phase contrast (Ph) microscopy. Lipid vesicles in liquid medium can be visualised and studied via Ph microscopy.

Figure 2. Incandescent light source of inverted microscope.

Figure 3. White LED light source with its power supply atop.

3.2. CellASIC microfluidic setup - modification

Phase contrast microscopy technique has been bundled with Onix CellASIC Millipore microfluidic system. Microfluidic (MF) setup as shown on figure 4 has been obtained for

Journal of Physics: Conference Series **682** (2016) 012029 doi:10.1088/1742-6596/682/1/012029

conducting experiments with cells or cell-like objects. In our study we only focus on cells' model membranes, the vesicles.

The CellASIC ONIX Microfluidic System uses microfluidic technology to enable continuous live imaging with or without media flow. The proprietary design allows objects of interest (cells, vesicles) to be exposed to different solutions and conditions via pressurised flow channels controlled by user-specified time intervals and flow rates.

The microfluidic plate can be used with typical inverted microscopes. The CellASIC ONIX Microfluidic System connects to the microfluidic plate via a pneumatic manifold (F84-DL3-015 regular manifold type) that uses pressurised air to pump objects of interest and liquids from the plate wells into the microfluidic cell culture chambers. A vacuum seal created between the manifold and the microfluidic plate ensures that each well is independent and that flow rates and fluid switching are accurate. A modification of the standard microfluidic setup is accomplished. A syringe is screwed to the vacuum outlet to ensure the proper working of the setup. Controls of the pressure appliance are managed through a USB connection to a computer on which a CellASIC software program is run [12]. MF setup uses special well plate (in our case M04S-03

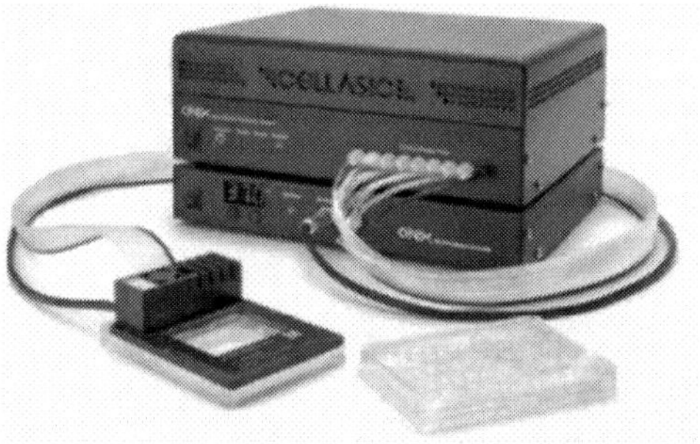

Figure 4. Picture of an original (non-modified) microfluidic setup.

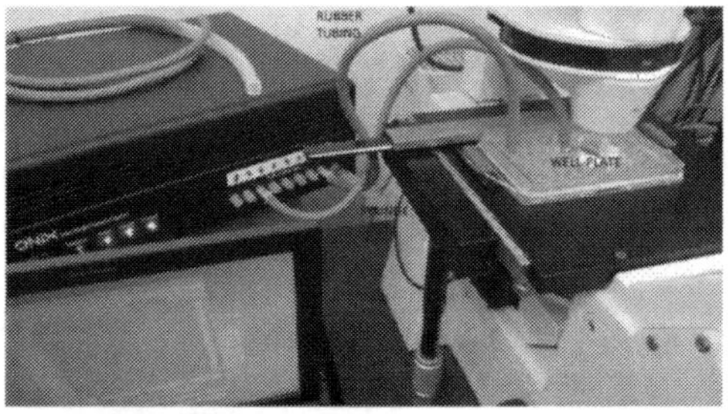

Figure 5. Modification of the microfluidics setup.

type microfluidic plate was used as shown on figure 6) which has microchannels stretching from inlet wells to culture chamber and later on connecting it to outlet wells. It is possible to load vesicles into observation chamber via inlets 1 or 6. For experiments we can supply up to four different solutions (inlets from 2 to 5) e.g. proteins, antibiotics, etc., to the culture chamber and observe the response of vesicles' membranes. The desired pressure is applied to microchannels

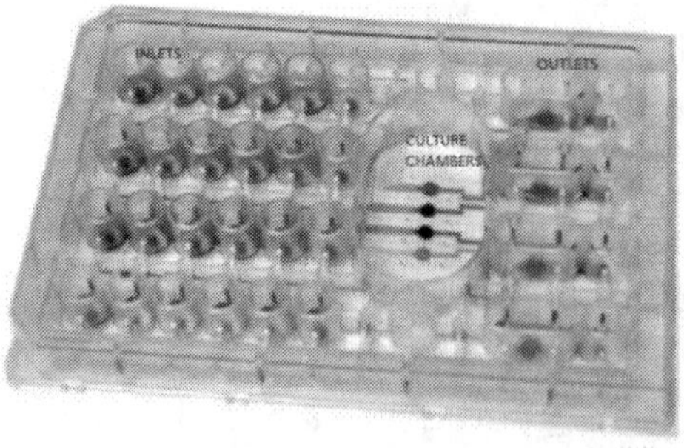

Figure 6. Picture of the well-plate (M04S-03) commonly used in our experiments.

Figure 7. Well-plate M04S-03 with modifications. Rubber/silicone tubing is used to apply pressure to the well-plate inlets. A syringe ensures vacuum for a proper working of the pressure appliance.

via setup's eight nozzles. It is possible to decouple nozzles 1 and 2 from the rest of them (nozzles from 3 to 8). Thus a possibility to apply two different pressures at the same time is achieved. Originally, for applying a desired pressure a special manifold is used. It is vacuum sealed to the well-plate every time the experiment starts.

For the application of pressure to the well-plate we used rubber and/or silicone tubings and vacuum was ensured by syringe screwed onto a vaccum nozzle (black colour) as shown on figure 4.

With such modifcation we achieved three main advantages of the modified microfluicid setup: there is no need to seal a manifold onto the well- plate, which is not so trivial task to accomplish; we can use only one row in well-plate when applying desired pressure to the well-inlets; there is no need to use the whole column. Also this realisation opens up many possibilities for custom microfluidic circuit design with two different pressure inlet options.

4. Results and Discussion

Using a modified CellASIC ONIX Microfluidic Platform morphological study of SOPC vesicles at asymmetrical presence of amphotericin B antibiotic was performed.The setup has been modified thus opening new opportunities for a variety of experimental realizations. A possibility of custom micro-chip design or Millicell μ-Migration Assay usage is achieved. The modified realization of the setup enables the application of two different pressure levels.

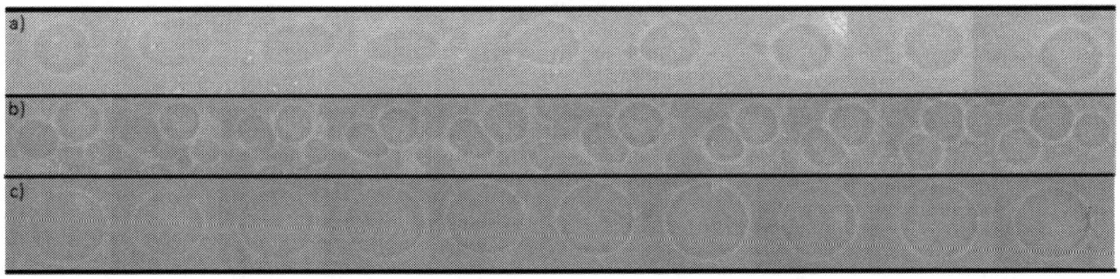

Figure 8. Morphology of lipid vesicles in addition of a) amphotericin B in concentration 10^{-5} g/l in the outer vesicle environment; b) amphotericin B in concentration 10^{-3} g/l in the outer vesicle environment; c) control with addition of glucose solution.

The visualization of the vesicles was carried out via inverted phase contrast microscopy. The illumination source was modified in a way that tungsten light bulb was replaced by 10 W white LED chip.

For achieving a high contrast images the vesicles for the experiment were formed in 180 mM sucrose solution of double distilled water and afterwards dissolved in glucose solution of the same osmolarity. At the beginning of the experiment all the compartments of the well-plate were thoroughly washed with 180 mM glucose solution. The giant lipid vesicles were added to the observation chamber. The amphotericin B dissolved in 180 mM glucose solution at different concentration (10^{-5} g/l and 10^{-3} g/l) was injected to the vesicle sample at pressure level of 0.25 psi for 15 seconds. The change of the shape of the lipid vesicle started right after the addition of the antibiotic. On figure 8 the course of vesicle shape changes, induced by the asymmetrical presence of the molecules of the antibiotic is shown in time. As it is seen on the figure (case a)) a strong and irreversible effect on the vesicle shape was observed at the presence of amphoteric in B in concentration 10^{-5} g/l in the outer for the liposomes membrane solution. At concentration 10^{-3} g/l AmB (case b)) the effect was less visible and in 15-20 minutes the vesicles regained its initial spherical shape. A control experiment was performed where we added glucose solution at the same pressure and time (case c) on figure 8) and no effect on the vesicles shape was observed in time. The possible explanation of the visualized effect on the morphology of lipid vesicle is the incorporation of antibiotic molecules in the structure of the outer layer of the liposome thus changing the local membrane curvature. At higher concentration of AmB the antibiotic molecules tend to agglomerate into bigger composites and do not manage to build into the lipid membrane structure.

5. Acknowledgement

This study was supported by the INERA project Research and Innovation Capacity Strengthening of ISSP-BAS in Multifunctional Nanostructures, Grant agreement no: 316309.

References

[1] Wolf B and Hartsel S 1995 *Biochimica et Biophysica Acta (BBA) - Biomembranes* **1238** 156
[2] Fujii G et al 1997 *Biochemistry* **36** 4959
[3] Gruszecki W I et al 2002 *FEBS Lett* **524** 92
[4] Baginski M et al 2006 *Chem Rec 6* 320
[5] Gabrielska J et al 2006 *FEBS Lett* **580** 2677
[6] Gagos M et al 2010 *Biochim Biophys Acta* **1798** 2124
[7] Angelova M I et al 1992 *Prog. Coll. Pol. Sci.* **89** 127
[8] Silk E 2002 *Microscope* **50** 101
[9] Cole R W and Turner J N 2008 *Microsc. Microanal.* **14** 1
[10] Aswani K et al 2012 *Microscopy Today* **20** 22
[11] Wessels J T et al 2012 *Cytometry Part* A **81A** 188197
[12] User Guide CellASIC ONIX Microfluidic Platform

Photo-controllable electro-optics of aerosil/7CB nanocomposite nematic doped with azo-bonded molecules

Y G Marinov[1,*], **G B Hadjichristov**[1], **A G Petrov**[1] **and S Krishna Prasad**[2]
[1] Georgi Nadjakov Institute of Solid State Physics, Bulgarian Academy of Sciences, 72 Tzarigradsko chaussee, Sofia 1784, Bulgaria
[2] Centre for Nano and Soft Matter Sciences, Jalahalli, Bangalore 560 013, India

E-mail: ymarinov@issp.bas.bg

Abstract. We demonstrate that the electro-optics of nanostructured nematic liquid crystal (LC) doped with a small amount of photoactive LC molecules can be efficiently controlled by light. In particular, the inclusion of 3 wt.% azobenzene LC 4-(4'-ethoxyphenylazo)phenyl hexanoate (EPH) into a gel nanocomposite material produced from nematic LC heptylcyanobiphenyl (7CB) and 3 wt.% hydrophilic silica nanoparticles of size ca. 7 nm (Aerosil 300) allows both the static (the transmittance versus the voltage) and the dynamic (amplitude-frequency electro-optic modulation) characteristics of thin films (25 μm) of such a complex LC system in an alternating-current electric field to be enhanced by UV light at a wavelength of 375 nm. This photo-effect that is reversed with white light is based on the photo-isomerization of the doped azobenzene molecules. The efficient photo control makes the considered EPH-doped Aerosil/7CB photo-responsive nematic nanocomposites attractive for specific electro-optic applications.

1. Introduction

Nanostructured materials with unique physical properties obtained by the doping of liquid crystals (LCs) with nanoparticles (NPs) have recently attracted much attention [1-4]. The significance of such promising nanocomposite materials comes from the fact that they can provide an improvement of LC device performance by the development of new switching modes and tuning the physical properties of the LC materials [5-7]. Since in most cases the NPs-induced effects are due to their special physical properties such as ferroelectric, magnetic or anisotropic shapes, it is of special interest to investigate dispersions of spherical NPs with no special properties, in particular, quartz nanospheres – aerosil NPs (ANPs) having a diameter of about 7 – 10 nm dispersed in LCs. The confined geometry leads to the introduction of disorder in such heterogeneous nanostructured LC systems [1,7-12]. By varying the concentration of such filled nematics, the random disorder can be created *in situ* [13,14]. Very recently, we have studied the electro-optics of thin (25 μm) films of gel nanocomposites produced from nematic LC heptylcyanobiphenyl (7CB) and 3 wt.% ANPs [15].

On the other hand, it is of a great practical interest to make the nanostructured materials photo-responsive. A simple way for this is to dope them with photoactive agents. Good candidates can be the azo-containing compounds that possess an efficient *trans-cis* photoisomerizm. For instance, azobenzene-doped nematics exhibit a photo-induced response [16,17] useful for a variety of

---

* To whom any correspondence should be addressed.

Journal of Physics: Conference Series **682** (2016) 012030 doi:10.1088/1742-6596/682/1/012030

applications [18-21]. Recently, we have investigated photo-induced effects in guest-host systems composed of a nematic LC (as a host) and a photoactive azobenzene-containing LC compound (as a guest) [22-24].

Here, we present an experimental study of the electro-optics of ANPs/7CB films (25 μm thick, 3 wt.% hydrophilic ANPs) doped with 3 wt.% azobenzene LC 4-(4'-ethoxyphenylazo)phenyl hexanoate (abbreviated hereafter as EPH). Our research is focused mainly on the photo-stimulation of the electro-optics of such photoactive nematic nanocomposite materials, relevant to their practical applicability.

2. Experimental

The preparation of EPH-doped ANPs/7CB nanocomposite films was similar to that described in [15]. Hydrophilic ANPs (Aerosil 300 nanospheres having a diameter of ~ 7 nm and capped with silanol groups) at a concentration of 3 wt.% were added to LC 7CB mixed with 3 wt.% LC EPH. EPH-doped ANPs/7CB films with a thickness of 25 μm were formed in cells assembled from two glass plates, to serve as electrodes, each coated inside by ultrathin transparent conductive layers of indium tin oxide (ITO) overcoated with a polyimide layer with a unidirectional rubbing. The cells were assembled with antiparallel orientation directions of the two substrates.

The experimental set up and the measurement configuration have been described in our previous paper [24]. The electro-optical response of EPH-doped ANPs/7CB films to an alternating-current (AC) electric field was investigated by use of an unpolarized laser diode (wavelength λ = 635 nm, 1 mW power) and computer-controlled lock-in amplifier (SR830, Stanford Research Systems). The intensity of the diode laser beam transmitted through the films (I_{out}) was measured with a photodiode. The transmittance of the films was calculated as $T = I_{out}/I_{in}$, where I_{in} was the intensity of the incoming laser beam. The interval between data acquisitions was equal to 8 s (the same applies to the recording of frequency spectra). The UV continuous light source was a LED (λ = 375 nm), and the UV light intensity on the sample was ~ 1.5 mW/cm^2. The measurements were performed at room temperature.

3. Results and Discussion

Figure 1 shows the texture of EPH-doped ANPs/7CB film as viewed by optical microscopy. Separate microscale regions of chaotically disordered texture of the LC containing ANPs agglomerates (cloudy texture) can be seen. The nematic phase of the ANPs/7CB nanocomposite is a weak gel [15] with an irregular and fragile gel network of hydrogen-bonded ANPs [25]. The changes induced by illumination with UV light (λ = 375 nm) appear as a brightening of the texture. This morphological modification occurs via UV-induced photoisomerization of the doped azobenzene EPH LC, i.e. light-driven molecular conformation change of the EPH molecules (the transition from rod-shaped *trans* isomers to bent-shaped *cis* isomers) [16,17,22-24]. No such textural change was observed in ANPs/7CB nanocomposite under the same experimental conditions and in identical cells but without EPH, i.e. the undoped ANPs/7CB nanocomposite itself was photo-insensitive, as further confirmed by other measurements.

At zero electric field, a strong light scattering by EPH-doped ANPs/7CB nematic films takes place due to their gelly structures. The mechanism of light scattering in filled nematics due to the gel network and aggregation structures is well known (see, e.g., [8]). Applying an AC electric field to the films, the low-density gel state can be electrically driven through the dielectric orientation of the nematic director, so that the light-scattering properties (and thereby, the light transmittance) of the films can be electrically controlled. Reasonably, the voltage-dependent light transmittance of the EPH-doped ANPs/7CB nematic films (figure 2) should follow the reorientation of the 7CB LC molecules. By increasing the electric field, the 7CB molecules are forced to orientate along the field direction due to the positive dielectric anisotropy of 7CB, and at higher fields they tend to the homeotropic alignment. This is why the optical scattering is reduced and the optical transmission of the EPH-doped

ANPs/7CB films is increased by the increasing AC field (like the undoped NPs/7CB nanocomposite gel films [15]).

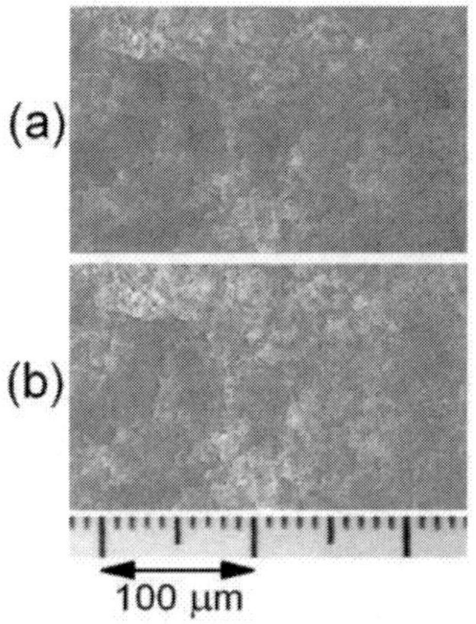

Figure 1. Micrographs of a film of the studied EPH-doped ANPs/7CB nanocomposite gel at an AC voltage of 30 V_{RMS}. No polarizers were used. The frequency of the electric field applied to the film was 300 Hz, and the temperature of the film was 29° C. Pictures taken without (a) or with (b) illumination by UV LED (λ = 375 nm).

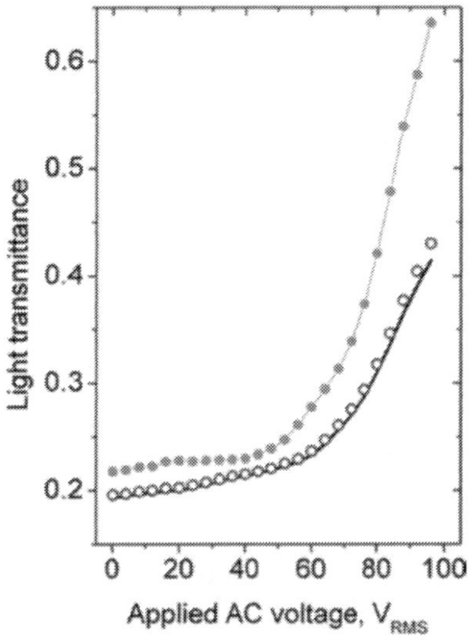

Figure 2. Laser beam (λ = 635 nm) transmission through EPH-doped ANPs/7CB film measured as a function of AC voltage applied to the film, either without (solid line) and with (solid circles) UV illumination at 375 nm, as well as by subsequent illumination by white light (open circles). The laser beam was chopped at 90 Hz, the frequency of the electric field applied to the film was 300 Hz, and the temperature of the film was 29°C. No polarizers were used.

By using UV light illumination, the transmittance of the considered EPH-doped ANPs/7CB films is enhanced (figure 2). The UV-produced change can be considered as a photo-stimulation of the film transmittance (up to ~ 60 % at 100 V, in the case shown in figure 2), and could have useful practical applications. After the UV illumination, by subsequent white light (from room lighting), the EPH-doped ANPs/7CB system tends to return back to its original state (figure 2, open circles) due to the *cis*-to-*trans* photoisomerization of the EPH azo-bonded molecules.

Our supplementary tests indicated that the polyimide layers that overcoat the ITO layers of the LC cell glass plates could play a role for the UV-induced change in the light transmittance of the studied EPH-doped ANPs/7CB gel films (the optical, electro-optical and electrical experimental proofs of the UV-induced photo-effect through the polyimide layers will be presented elsewhere). However, in the present case the contribution of the surface and bulk photoisomerization of EPH is considerably stronger, and thereby the UV-induced effect due to photoactive EPH dopants is much more pronounced than the UV-induced effect from the polyimide layers themselves.

The presence of the photoactive agent EPH in the ANPs/7CB nanocomposite films also strongly modifies their electro-optical modulation response which is rather an interesting feature. Thus, the electro-optic modulation characteristics of the azo-doped nanostructured nematic system studied here can be efficiently photo-controlled. As an example, figure 4 shows frequency spectra of the electro-optic modulation of transmitted laser light by EPH-doped ANPs/7CB films at a 100 V AC driving voltage. The relatively flat and rather extended electro-optical frequency-modulation response of the examined nanocomposite towards the higher frequencies (except for the first harmonic – the modulation at the fundamental frequency) results from the enhanced role of the very small structural domains formed in this nanostructured nematic [15]. The UV-induced modulation changes were reversible (a white-light illumination restores the amplitude-frequency electro-optical response of EPH-doped ANPs/7CB system (figure 4) that is of great importance for the practical application of photo-responsive azo-doped ANPs/7CB nanocomposite films considered here.

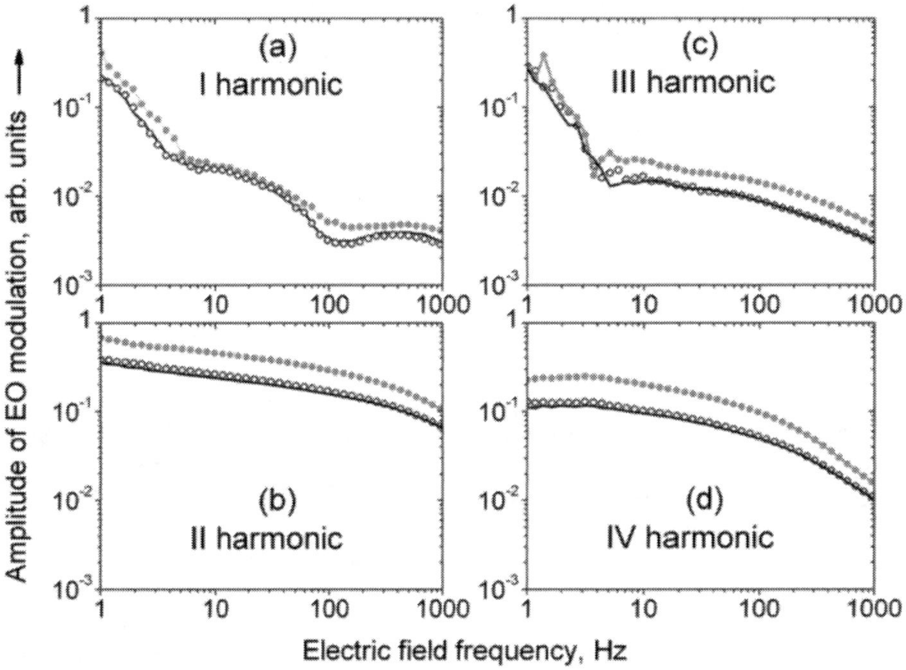

Figure 4. Frequency spectra of the amplitude of electro-optic modulation by EPH-doped ANPs/7CB film measured with a laser beam ($\lambda = 635$ nm) transmitted through the film at a 100 V_{RMS} driving voltage. Data obtained for the first (a); the second (b); the third (c) and the fourth (d) harmonic of the modulated signal versus the applied electric field frequency. No polarizers were used. The temperature of the film was 30° C.

4. Conclusion

We have shown that UV light illumination of aerosil/nematic nanocomposite films doped with photoactive azobenzene (in our case, the LC EPH at concentration of 3 wt.%) can enhance their electro-optical voltage-transmittance behavior and amplitude-frequency electro-optic modulation. Based on the light-driven molecular conformation change of the azo-bonded EPH molecules, the UV-induced stimulation of the electro-optics of EPH-doped aerosil/nematic nanocomposites is reversible (the initial state of the system is restored by white-light illumination). The observed photo-effect and the photo-responsive room-temperature filled nematic considered here could find various photo-controllable electro-optic applications for control and modulation.

Acknowledgments

The authors acknowledge the INERA EU project Research Potential "Research and Innovation Capacity Strengthening of ISSP-BAS in Multifunctional Nanostructures" (FP7 REGPOT-2012-2013-1) support for their participation at the LNN2015 conference. Work supported by the research project DFNI-TO2/18 "Nanostructured liquid crystals for tunable photonic devices" by the Ministry of Education and Science, National Science Fund of Bulgaria. This study was carried out as a part of Indo-Bulgarian joint research project DNTS/In-01/4/2013 "Investigation of photo-stimulation effects in nano-structured liquid crystalline systems"supported by the Department of Science and Technology, New Delhi and the Ministry of Education and Science, National Science Fund of Bulgaria.

References

[1] Iannacchione G S 2004 *Fluid Phase Equlibria* **222-223** 177
[2] Zakharov A V and Thoen J 2004 *Phys. Rev. E* **69** 011704
[3] Cloutier S G, Eakin J N, Guico R S, Sousa M E, Crawford G P and Xu J M 2006 *Phys. Rev. E* **73** 051703
[4] Prasad S K, Rao D S S and Nair G G 2009 *J. Indian Inst. Sci.* **89** 211
[5] Ouskova E, Buchnev O, Reshetnyak V, Reznikov Y and Kresse H 2003 *Liq. Cryst.* **30** 1235
[6] Li F, Buchnev O, Cheon C, Glushchenko A, Reshetnyak V, Reznikov Y, Sluckin T and West J 2006 *Phys. Rev. Lett.* **97** 147801
[7] Li Q 2012 *Liquid Crystals Beyond Displays: Chemistry, Physics, and Applications* (Hoboken, NJ: Wiley & Sons)
[8] Kreuzer M and Eidenschink R 1996 Filled Nematics *Liquid Crystals in Complex Geomteries Formed by Polymer and Porous Networks* eds G P Crawford and S Zumer (London: Taylor & Francis) chapter 15 pp 307–324
[9] Mercuri F, Ghosh A K and Marinelli M 1999 *Phys. Rev.* E **60** R6309
[10] Bellini T, Buscaglia M, Chiccoli C, Mantegazza F, Pasini P and Zannoni C 2000 *Phys. Rev. Lett.* **85** 1008
[11] Marinelli M, Ghosh A K and Mercuri F 2001 *Phys. Rev. E* **63** 061713
[12] Jin T and Finotello D 2001 *Phys. Rev. Lett.* **86** 818
[13] Lobo C V, Prasad S K, Yelamaggad C V 2006 *J. Phys.: Condens. Matter* **18** 767
[14] Bapat P N, Rao D S S, Prasad S K and Yelamaggad C V 2010 *J. Phys. Chem. B* **114** 12825
[15] Kumar M V, Prasad S K, Marinov Y, Todorova L and Petrov A G 2015 *Mol. Cryst. Liq. Cryst.* **610** 51
[16] Prasad S K, Nair G G, Hegde G, Sandhya K L, Rao D S S, Lobo C V and Yelamaggad C V 2005 *Phase Transitions* **78** 443
[17] Yelamaggad C V, Prasad S K and Li Q 2012 Photo-Stimulated Phase Transformations in Liquid Crystals and Their Non-Display Applications *Liquid Crystals Beyond Displays: Chemistry, Physics, and Applications* ed Q Li (Hoboken, NJ: Wiley & Sons) chapter 5 p 157
[18] Tsutsumi O and Ikeda T 2002 *Mater. Sci.* **6** 563
[19] Urbas A, Klosterman J, Tondiglia V, Natarajan L, Sutherland R, Tsutsumi O, Ikeda T and Bunning T 2004 *Adv. Mater.* **16** 1453
[20] Barrett C J, Yager K G, Mamiya J and Ikeda T 2007 *Soft Matter* **3** 1249
[21] Yager K G and Barrett C J 2009 Azobenzene Chromophores in Photo-Reversible Materials *Smart Light-Responsive Materials: Azobenzene-Containing Polymers and Liquid Crystals* eds Y Zhao and T Ikeda (Boca Raton, Florida: John Wiley and Sons) chapter 1 p 1
[22] Petrov A G, Marinov Y G, Hadjichristov G B, Sridevi S, Hiremath U S, Yelamaggad C V and Prasad S K 2011 *Mol. Cryst. Liq. Cryst.* **544** 3/[991]
[23] Sridevi S, Hiremath U S, Yelamaggad C V, Prasad S K, Marinov Y G, Hadjichristov G B and Petrov AG 2011 *Mater. Chem. Phys.* **130** 1329
[24] Hadjichristov G B, Marinov Y G and Yelamaggad C V 2014 *J. Phys.: Conf. Ser.* **558** 012026
[25] Jayalakshmi V, Nair G G and Prasad S K 2007 *J. Phys.: Condens. Matter* **19** 226213

New optical method for measuring the bending elasticity of lipid bilayers

C Minetti[1], V Vitkova[2,4], F Dubois[1], and I Bivas[2]

[1] Université libre de Bruxelles, Service de Chimie Physique EP CP 165/62, 50 Avenue Franklin Roosevelt, 1050 Bruxelles, Belgium
[2] G. Nadjakov Institute of Solid State Physics, Bulgarian Academy of Sciences, 72 Tzarigradsko chaussee blvd., Sofia 1784, Bulgaria

E-mail: victoria@issp.bas.bg

Abstract. The knowledge of the elasticity of lipid bilayer structures is fundamental for new developments in biophysics, pharmacology and biomedical research. Lipid vesicles are readily prepared in laboratory conditions and employed for studying the physical properties of lipid membranes. The thermal fluctuation analysis of the shape of lipid vesicles (or flicker spectroscopy) is one of the experimental methods widely used for the measurement of the bending modulus of lipid bilayers. We present direct phase measurements performed on dilute vesicular suspensions by means of a new optical method exploiting holographic microscopy. For the bending constant of phosphatidylcholine bilayers we report the value of $23k_BT$ in agreement with values previously measured by micropipette aspiration, electrodeformation and flicker spectroscopy of giant lipid vesicles. The application of this novel approach for the evaluation of the bending elasticity of lipid membranes opens the way to future developments in the phase measurements on lipid vesicles for the evaluation of their mechanical constants.

1. Introduction

The bending rigidity of lipid membranes is studied in order to quantify their ability to deform, which is relevant to many biomedical applications [1-4]. The deformability of red blood cells, for example, can be affected by various pathologies. It is of major importance in blood circulation and is closely related to the mechanical properties of the cells. In water, lipid bilayers spontaneously form closed structures, which in the case of unilamellar objects with characteristic sizes of the order of tens of micrometers are called giant unilamellar vesicles (GUVs). During the last decades GUVs have been used to probe the membrane properties [5] and cell hydrodynamics in flows [6]. By the application of elaborated experimental protocols, GUVs are reproducibly prepared in laboratory conditions with a good control of the chemical composition of membranes and bathing solutions. Due to the considerable progress so far, the application of advanced experimental methods allows quantifying the membrane bending rigidity [4, 7, 8]. Several decades ago, Schneider et al. [9] have proposed a method for the measurement of the membrane bending constant based on the analysis of the thermal shape fluctuation analysis (or flicker spectroscopy) applied on GUVs. They registered and analyzed the fluctuations of the vesicle diameter in two mutually perpendicular directions. The progress in this direction continued

[4] To whom any correspondence should be addressed.

Journal of Physics: Conference Series **682** (2016) 012031 doi:10.1088/1742-6596/682/1/012031

with the contributions by Helfrich [10] and Milner and Safran [11], followed by the development of an approach based on the Fourier decomposition of the thermal fluctuations of the vesicle radius [12, 13]. The fluctuation spectroscopy method has been further refined with the introduction of Legendre analysis of the autocorrelation function of the vesicle contour and accounting for the white noise contribution to the calculation of the membrane bending modulus [14]. In the fluctuation spectroscopy, a particularly important step in the image treatment is precisely locating the position of the vesicle contour representing the equatorial cross section of the vesicle with the focal plane of the microscope in each image from the recorded sequence of snapshots [7, 8]. Optical techniques providing information about the fluctuations of vesicle membranes are expected to increase the precision of the bending rigidity measurements. The work presents an experimental evaluation of the bending elasticity modulus of stearoyl-oleoyl phosphatidylcholine membranes by means of a holographic tool that allows performing quantitative phase measurements and reconstituting the vesicle shape.

2. Materials and experimental set-up

Giant unilamellar vesicles (GUVs) were prepared from the lipid 1-stearoyl-2-oleoyl-sn-glycero-3-phospho-choline (Avanti Polar Lipids Inc., AL, USA). A high yield of GUVs with radii of tens of micrometers and without visible defects was obtained by the electroformation method [15, 16]. Methanol and chloroform (*for analysis* grade) used for dissolving the lipid were provided from Fluka Inc. (Germany).

Prior to the bending elasticity measurements, vesicular suspensions were prepared in the manner that the vesicle membranes enclose 0.2 mol/l of sucrose (Sigma Ultra) in water, while the suspending medium was the iso-osmolar aqueous solution of glucose containing 1 wt% concentration of the biocompatible polymer of glucose, dextran. Sucrose and dextran were provided by Sigma-Aldrich Chemie (Gremany), while glucose was purchased from Merck (Germany). Double distilled water with pH~5.5 was used for the preparation of sucrose and glucose solutions. The measurements of the membrane bending rigidity were performed at room temperature (~22°C).

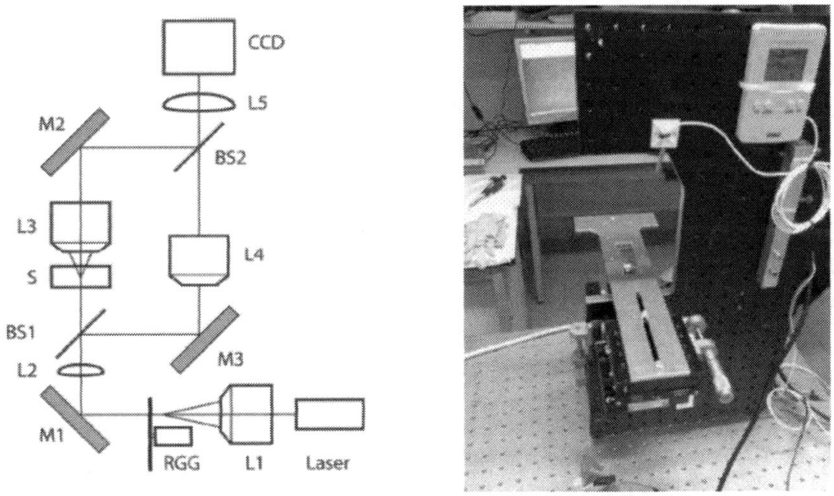

Figure 1. Schematic view and a photograph of the experimental set-up for the observation of vesicular suspensions: M1, M2 and M3 – mirrors; L1 and L4 – objectives; L2, L3 and L5 – optical lenses; RGG – rotating ground glass; S – sample (for further detail see [17]).

Vesicle fluctuations were observed and registered using a microscopic holographic system described in details elsewhere [17]. Its schematic presentation and photograph are shown in figure 1.

Objectives (Olympus, Japan) with magnification ×20 and numerical aperture NA 0.3 were implemented in the set-up in order to achieve the appropriate optical resolution demanded for the purposes of our study. The experimental parameters of our set-up included as follows: laser wavelength of 633 nm; pixel size of 208 nm; frame rate of 24 frames per second with exposure time of 200 μs.

3. Results and discussion

Digital holographic microscopy [17] takes benefit of the knowledge of the optical complex amplitude (amplitude and phase) to refocus numerically each slide of an experimental volume. In the experimental protocol, the vesicles were slightly defocused for a correct visualization. Consequently, the optical phase map of the vesicles might be distorted because they don't correspond to the focus plane of the vesicle. The first step of the processing is thus the determination of the focus plane of each vesicle.

The focus plane of a vesicle was determined by scanning, in depth, a region of interest around the object under investigation and by computing for each plane the gradient of the amplitude modulus. The figure 2 shows an example of the focus curve. Once the focus plane for the object determined, we computed the amplitude and the phase map in that plane.

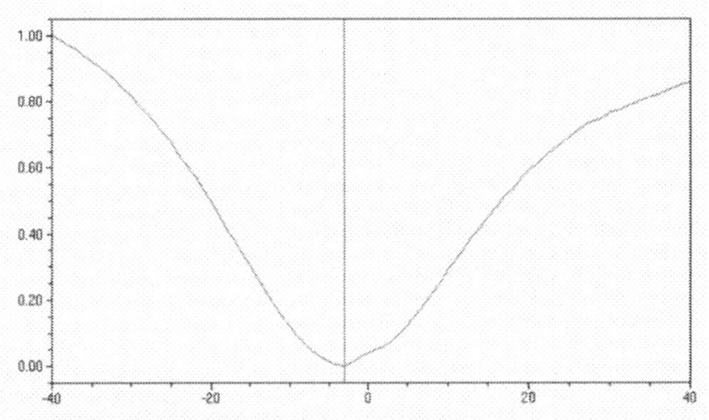

Figure 2. Focus determination; amplitude modulus gradient (arbitrary unit) vs. depth reconstruction distance (μm).

Phase maps were unwrapped in order to remove the discontinuities (phase jumps) with a period of 2π. Examples of wrapped and unwrapped phase maps are shown in figure 3. The background phase (phase obtained when no object is present in the field of view) was approximated by a bi-dimensional quadratic phase map. For every studied vesicle the background phase map was fitted on the basis of the phase map in a neighborhood region where no objects were present. The fitted map is then subtracted from the phase image resulting in a flat and constant background.

The measured phase is related to the vesicle diameter and the refractive index difference between the fluid, enclosed by the vesicle membrane and the surrounding aqueous solution. In the present study only the fluctuations of the vesicle diameter laying on the optical axis of the microscope were extracted and analyzed. Therefore, a segmentation process is performed on the vesicle to extract its contour and determine its center. The value of the phase in the center of the vesicle, proportional to the vesicle radius, is then extracted. Its time autocorrelation function is related to the bending modulus and the lateral tension of vesicle membrane. The experimentally acquired data for the fluctuations of the vesicle diameter with time was fitted with the theoretical expression for the autocorrelation function. The fit yielded the values of the two free parameters, namely the bending modulus and the lateral membrane tension for every analyzed vesicle.

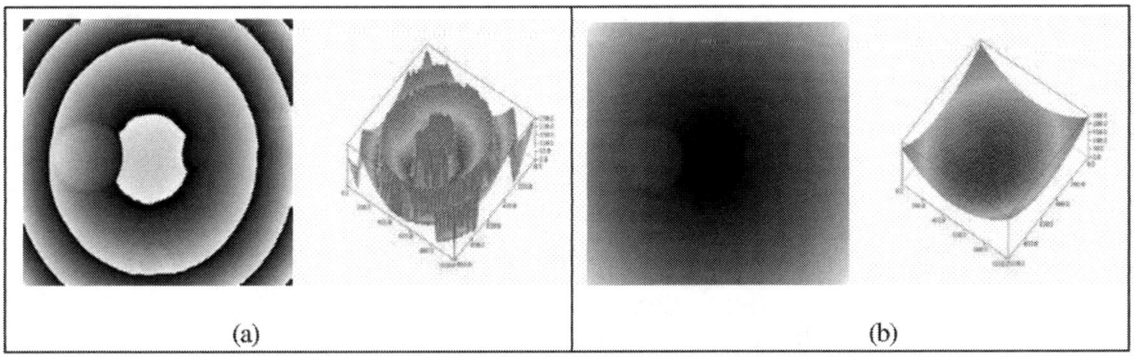

Figure 3. Examples of (a) wrapped and (b) unwrapped phase map.

The ensemble of vesicles with goodness-of-fit (see e.g. [18], Chapter 15) higher than 0.1, used for the evaluation of the membrane bending modulus, contained 9 vesicles out of more than 140 recorded and processed. On the basis of this ensemble we calculated for the membrane bending constant and its error the value of $(23\pm1)k_BT$, where k_B is the Boltzmann constant and T is the absolute temperature as described in detail in [19]. This experimental finding is in quantitative agreement with our previous results from micromanipulation experiments for the bending elasticity of stearoyl-oleoyl phosphatidylcholine bilayers in aqueous solutions with comparable concentrations of sucrose [20]. It corresponds also to the values of the bending constant of phosphatidylcholine membranes published in the literature so far [21, 22].

4. Concluding remarks

We reported on the determination of the bending rigidity of phosphatidylcholine lipid membranes in aqueous solutions containing carbohydrate molecules by means of a new experimental set-up. It is based on quantitative phase measurements performed on fluctuating giant unilamellar vesicles and the subsequent analysis of the fluctuations of the vesicle diameter along the optical axis of the microscopic imaging system. The complete list describing the advantages of the above-presented approach for the determination of the bending constant for lipid bilayers goes beyond the fully automated data treatment and the potential to reconstitute the tri-dimensional shape of lipid vesicles. Further experimental investigations will shed light on the applicability of the method with other lipid compositions and various aqueous media. Its development and application would be of use in studying the relevance of the physicochemical parameters of the aqueous surroundings to the membrane mechanics.

Acknowledgments

This publication has been made possible thanks to the Agreement between WBI, FRS-FNRS and the Bulgarian Academy of Sciences. The contribution of the Bulgarian Science Fund (Grant DMU03-80/2011) is acknowledged.

References

[1] Seifert U 1997 *Adv. Phys.* **46** 13
[2] Safran S 1999 *Adv. Phys.* **48** 395
[3] Nagle J F 2013 *Faraday Discuss. Chem. Soc.* **161** 11
[4] Dimova R 2014 *Adv. Colloid Interface Sci.* **208** 225
[5] Dimova R, Aranda S, Bezlyepkina N, Nikolov V, Riske K A and Lipowsky R 2006 *J. Phys. Condens. Matter* **18** S1151
[6] Vlahovska P M, Podgorski T and Misbah C 2009 *C. R. Physique* **10** 775
[7] Genova J, Vitkova V and Bivas I 2013 *Phys. Rev. E* **88** 022707

[8] Vitkova V and Petrov A G 2013 *Lipid Bilayers and Membranes: Material Properties* (Advances in Planar Lipid Bilayers and Liposomes vol 17) eds A Iglic and J Genova (Burlington: Academic Press) chapter 5 pp 89-138
[9] Schneider M B, Jenkins J T and Webb W W 1984 *J. Physique* **45** 1457
[10] Helfrich W 1986 *J. Physique* **47** 321
[11] Milner S T and S. A. Safran S A 1987 *Phys. Rev. A* **36** 4371
[12] Fricke K, Wirthensohn K, Laxhuber R and Sackmann E 1986 *Eur. Biophys. J.* **14** 67
[13] Sackmann E, Duwe H P and Engelhardt H 1986 *Faraday Discuss. Chem. Soc.* **81** 281
[14] Bivas I, Hanusse P, Bothorel P, Lalanne J and Aguerre-Chariol O. 1987 *J. Physique* **48** 855
[15] Angelova M and Dimitrov D S 1988 *Progr. Colloid Polym. Sci.* **76** 59
[16] Vitkova V, Antonova K, Popkirov G, Mitov M D, Ermakov Y A and Bivas I 2010 *J. Phys.: Conf. Ser.* **253** 012059
[17] Minetti C, Podgorski T, Coupier G and Dubois F 2012 *Proc. of SPIE* **8429** 84291I
[18] Press W H, Flannery B P, Teukolsky S A and Vetterling W T 1992 *Numerical Recipes in Fortran 77: The Art of Scientific Computing* (Cambridge, Cambridge University Press)
[19] Minetti C, Vitkova V, Dubois F and Bivas I 2015 Digital holographic miscroscopy as a tool to study the thermal shape fluctuations of lipid vesicles (submitted)
[20] Vitkova V, Genova J, Mitov M D and Bivas I 2006 *Mol. Cryst. Liq. Cryst.* **449** 95
[21] Marsh D 2006 *Chem. Phys. Lipids* **144** 146
[22] Evans E and Needham D 1987 *J. Phys. Chem.* **91** 4219

Interaction of a single mode field cavity with the 1D XY model: Energy spectrum

H Tonchev, A A Donkov, H Chamati

Institute of Solid State Physics, Bulgarian Academy of Sciences,
72 Tzarigradsko Chaussée, 1784 Sofia, Bulgaria

E-mail: htonchev@issp.bas.bg, aadonkov@issp.bas.bg, hassan.chamati@issp.bas.bg

Abstract. In this work we use the fundamental in quantum optics Jaynes-Cummings model to study the response of spin $\frac{1}{2}$ chain to a single mode of a laser light falling on one of the spins, a focused interaction model between the light and the spin chain. For the spin-spin interaction along the chain we use the XY model. We report here the exact analytical results, obtained with the help of a computer algebra system, for the energy spectrum in this model for chains of up to 4 spins with nearest neighbors interactions, either for open or cyclic chain configurations. Varying the sign and magnitude of the spin exchange coupling relative to the light-spin interaction we have investigated both cases of ferromagnetic or antiferromagnetic spin chains.

1. Introduction

The Jaynes-Cummings [1] model involving the interaction of an atom with a quantized electromagnetic field describes electron transitions between the atom levels induced by the modes of the field. Since its introduction this interaction have found many applications. It has been suggested that an ion trap of atoms or ions could be a realization of a quantum computer. In this context the Jaynes-Cummings model was used in [2] to investigate atom- or ion-field interactions in an ion trap and in [3] in quantum dots. Other uses of this model are for one photon lasers [4] and for single-photon photo detectors [5]. For some pedagogical texts describing the model see, for example [6, 7].

In solid state physics there are many models used to quantum-mechanically describe the interaction between the spins in a spin chain. Commonly used models are the quantum Ising [8] and the Heisenberg [9] ones. In the Ising model the spins are allowed the fewest degrees of freedom – only one. If the spins have all the degrees of freedom – three, then the Heisenberg model is used. In the intermediate XY model [10] the spins interact through two degrees of freedom. The XY model is used to study entanglement [10], spin glasses [11], phase transitions [11], and the W state in quantum computing [12].

In the literature, there exist several models describing the collective interaction between the electromagnetic field and a spin chain. One such model is the Dicke model, often used to study the superadiance phase transitions [13]. Another recent model [14] describes the collective interaction between classical monochromatic circularly polarized light and the spin chain. Here, we report a novel use of the Jaynes-Cummings model, namely, the study of the response of a spin $\frac{1}{2}$ chain to a single quantum mode of a laser light focused on a particlar spin.

The description of the XY-Jaynes-Cummings model for the interaction Hamiltonian follows in the next Section 2, after that we report results obtained by solving the model for the particular case of a finite chain of four spins in Section 3, followed by concluding remarks.

2. The Hamiltonian

The object of our study is a spin $\frac{1}{2}$ chain of N sites and one mode of a laser field interacting with one of the spins in the chain. The spins form either an open or a closed, cyclic chain, for which the last spin interacts with the first, see Figure 1. We consider chains with nearest neighbor interactions J, in the XY Model, which couples the S^x and S^y components of the neighboring i,j spins operators $S_{i,j} = (S^x_{i,j}, S^y_{i,j}, S^z_{i,j})$, and the field is coupled to the k−th spin.

The interaction Hamiltonian $\widehat{H}$ reads [15]:

$$\widehat{H} = G\left(aS_k^+ + a^\dagger S_k^-\right) - 2J\sum_{i=1}^{N-1}\left(S_i^x S_{i+1}^x + S_i^y S_{i+1}^y\right), \tag{1}$$

where the first term is nothing but the Jaynes-Cummings model. Here G incorporates information about the interaction between the mode of the field and the k-th spin, and S^+ (S^-) lift (lower) the spin projection, respectively, and are given by $S^\pm = S^x \pm \mathrm{i}S^y$, with "i" the imaginary unit. They would correspond to the transitions between the lower and the upper atomic levels in the commonly used formulations of the Jaynes-Cummings model. The operators a and $a^\dagger$ are the annihilation and creation operators of the mode of the field. The expression with the spin exchange interaction J between the spin dipoles is the XY model. With the "−" sign in mind, in the classical limit of spin vectors, $J > 0$ would mimic a mostly ferromagnetic (FM) alignment in the chains (the S_z components being arbitrary), $J < 0$ an antiferomagnetic (AFM) alignment, and $J = 0$ would describe independent spins along the chain.

Figure 1. Theorist's rendering of open (a), or cyclic (b) configurations of four spin chains. The wiggle line represents the photon.

For the calculations we choose a basis set of states by first ordering the field states, ($|0\rangle, |1\rangle$), where 0 and 1 stand for the number of photons, then for each spin site we have the spin states ($|\uparrow\rangle, |\downarrow\rangle$). This gives a set of 2^{N+1} basis states. The spin operators in this basis are expressed through the Pauli matrices by the standard relations, $S^{x,y,z} = \frac{1}{2}\hbar\sigma^{x,y,z}$. Since the spin-spin term $S_i^x S_{i+1}^x + S_i^y S_{i+1}^y$ can be written as $\frac{1}{2}(S_i^+ S_{i+1}^- + S_i^- S_{i+1}^+)$, it useful to define additionally $\sigma^\pm = \frac{1}{2}(\sigma^x \pm \mathrm{i}\sigma^y)$, so by definition $S^\pm = \hbar\sigma^\pm$. To summarize for spin $\frac{1}{2}$ we have the expressions:

$$\sigma^x = \begin{pmatrix} 0 & 1 \\ 1 & 0 \end{pmatrix}, \quad \sigma^y = \begin{pmatrix} 0 & -\mathrm{i} \\ \mathrm{i} & 0 \end{pmatrix}, \quad \sigma^z = \begin{pmatrix} 1 & 0 \\ 0 & -1 \end{pmatrix}, \quad \text{and} \quad \mathbb{I} = \begin{pmatrix} 1 & 0 \\ 0 & 1 \end{pmatrix}, \tag{2a}$$

$$|\uparrow\rangle = \begin{pmatrix} 1 \\ 0 \end{pmatrix}, \quad |\downarrow\rangle = \begin{pmatrix} 0 \\ 1 \end{pmatrix}, \tag{2b}$$

and for the photon we have

$$a = \begin{pmatrix} 0 & 1 \\ 0 & 0 \end{pmatrix}, \quad a^\dagger = \begin{pmatrix} 0 & 0 \\ 1 & 0 \end{pmatrix}, \tag{3a}$$

$$|0\rangle = \begin{pmatrix}1\\0\end{pmatrix}, \ |1\rangle = \begin{pmatrix}0\\1\end{pmatrix}. \tag{3b}$$

Since the field and the spin operators on different sites act independently, the matrix form for the subscripted σ_i is a tensor product of $N+1$, 2×2 matrices as follows (e.g. §2.5 of [16]):

$$\sigma_i^x = \underbrace{\mathbb{I}}_{\text{Field mode}} \otimes \underbrace{\mathbb{I}\otimes\mathbb{I}\otimes\cdots\otimes\mathbb{I}}_{i-1 \text{ terms}} \otimes \underbrace{\sigma^x}_{i\text{th spin}} \otimes \underbrace{\mathbb{I}\otimes\mathbb{I}\otimes\cdots\otimes\mathbb{I}}_{N-i \text{ terms}}, \tag{4}$$

and similarly for $\sigma_i^y, \sigma_i^{\pm}$. This gives for the field-spin parts of the Hamiltonian:

$$a^\dagger\sigma_j^y = \underbrace{a^\dagger}_{\text{Field mode}} \otimes \underbrace{\mathbb{I}\otimes\mathbb{I}\otimes\ldots\otimes\mathbb{I}}_{j-1 \text{ terms}} \otimes \underbrace{\sigma^y}_{j\text{th spin}} \otimes \underbrace{\mathbb{I}\otimes\mathbb{I}\otimes\ldots\otimes\mathbb{I}}_{N-j \text{ terms}}, \tag{5}$$

and for the spin-spin terms:

$$\sigma_i^x\sigma_j^y = \underbrace{\mathbb{I}}_{\text{Field mode}} \otimes \underbrace{\mathbb{I}\otimes\cdots\otimes\mathbb{I}}_{i-1 \text{ terms}} \otimes \underbrace{\sigma^x}_{ith \text{ spin}} \otimes \underbrace{\mathbb{I}\otimes\cdots\otimes\mathbb{I}}_{j-i-1 \text{ terms}} \otimes \underbrace{\sigma^y}_{\text{jth spin}} \otimes \underbrace{\mathbb{I}\otimes\cdots\otimes\mathbb{I}}_{N-j \text{ terms}}. \tag{6}$$

3. The energy spectra for 4 spin chain

We diagonalize equation (1), expressed in matrix form using (5-6) for the case of four spins, $N=4$, to obtain the exact solutions for the energy levels. To this end we solve the characteristic equation for E i.e.

$$\det\left(\widehat{H} - E\,\mathbb{I}_{32\times 32}\right) = 0.$$

The Hamiltonian is a 32×32 matrix, and accordingly, there are 32 (some multiply degenerate) energy levels, which we found with help of a computer algebra system. For the open spin chain, the results depend on whether the photon falls on the first or the second spin, while the addition of the cyclic interactions term would appear to make this distinction irrelevant for the closed chain, as the results show as well. In order to check the computer derived formulae, it is useful to compare the limit $J=0$ with the known solutions of the two level Jaynes-Cummings model, which are $\pm G$ and 0 in units of $\hbar$.

3.1. Open chain – edge coupled photon

The spectrum in this case consists of 9 distinct energy branches shown in Figure 2. The analytic expressions , where the superscripts D, Q, or O stand for the degeneracy of each level 2, 4, or 8 times, listed in decreasing order at $G=0$ are:

$$E_1^D(G,J) = \sqrt{G^2+3J^2+2\sqrt{2G^2J^2+J^4}}, \tag{7a}$$

$$E_2^Q(G,J) = \frac{\sqrt{G^2+3J^2+\sqrt{G^4-2G^2J^2+5J^4}}}{\sqrt{2}}, \tag{7b}$$

$$E_3^D(G,J) = \sqrt{G^2+3J^2-2\sqrt{2G^2J^2+J^4}}, \tag{7c}$$

$$E_4^Q(G,J) = \frac{\sqrt{G^2+3J^2-\sqrt{G^4-2G^2J^2+5J^4}}}{\sqrt{2}}, \tag{7d}$$

$$E_5^O(G,J) = 0, \tag{7e}$$

and $E_6^Q = -E_4^Q$, $E_7^D = -E_3^D$, $E_8^Q = -E_2^Q$, $E_9^D = -E_1^D$. The (at least) double degeneracy of the levels follows from the Kramers theorem (we do not have an explicit magnetic field in the Hamiltonian), where the photon acts as an additional spin in the chain for the purposes of the theorem (see e.g. § 60 of [17]). To better show the behavior of the levels with predominantly light-spin interaction ($J \ll G$) or predominantly

Journal of Physics: Conference Series **682** (2016) 012032 doi:10.1088/1742-6596/682/1/012032

spin-spin interaction ($G \ll J$) we have used the substitution $G = C\cos\varphi, J = C\sin\varphi$, where in the Figure 2(b), $C = 1$. With this notation, when $\varphi \in (-\frac{\pi}{2}, 0)$ the chain is an AFM chain, and when $\varphi \in (0, \frac{\pi}{2})$ it is a FM chain. Additionally shown is the region $\varphi > \frac{\pi}{2}$, where G formally takes negative values. This would mean an overall change of the sign of the Hamiltonian, and accordingly, we see an energy spectrum that consists of the same levels, just arranged with opposite signs, so solid lines go to the corresponding solid lines at the line at $\varphi = \frac{\pi}{2}$.

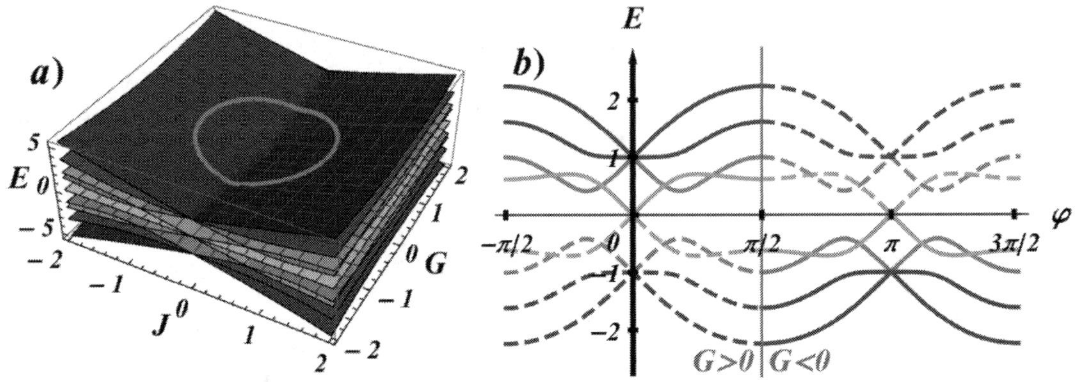

Figure 2. (Color online) The energy spectra given by equations (7) for a 4 spin chain coupled to a photon at the edge of the chain. (a) Energy surfaces at various finite strengths G and J. (b) The cross sections of the energy surfaces at the red (solid line) circle shown in (a). See text for further discussion of the notations and the symmetries of the spectrum. Close to $\varphi = 0$, $J \ll G$, and close to $\varphi = \frac{\pi}{2}$, $G \ll J$.

3.2. Open chain – photon coupled to the second spin in the chain

Due to the extra neighbor to the spin impacted by light, the qualitative change in the spectrum, is that the quadruply degenerate bands in (7) split into doubly degenerate bands, and thus all levels, except the zero energy, are now doubly degenerate. The eigenvalues in explicit form look rather cumbersome, so we first give the characteristic equation. It could be written in factored form as:

$$\begin{aligned}\det\left(\widehat{H} - E\,\mathbb{I}_{32\times32}\right) = E^8\left[E^4 - (G^2+3J^2)E^2 + (G^2+J^2)J^2\right]^2 \\ \times\left\{E^8 - (3G^2+9J^2)E^6 + (3G^4+15G^2J^2+24J^4)E^4\right. \\ -(G^6+7G^4J^2+23G^2J^4+21J^6)E^2 + G^6J^2 + 7G^4J^4 \\ \left.+\,3G^2J^6+5J^8\right\}^2 \\ = 0. \end{aligned} \tag{8}$$

The solutions of this equation are shown in Figure 3, and there are 6 branches in the upper half-plane, with 13 overall distinct solutions. From the first factor we get the octuply degenerate zero level,

$$E^O(G,J) = 0, \tag{9}$$

from the second factor (expression in the square brakets) and the third factor (curly brackets) the remaining doubly degenerated eigenvalues. The expressions from the second factor:

$$E^D_{a,b}(G,J) = -E^D_{c,d}(G,J) = \frac{\sqrt{G^2+3J^2 \pm \sqrt{G^4+2G^2J^2+5J^4}}}{\sqrt{2}}, \tag{10}$$

look similar to the $E^Q_{2,4}(G,J)$ (and $E^Q_{6,8}(G,J)$) in equation (7), with the only change in sign in the term in the second square root: $-2G^2J^2 \to +2G^2J^2$. The expressions for those from the third factor (the eight remaining lines on the figure), as solutions of fourth order equation for E^2, look involved and we only note that the topmost band, which is given by one of the roots of this fourth order equation, looks similar, with a slight decrease in the initial slope at $J \to 0$, to the topmost band E^D_1 in (7).

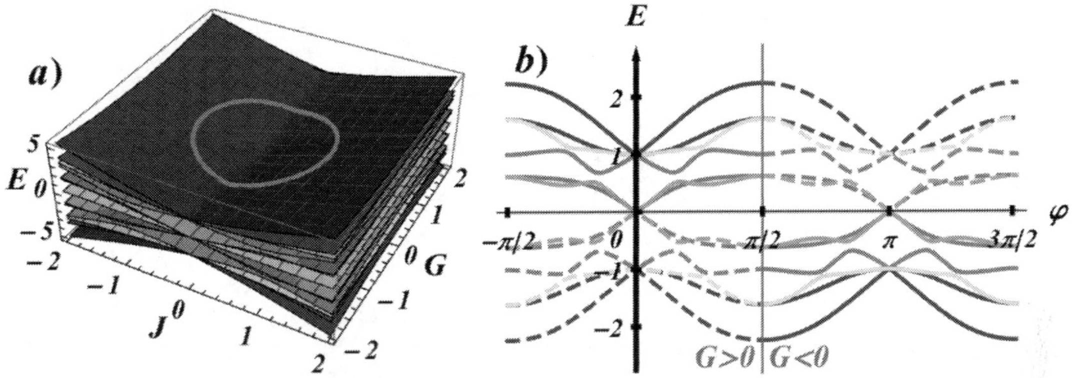

Figure 3. (Color online) The energy spectra given by equations (8) for a 4 spin chain coupled to a photon at second spin in the chain. Notations are similar to Figure 2.

3.3. Closed chain

The 4-spin cyclic configuration is described by the following characteristic equation:

$$\begin{aligned}\det\left(\widehat{H}_C - E\,\mathbb{I}_{32\times 32}\right) &= E^8\left(E^2-G^2\right)^2\left[E^4-(G^2+4J^2)E^2+2G^2J^2\right]^2\\ &\times\left\{E^6-(2G^2+12J^2)E^4+(G^4+6G^2J^2+32J^4)E^2-2G^4J^2-8G^2J^4\right\}^2\\ &=0, \qquad (11)\end{aligned}$$

whose 13 distinct solutions are shown in the Figure 4. $\widehat{H}_C$ compared to (1), has an additional spin-spin interaction term between the spins on sites N and 1; thus closing the chain. Except for the octuply degenerate zero level,

$$E^O(G,J) = 0, \qquad (12)$$

the energy spectrum consists of doubly degenerate levels. The second factor gives:

$$E^D_{A,B}(G,J) = \pm G, \qquad (13)$$

which is independent of J. In Figure 4(b) it appears as a cosine (cyan) line. We have checked, that for a 3-spin chain there also is an energy level that is linear in G, but has an additional, also linear, dependence on J. Here such levels come from a mixture of different basic spin configurations, for which the magnetic part of the Hamiltonian (1) produces canceling contributions. The third, quadratic in E^2, factor gives:

$$E^D_{a,b}(G,J) = -E^D_{c,d}(G,J) = \frac{\sqrt{G^2+4J^2\pm\sqrt{G^4+16J^4}}}{\sqrt{2}}. \qquad (14)$$

The last factor, curly brackets in (11), is cubic in E^2 and gives the remaining 12 roots (counting the degeneracy), or 6 lines on the figure, and the topmost band is one of these roots. The explicit formulae are again somewhat cumbersome, and we skip writing them explicitly.

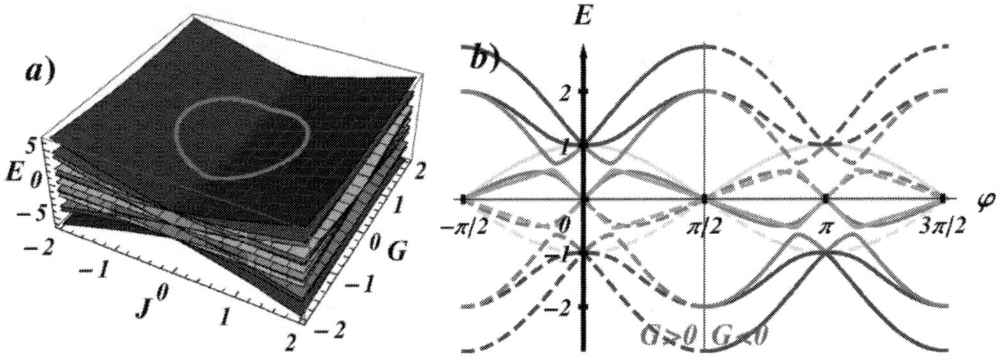

Figure 4. (Color online) The energy spectra from equation (11) for a 4 spin cyclic chain coupled to a photon. Notations are similar to Figure 2.

4. Conclusions

In this report we have discussed some analytical results of applying the Jaynes-Cummings model to an XY- AFM or FM spin $\frac{1}{2}$ chain. Because we limited ourselves to a single mode photon field, the resulting Hamiltonian in matrix form resembles the case of a pure spin chain with one extra spin site inducing anisotropy. Nevertheless the light-spin interaction term does bring a novel effect: it does not commute with the total spin $S_z^{\text{total}} = \frac{1}{2}\sum_{i=1}^{N} \sigma_i^z$. This means that the eigenstates of the XY-Jaynes-Cummings Hamiltonian consist of some mixtures of the product basis states, which have definite spin moment value. For that reason only for $G = 0$ ($\varphi = \frac{\pi}{2}$ point in figures 2, 3, 4) it is possible to assign definite spin moment to the energy levels. As can be seen from the equations, the 4-spin chain seems to be the limit for the analytical formulae, since more sites would require solving polynomial equations of more than 5-th power, which may not have a closed form. It was possible to obtain numerical data for the energy levels of a chain made of more than 4 spins, this would be reported in some future work [18].

Acknowledgments

This work was supported by EU FP7 INERA project grant agreement number 316309. We thank N. Tonchev for bringing our attention to [19].

References

[1] Jaynes E and Cummings F 1963 *Proceedings of the IEEE* **51** 89–109
[2] Vogel W and de Matos Filho R L 1995 *Phys. Rev.* A **52** 4214–4217
[3] You J Q and Nori F 2003 *Phys. Rev.* B **68** 064509
[4] Rempe G, Walther H and Klein N 1987 *Phys. Rev. Lett.* **58** 353–356
[5] Dodonov A V, Mizrahi S S and Dodonov V V 2006 *Phys. Rev.* A **74** 033823
[6] Gerry C and Knight P 2004 *Introductory quantum optics* (Cambridge: Cambridge University Press)
[7] Scully M O and Zubairy M S 1997 *Quantum Optics* (Cambridge: Cambridge University Press)
[8] Dziarmaga J 2005 *Phys. Rev. Lett.* **95** 245701
[9] Arnesen M C, Bose S and Vedral V 2001 *Phys. Rev. Lett.* **87** 017901
[10] Batle J and Casas M 2010 *Phys. Rev.* A **82** 062101
[11] Maucourt J and Grempel D R 1998 *Phys. Rev. Lett.* **80** 770–773
[12] Wang X 2001 *Phys. Rev.* A **64** 012313
[13] Wang Y K and Hioe F T 1973 *Phys. Rev.* A **7** 831–836
[14] Takayoshi S, Aoki H and Oka T 2014 *Phys. Rev.* B **90** 085150
[15] The invariant terms in the total Hamiltonian $\sim (a^\dagger a + \sum_i S_i^z)$ have been rotated away with a unitary transformation, cf. [19]
[16] Rumer Y and Fet A 1970 *The theory of unitary symmetry* (Moscow: Nauka)
[17] Landau L D and Lifshitz E M 1989 *Quantum Mechanics: Non-relativistic Theory (Vol 3)* 3rd ed (Oxford: Pergamon Press)
[18] Tonchev H, Donkov A A and Chamati H (in preparation)
[19] Juárez-Amaro R, Zúñiga-Segundo A and Moya-Cessa H M 2015 *Appl. Math. Inf. Sci.* **9** 299–303

Ultraviolet laser treatment of titanium surface

Ivaylo Balchev[1], Nikolai Minkovski[2], Krasimir Dimitrov[3], Maria Shipochka[4], Robert Barbucha[5]

[1]Institute of Electronics, BAS, 72 Tsarigradsko chaussee blvd., Sofia 1784, Bulgaria
[2]University of Forestry, 10 Kliment Ohridsky blvd., Sofia 1756, Bulgaria
[3]Institute of Solid State Physics, BAS, 72 Tsarigradsko chaussee blvd., Sofia 1784, Bulgaria
[4]Institute of General and Inorganic Chemistry, Bulgarian Academy of Sciences, Acad. G. Bonchev str., bl. 11, 1113 Sofia, Bulgaria
[5] Institute of Fluid flow machinery, PAS, 14 J. Fiszera str., Gdansk 80-231, Poland

E-mail: ivobalchev@gmail.com

Abstract. Interaction of a third harmonic of DPSS laser, wavelength 355 nm and pulse duration of 30 ns with titanium wafers was studied. It was investigated the structure of laser ablated titanium surface, depending on the laser beam scanning speed, and laser pulse frequency. The titanium surface modification was studied by scanning electron microscopy (SEM) and XPS (X-ray Photoelectron Spectroscopy). Nanosecond irradiation with ultraviolet light of Ti plate led to the formation of high porous granular structures consisting of agglomerated micro- and submicro- particles.

1. Introduction

Titanium oxide coatings exhibit a wide variety of physical properties that allow using them in many fields such as photocatalysis, gas sensors, medical implants and optical coatings. Anatase and rutile are used as catalysts, water splitting systems, as well as gas and humidity sensors. The adsorption and desorption characteristics of TiO_2 sensing films can be improved by maximizing the pore sizes on the surface. Titanium is also a widely used biomaterial, especially for bone-tissue implants. The good biocompatibility of titanium is to a large extent depends on its chemical surface properties and surface topography at mezoscopic level [1].

Surface oxidation by laser treatment of metals and semiconductors in air or in chemically active atmospheres promise approach for producing of metal oxide structures with large active surface area and open pore structure. In this work we explored the possibility to obtain titanium oxide coatings by laser irradiation of titanium wafers in air.

SEM and XPS methods were used to investigate the surface morphology, composition and chemical state of Titanium surface before and after laser ablation with ultraviolet light.

2. Experiment

The third harmonic of a Nd:YAG laser (DPSS MATRIX 355; λ= 355nm, pulse duration ~ 30ns, Average power – 5W, beam divergence – 2mrad) operating at a repetition rate of 50kHz was focused by 100mm lens on the Titanium plate. Movement of laser beam in parallel lines with pitch 80μm was achieved by Scanlab scanner head. The treatment was done at fluence 25J/cm^2 (focal spot d = 20μm) and flowing scanning beam velocities: 5mm/s, 40mm/s, 160mm/s, 320mm/s, 640mm/s, 1280 mm/s. The experiments were performed at room temperature in the air.

During laser interaction with metals, including titanium, the laser energy is absorbed primarily by free electrons [4, 5]. The absorbed radiation energy in the skin layer involves thermalization within the electron sub-system.

Due to the electron thermal diffusion, energy transfers to the lattice and heat is transported into metal. For the laser pulses with laser fluence higher than the damages threshold the thermal diffusivity cannot be neglected. Generally say, a series of effects such as melting, vaporization of the molten materials, dissociation and ionization of the vaporized material, etc., can be generated on the target.

For this laser source with relatively short pulses (t_p= 30ns) the heat affected zone is small. It can be estimate after one laser pulse using the relation:

$$L_t = 2\sqrt{D.t_p},$$

where D is thermal diffusivity coefficient, $0.86.10^{-5}m^2.s^{-1}$ for titanium. In this case the length for significant effect on heat wave is L_t=1μm which is very small compare to the focal laser spot of 20μm.

The composition and the chemical properties of the films were analyzed by X-ray photoelectron spectroscopy (XPS). The measurements were carried out on AXIS Supra electron- spectrometer (Kratos Analitycal Ltd.) using monochromatic AlK_α radiation with photon energy of 1486.6eV. The energy calibration was performing by normalizing the C1s line of adsorbed adventitious hydrocarbons to 285.0eV. The binding energies (BE) were determined with an accuracy of ±0.1eV. The chemical compositions of the films were determined monitoring the areas and binding energies of C1s, O1s, Ti2p and N1s photoelectron peaks. Using the commercial data-processing software of Kratos Analytical Ltd. the concentrations of the different chemical elements (in atomic %) were calculated by normalizing the areas of the photoelectron peaks to their relative sensitivity factors.

The morphology of the titanium surface irradiated by the third harmonic of Nd:YAG laser was observed by scanning electron microscopy (SEM) using a Philips SEM-515 electron microscope.

3. Results

3.1. SEM results

Figure 1 shows SEM images of the surface of the irradiated Ti targets with the nanosecond laser pulses under atmospheric pressure in air. The laser irradiation causes a change in the surface morphology due to the processes of melting and subsequent solidification. SEM studies showed that laser beam velocity determine the Ti morphology evolution under pulsed laser irradiation. In air, the vaporized titanium particles from the plume react with O_2 and water vapor, this resulting in formation of oxide nanoclusters. These clusters then are deposited on the surface, which becomes less flat at a submicron level.

Journal of Physics: Conference Series **682** (2016) 012033 doi:10.1088/1742-6596/682/1/012033

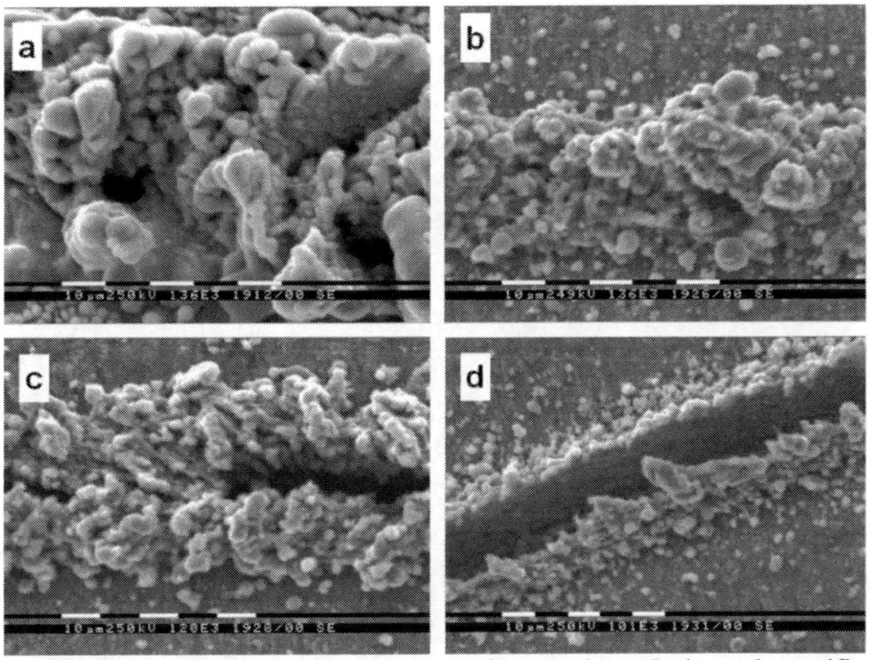

Figure 1. SEM micrographs after laser ablation with scanning velocity and magnification: a) 320mm/s; 1360, b) 80mm/s;1360 c) 40mm/s;1200, d) 5mm/s; 1010.

Morphology of the ablated surface depends strongly of number of pulses over focal spot. This can be calculated by N = d.f/V and varied from 3 for 320mm/s to 200 for 5mm/s. It is visible on micrographs that the laser ablation increases when N is increasing, that means for slower scanning velocity (see figure 1d). In air, the vaporized titanium particles from the plume react with O_2 and water vapor, this resulting in formation of oxide micro and submicro clusters. But main part of microclusters are formed by solidification of melted titanium. The result of this process is visible mainly on figure 1a, b achieved at higher scanning velocity.

3.2. XPS studies

The surface composition and chemical state of the titanium wafers exposed to laser irradiation have been investigated by XPS. The analysis shows a presence of C, O, Ti and less amount of N on the surface of plates. The XPS spectra of titanium is well known in literature. Typical peaks and the values for binding energy are available in table 1.

Core level	**Binding energy (eV)**
Ti2s	561.4
Ti2p$_{1/2}$	460.8
Ti2p3$_{/2}$	454.7
Ti3s	59.1
Ti3p	34.0
O1s	531.4
O2s	23.5

Table 1. XPS binding energies of untreated titanium.

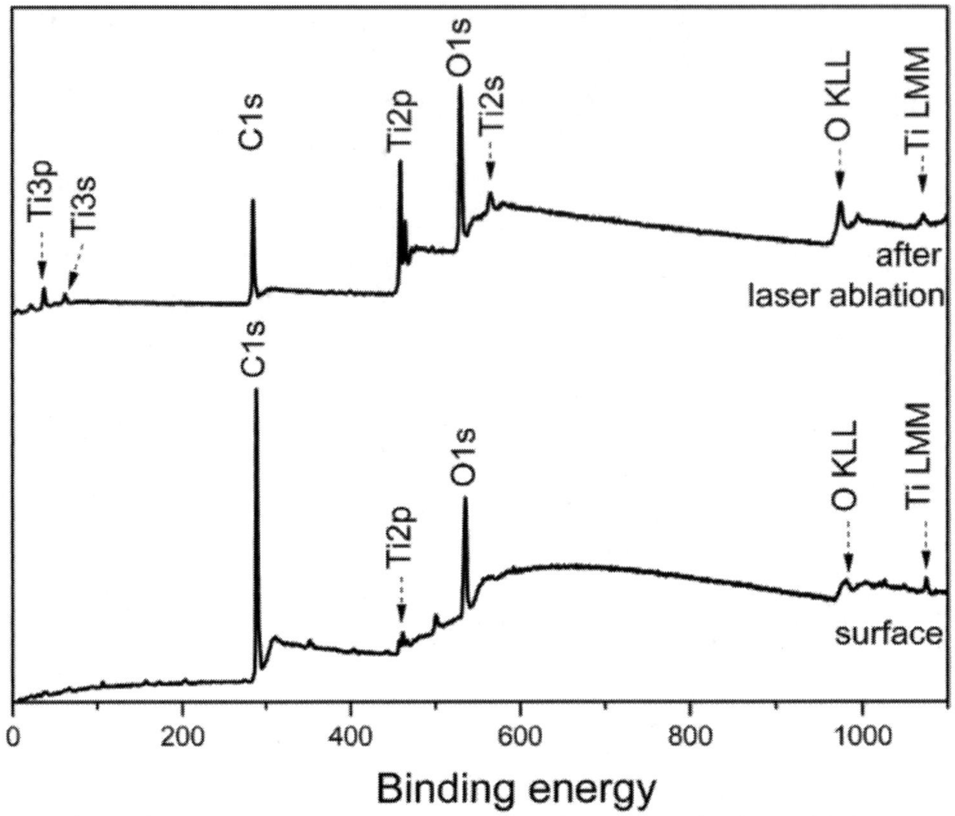

Figure 2. XPS survey of the titanium surface before and after laser ablation.

One typical spectra for treated Ti surface and the same for untreated we presented on figure 2. The XPS analysis indicated that the O/Ti atomic ratio is in the range of (2.6-3.1), which was calculated by the peak areas of the O1s and $Ti2p_{3/2}$, normalized by the corresponding photoionization cross–sections. The value of O/Ti ratio increase with the increasing of the scanning velocity of the laser beam.

scanning velocity	C [at. %]	O [at. %]	Ti [at. %]	N [at. %]	O/Ti
5mm/s	46.1	38.6	15.0	0.3	2.6
40mm/s	48.4	36.7	14.4	0.5	2.6
80mm/s	51.4	34.4	13.4	0.8	2.6
160mm/s	51.1	34.9	13.0	1.0	2.7
320mm/s	53.5	33.5	12.1	0.9	2.8
640mm/s	61.6	28.1	9.7	0.6	2.9
1280mm/s	67.1	24.4	7.9	0.6	3.1

Table 2. The calculated concentrations of chemical elements on the surface after laser treatment with scanning velocity from 5mm/s to 1280mm/s.

We recorded the Ti2p and O1s core level spectra. The binding energies of the Ti2p and O1s lines give information about the oxidation state of the titanium targets. Figure 3 shows the O1s core level spectra of titanium wafers. The O1s peaks are wide with an asymmetric shoulder. They are deconvoluted by Lorentzian–Gaussian curve fitting into two components at 530.0eV and 531.8eV, respectively (figure 4). The first component with the lower binding energy is attributed to O^{2-} ions in the TiO_2 lattice and the second one with the higher binding energy is ascribed to oxygen atoms in hydroxyl groups [5].

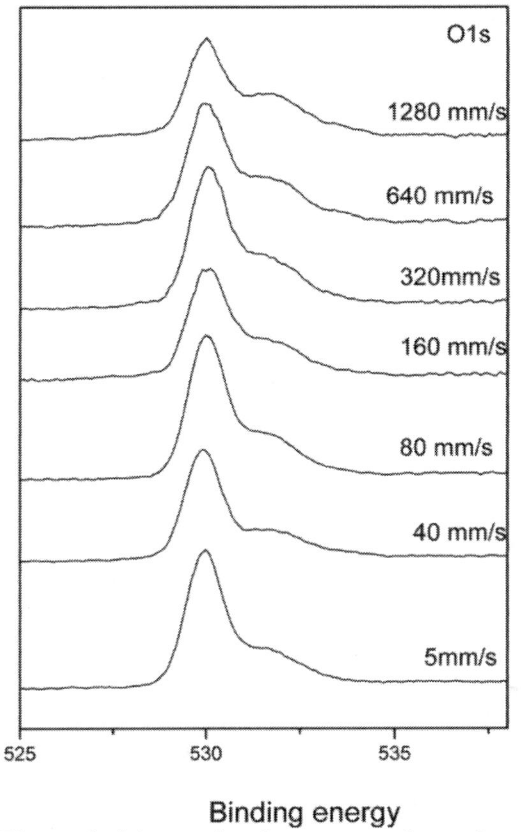

Figure 3. O1s core level spectra on the surface of the titanium wafers.

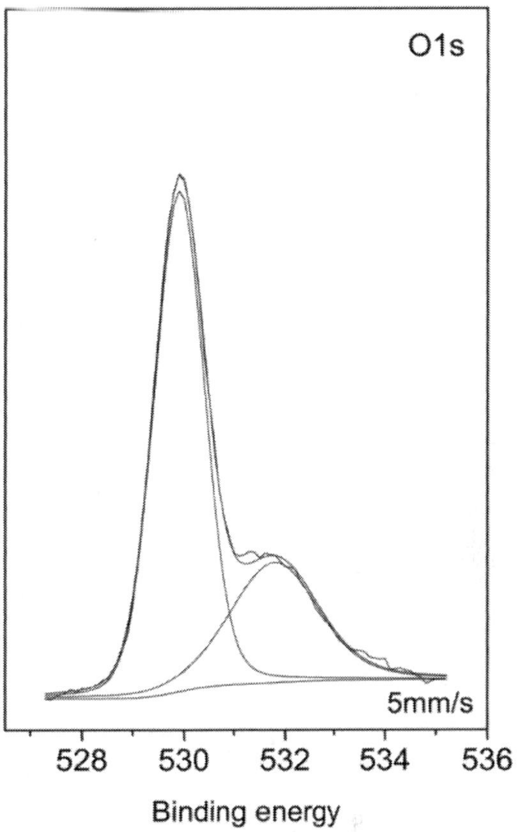

Figure 4. Deconvolution of O1s peak treated with beam velocity 5mm/s.

Figure 5 shows the Ti2p core level spectra of titanium plates prepared with different velocities of the laser beam. The Ti2p spectra are symmetric with a maximum at 458.6eV. Ti2p peak has significantly split spin-orbit components (Δmetal=6.1eV). Splitting-value varies with chemical state (Δnitride=6.0eV, Δoxide=5.7eV). Typically FWHM for each spin-orbit component is the same, but for Ti2p the $Ti2p_{1/2}$ component is much broader than the $Ti2p_{3/2}$ peak. Post-ionization, $Ti2p_{1/2}$ state is very short lived compared to $Ti2p_{3/2}$ state.

The observed peak position, the doublet separation between the $2p_{1/2}$ and $2p_{3/2}$ peaks of ~5.7eV and the satellite at 14eV from the main peak are characteristic of TiO_2.

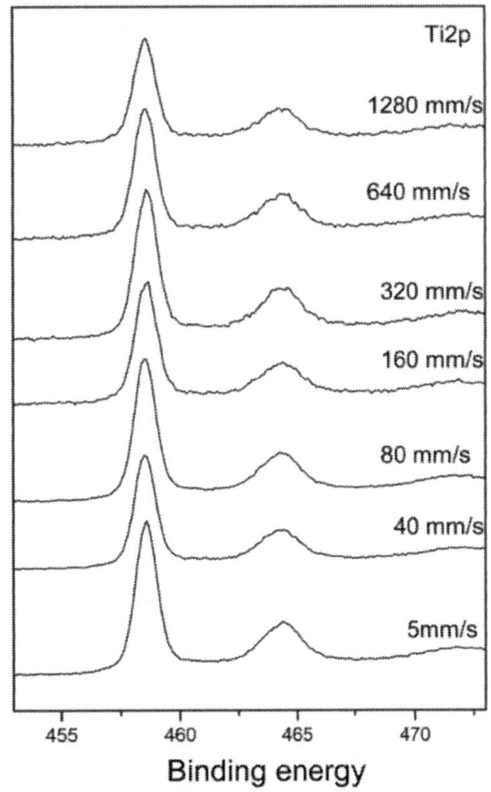

Figure 5. Ti2p core level spectra on the surface of the titanium wafers.

4. Conclusion

XPS analysis indicated that the treatment with UV laser leads to formation of oxide layer on the surface of the plates. The oxide layer covered by hydroxyl species is formed by O_2 and water vapor present in air. The TiO_2 layer is nonstoichiometric which was proved by the calculation of the O/Ti atomic ratio.

References

[1] Diebold U 2002 The surface science of titanium dioxide *Surface Sci. Rep.* **48** 53-229

[2] Fasasia A Y Mwenifumbob S Rahbarb N Chenb J Lid M Beyeb A C Arnoldb C B and Soboyejo W O 2009 Nano-second UV laser processed micro-grooves on Ti6Al4V for biomedical applications *Materials Science and Engineering* **29** 5–13

[3] Pérez del Pino A Serra P and Morenza J L 2002 Coloring of titanium by pulsed laser processing in air *Thin Solid Films* **415** 201–205

[4] Bauerle D 2003 *Laser Processing and Chemistry* (Berlin: Springer Verlag) Chapter 2

[5] Korotin D Bartkowski S Kurmaev E Meumann M Yakushina E Valiev R and Cholakh S 2012 Surface Characterization of Titanium Implants Treated in Hydrofluoric Acid *J. of Biomat. and Nanobiotech* **3** 87-91

Acknowledgements

The authors are indebted to the National Science Fund of Bulgaria (Contract No. DFNI-B02/9/2014) for the financial support and this work was supported by Program for cooperation between the ministry of education and science of the republic of Bulgaria and the minister of science and higher education of the Republic of Poland.

Journal of Physics: Conference Series **682** (2016) 012034 doi:10.1088/1742-6596/682/1/012034

Studying complex chemistries using PLASIMO's global model

PMJ Koelman, S Tadayon Mousavi, R Perillo, WAAD Graef, DB Mihailova and J van Dijk

Department of Applied Physics, Eindhoven University of Technology, P.O. Box 513, 5600 MB Eindhoven, The Netherlands

E-mail: p.m.j.koelman@tue.nl

Abstract. The Plasimo simulation software is used to construct a Global Model of a CO_2 plasma. A DBD plasma between two coaxial cylinders is considered, which is driven by a triangular input power pulse. The plasma chemistry is studied during this power pulse and in the afterglow. The model consists of 71 species that interact in 3500 reactions. Preliminary results from the model are presented. The model has been validated by comparing its results with those presented in Kozák *et al.* (Plasma Sources Science and Technology **23**(4) p. 045004, 2014). A good qualitative agreement has been reached; potential sources of remaining discrepancies are extensively discussed.

1. Introduction

Numerical plasma simulations are a powerful tool for studying the characteristics and essential parameters of a plasma source and have enough predictive value to inspire the development of new plasma technologies. In modern plasma applications, such as plasma surface treatment, biomedical applications and environmental technologies, plasmas are typically created in a mixture of gases. As a consequence of the plasma activities, the filling gas is transformed into a myriad of excited atoms, molecules, radicals and various types of ions.

Two or three dimensional modeling of such a large number of plasma species results in a significant calculation time, from several days to weeks and even more. Global Models are only time resolved, and hence require less computational time than spatially resolved models. Therefore Global Models can be favourable tool to study the plasma processes for chemically complex plasmas.

The overall goal of this study is the efficient transformation of CO_2 in valuable fuels. This process starts with the dissociation of CO_2 in CO and $\frac{1}{2}O_2$, and the fuels are produced in subsequent steps. To that end we study the chemistry of a CO_2 plasma with particular interest in the vibrational levels of CO_2, which provide an energy efficient path for dissociation.

In this work we present a Global Model of CO_2, which is constructed using Plasimo simulation software [1, 5]. We use the same chemistry as Kozak *et al.* [2], who modeled a CO_2 plasma in a cylindrical concentric reactor using GlobalKin [3]. We make a detailed comparison between the results of our model and the results presented in [2], and the influence of the input power density on the CO_2 vibrational modes is studied.

2. The CO_2 Model

First the equations, which form the Global Model, will be introduced breefly. In the PhD thesis of W. Graef [5] an extended description of the Global Model can be found. Thereafter the model set-up and the used set of species and input data will be introduced.

2.1. Global Model

The Global Model solves the time evolution of the species densities n_s of the species s that occur in the plasma,

$$\frac{\mathrm{d}n_s}{\mathrm{d}t} = \sum_{i=1}^{j} (\beta_{i,s} - \alpha_{i,s}) R_i, \tag{1}$$

with j the number of reactions affecting n_s and R_i the rate of reaction i, which is given by:

$$R_i = k_i \prod_z n_z^{\alpha_{i,z}}. \tag{2}$$

Here k_i is the rate coefficient, n_z is the density of particle z and $\alpha_{i,z}$ and $\beta_{i,z}$ the numbers of particles of type z at the left and right hand sides of reaction i. The rate coefficient of reaction i follows from the cross sections using the relation:

$$k_i = \int_{E_{th}}^{\infty} v(\epsilon)\sigma_i(v) f(\epsilon)\,\mathrm{d}\epsilon, \tag{3}$$

with ϵ the energy, $v(\epsilon)$ the velocity of the electrons, $\sigma(v)_i$ the cross section of collision i, $f(\epsilon)$ the electron energy distribution function (eedf) and E_{th} the threshold energy of the collision. For the calculation of the eedf we use the free-ware Boltzmann solver BOLSIG+ [4].

Together with the particle balance, equation (1), the electron energy density balance is solved by:

$$\frac{\mathrm{d}E}{\mathrm{d}t} = \mathrm{P} - Q_{elas} - Q_{inelas}, \tag{4}$$

where P is the input power density, Q_{elas} the energy loss due to elastic collisions and Q_{inelas} the nett energy loss due to inelastic processes.

2.2. Model set-up

A dielectric barier discharge plasma between two coaxial cylinders is considered. The plasma has an inner radius of 11 mm and the outer radius at 13 mm. The length of the plasma cylinder is 90 mm.

Three different input power densities are used in the model: $1.06\ 10^7\ \mathrm{W/m^3}$, $1.06\ 10^8\ \mathrm{W/m^3}$ and $1.06\ 10^9\ \mathrm{W/m^3}$. Figure 1 shows the shape of the input power pulse during the first 100 ns. The power rises in 15 ns towards the maximum value. Thereafter it lowers to 0 in the same period of time, resulting in a pulse duration of 30 ns. The total duration of the cycle is $2.9\ 10^5$ ns.

The pressure is an input parameter which is set at 10^5 Pa. From the pressure the initial densities are calculated, resulting in a density of $1.9\ 10^{25}\ \mathrm{m^{-3}}$ for CO_2. Other species in the ground state and the ions have a density of $1.9\ 10^{15}\ \mathrm{m^{-3}}$. The vibrational states are initially in Boltzmann equilibrium with the species in the ground state. The gas temperature is fixed at 300 K.

Journal of Physics: Conference Series **682** (2016) 012034 doi:10.1088/1742-6596/682/1/012034

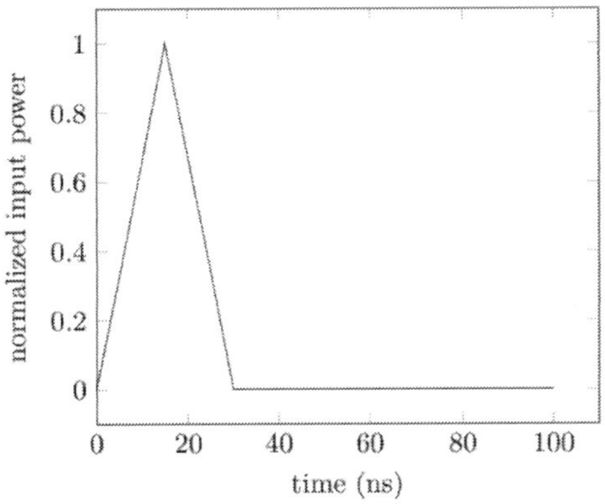

Figure 1. The shape of the input power as used in the model. Since the input power density in the model is one of the studied parameters, the normalized input power density is plotted. Here only the first 100 ns are depicted. The total duration of the cycle is 2.9 10^5 ns .

Table 1. The list of species included in the model.

Neutrals	CO_2	CO	C_2O	C	C_2	O	O_2	O_3
Pos. ions	CO_2^+	CO_4^+	CO^+	$C_2O_2^+$	$C_2O_3^+$	$C_2O_4^+$		
	C^+	C_2^+	O^+	O_2^+	O_4^+			
Neg. ions	CO_3^-	CO_4^-	O^-	O_2^-	O_3^-	O_4^-		
Elec. excited	CO_2e_1	CO_2e_2	COe_1	COe_2	COe_3	COe_4	O_2e_1	O_2e_2
Vib. excited	CO_2v_a	CO_2v_b	CO_2v_c	CO_2v_d	$CO_2v_{1...21}$			
	$COv_{1...9}$	O_2v_1	O_2v_2	O_2v_3				

2.3. Chemistry

The model consists of 71 species, which form a set of 3500 reactions. The species are listed in table 1, where the excitation is denoted by e and v for the electronicly and vibrationally excited states respectively. Cross sections or rate coefficients are available in literature for most reactions in which neutrals or ions are involved. There is a lack of cross sections and rate coefficients for higher excited vibrational level. For that reason it is suggested in [2] to make use of scaling laws, which are based on reaction types. The reaction types that are considered in the model, are listed below. Two of the reaction types scaling laws are involved, which are described concisely.

- **Electron impact reactions**
  The cross sections of the vibrationally excited levels are obtained by scaling of the ground state species cross section by the so called Fridman approximation: scaling of the cross section by a shift in energy and a modification of the absolute value of the cross section.
- **Electron attachment and electron-ion recombination reactions**
- **Neutral-neutral reactions**
- **Ion-heavy particle reactions**
- **Vibrational energy transfer reactions**
  The rate coefficients of these reactions are scaled using the SSH (Schawartz, Slawski and Hertzfeld) theory [6]. The rate coefficients scale strongly with the vibrational mode number and the transition energy of the reaction.

At [2] an extensive description of the species and chemistry can be found.

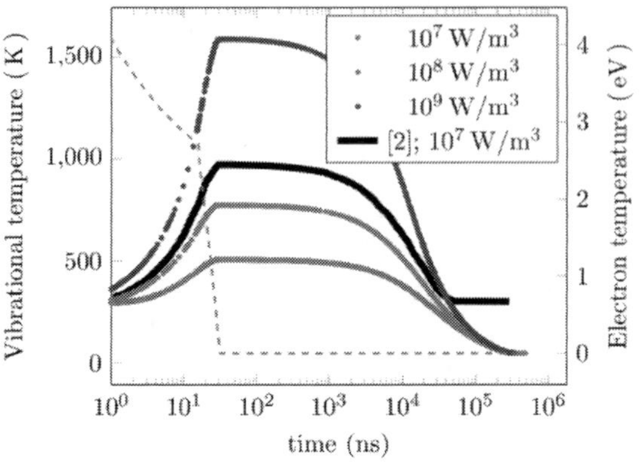

Figure 2. The vibrational temperature as a function of time at different values of the input power density, with logarithmic x-axis. The electron temperature is depicted using the red dashed line. The accompanying label is at the secondary y-axis at the right. Results of [2] are plotted as well for comparison.

3. Results and discussion

First, the results of the model will be analysed using the vibrational temperature of CO_2. Thereafter, the population of the vibrational levels will be the topic of discussion.

3.1. Vibrational temperature

Figure 2 shows the vibrational temperature as a function of time. During the power pulse, the first 30 ns, the vibrational temperature increases towards a maximum value. The electron temperature shoots up at the start of the pulse to a high value, followed by a decrease over time. The peak of the input power pulse can be recognized in the electron temperature by the elbow in the trend of the electron temperature at 15 ns. As the power pulse ends, the electron temperature decreases towards zero.

The rise in vibrational temperature stops at the moment when the electron temperature reaches zero. This result shows that the increase of vibrational temperature results from the electron impact collisions, which typically show an electron temperature dependence. The vibrational temperature starts to decrease significantly around 1000 ns. Within one millisecond the system is completely relaxed towards the equilibrium temperature.

The equilibrium temperature of the model is 0 K, while the vibrational temperature in a physical experiment would relax towards the temperature of the heavy particles. This difference in temperature is a result of omitting collisions of heavy particles which stimulate the population of vibrational levels from the ground state of CO_2.

Trends in the results are independent on the input power density. The input power density only influences the maximum vibrational temperature, which shows a non-linear dependence.

Agreement in the trend is observed when comparing the results with the results of [2]. There are, however, also differences between the results of the models. The vibrational temperature in the results of [2] reaches 950 K, which is significantly higher than the 500 K, which is observed for the same input power. Furthermore the system relaxes towards an equilibrium temperature of 300 K. The decay time of the vibrational temperature towards equilibrium is, with 0.1 ms, the same in both models.

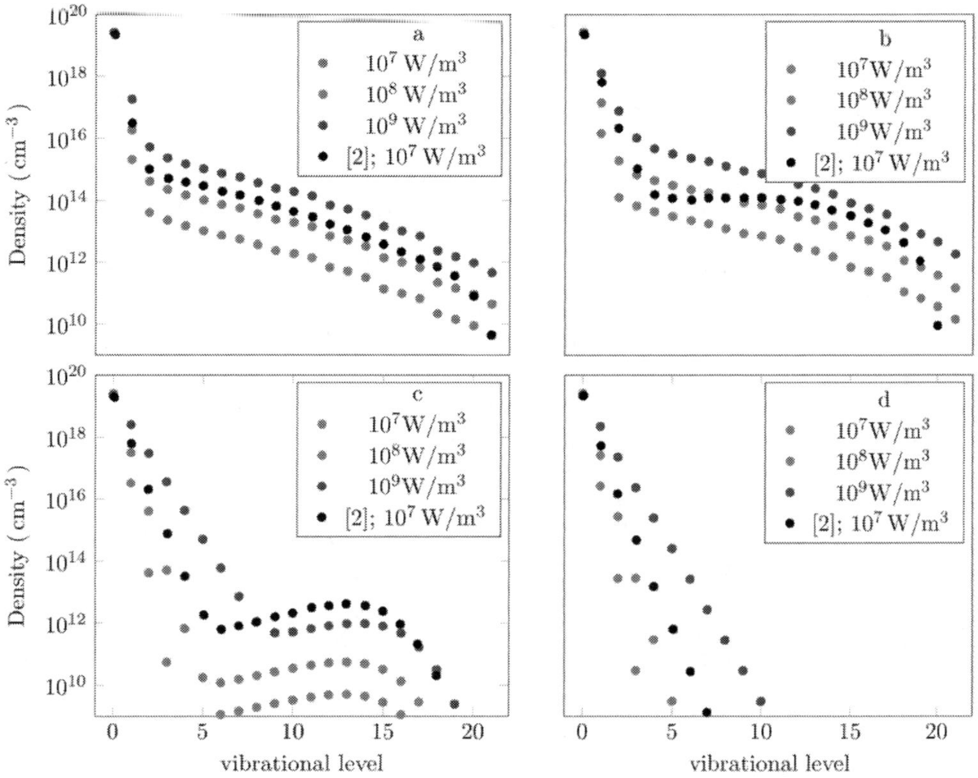

Figure 3. The distribution of vibrational levels after 6 ns, 30 ns, 100 ns and 1000 ns for the top left and right and bottom left and right respectively. The black markers show the results as obtained at [2]. The results are plotted in semi-logarithmic scale.

3.2. Vibrational level population

Figure 3 presents the population of the vibrational levels at four moments in time: at the start of the input power pulse 6 ns, at the end of the input power pulse 30 ns, at the early part of the afterglow 100 ns and further on during the afterglow, after 1000 ns.

The figures show the densities as a function of the vibrational level in logarithmic scale. During the start of the input power pulse the densities decrease linearly with the vibrational level, for vibrational levels above the first level. This trend is independent of the input power density. Increasing the input power density leads to an increased degree of ionization, while the electron temperature does not differ. This higher number of electrons does increase the rate for electron impact collisions, which is a linear dependence (cf. equation (2)).

At the end of the power pulse (at 30 ns, figure b) the distribution of the vibrational level densities do not differ significantly from the results at 6 ns, besides an increase in the densities. At this time the population of the vibrational levels increases, while the gradient remains unchanged.

In the early stage of the afterglow the results show that the vibrational levels, up to level 4-7, the densities exhibit a Boltzmann distribution. For the higher vibrational levels a more salient region can be defined. Species within this region show an inverse trend: increasing population for increasing vibrational level. This region of inverse trend reaches up to the 14th vibrational level. Further on during the afterglow, the results manifest a Boltzmann distribution.

During the afterglow the input power does show an influence on the scaling. During the early part of the afterglow, when the region of inverse scaling is present, the number of vibrational levels in this region are dependent on the input power density. For low input power densities the number of species forming this region is larger than for high input power densities.

The trends of [2] are all in close agreement with the results which are presented here. The difference in absolute densities is likely to be a result of the differences in the models. In the analysis of figure 2 is pointed out that it is likely that the chemistry still needs a modification to balance the chemistry in the afterglow towards a Boltzmann equilibrium at 300 K. It is hard to oversee the consequence of such an addition. It is however well possible that such an addition would stimulate the population of the vibrational levels and that the results will be in even closer agreement.

4. Conclusion and outlook

The Plasimo simulation software is used to construct a Global Model of CO_2 plasma. A dielectric barier discharge plasma is considered, situated between two coaxial cylinders. The plasma, consisting of 71 species which form 3500 reactions, is driven by a triangular shaped power pulse. The vibrational levels of the plasma are studied during the power pulse and the afterglow.

The preliminary results show a trend in vibrational temperature which is independent of the input power density. The population of the vibrational levels show a similar scaling during the time the input power pulse is active. During the early stage of the afterglow there is, however, an input power density dependence observed. A region is defined for which the trend of the vibrational level population exhibits an inverse trend. The size of this region is input power density dependent, with an increased number of vibrational levels within this region for decreased input power density. Further on in the afterglow the temperature dependence in the scaling is not present anymore.

The relaxation of the vibrational temperature in the model is towards 0 K. This is due to the absence of reactions which populate the vibrational levels from the CO_2 ground state via heavy-heavy particle collisions. Extending the chemistry with these reactions is needed.

The scaling in the presented results is in close agreement with results from literature. The major difference between the results of the model and literature is in the absolute value for the vibrational densities, for the same input power density.

In the near future the aim is to complete the chemistry so that the system reaches a Boltzmann distribution at the heavy particle temperature. When the model is further validated based on results of other models, the chemistry will be expanded by adding more vibrational levels of O_2.

Acknowledgements

This research is supported by the Dutch Technology Foundation STW, which is part of the Netherlands Organisation for Scientific Research (NWO), and which is partly funded by the Ministry of Economic Affairs.

References

[1] van Dijk J, Peerenboom K, Jimenez M, Mihailova D and van der Mullen J 2009 *Journal of Physics D: Applied Physics* **42** 194012 URL `http://stacks.iop.org/0022-3727/42/i=19/a=194012`
[2] Kozk T and Bogaerts A 2014 *Plasma Sources Science and Technology* **23** 045004 URL `http://stacks.iop.org/0963-0252/23/i=4/a=045004`
[3] Dorai R 2002 *Modelling of atmospheric pressure plasma processing of gasses and surfaces* Ph.D. thesis University of Illinois at Urban-Champaign
[4] Hagelaar G J M and Pitchford L C 2005 *Plasma Sources Science and Technology* **14** 722 URL `http://stacks.iop.org/0963-0252/14/i=4/a=011`
[5] Graef W A A D 2012 *Zero-dimensional models for plasma chemistry* Ph.D. thesis Eindhoven University of Technology, The Netherlands
[6] Fridman A 2008 *Plasma chemistry* (Cambridge New York: Cambridge University Press) ISBN 9780521847353

Journal of Physics: Conference Series **682** (2016) 012035 doi:10.1088/1742-6596/682/1/012035

Simplifying plasma chemistry via ILDM

T Rehman[1] , E Kemaneci[2], W Graef[1] and J van Dijk[1]
[1]Department of Applied Physics, Eindhoven University of Technology, Eindhoven, The Netherlands
[2]Ruhr University Bochum, Theoretical Electrical Engineering, Bochum, Germany

E-mail: t.rehman@tue.nl

Abstract. A plasma fluid model containing a large number of chemical species and reactions yields a high computational load. One of the methods to overcome this difficulty is to apply Chemical Reduction Techniques as used in combustion engineering. The chemical reduction technique that we study here is ILDM (Intrinsic Lower Dimensional Manifold). The ILDM method is used to simplify an argon plasma model and then a comparison is made with a CRM (Collisional Radiative Model).

1. Introduction

Plasma sources are used extensively for the purpose of surface treatment of materials[1], semiconductor manufacturing[2], fluorescent lamps for lighting[3], growth of thin film solar cells[4], and biomedical applications like skin treatment[5]. These plasma sources contain a large number of chemical species and reactions. A numerical description of these plasma models is given by solving the coupled set of partial differential equation. These equations are spatio-temporal continuity equations, namely mass, momentum, and energy conservation. A complete simulation to study the behaviour of the system involves the solution of all the equations for each species present in the system simultaneously. The numerical solution becomes challenging if the model has a large number of species and reactions. An example of such a plasma fluid model is the conversion of methane into higher hydrocarbons containing 36 species and 367 gas phase reactions[6]. These types of plasma models containing a large number of species and reactions makes the computation expensive. In addition to the high computational cost and time the data generated from these simulations is massive and the interpretation of results becomes difficult. A branch of science that deals with the same problem of chemical complexity alongside plasma science is combustion research. One of the examples from combustion research is the modelling of a 1-D laminar flame[7]. To overcome the difficulty associated with chemical complexity, the combustion community employs various chemical reduction techniques. Some examples of chemical reduction techniques are ILDM (Intrinsic Lower Dimensional Manifold) [7], TGLDM (Trajectory Generated Lower Dimensional Manifold) [8], FGM (Flamelet Generated Manifold) [9], and PCA (Principle Component Analysis) [10].

In this study we apply the ILDM technique to simplify plasma chemistry. The ILDM technique uses the fact that a reaction system containing a large number of species has widely varying time scales for the evolution of different species in the system. A system of chemical reactions can be studied by taking a fewer number of slow reactions, from a complete set of reactions occurring in a system. Other reactions are so fast that the change in the system due to the fast reactions occurs quickly.

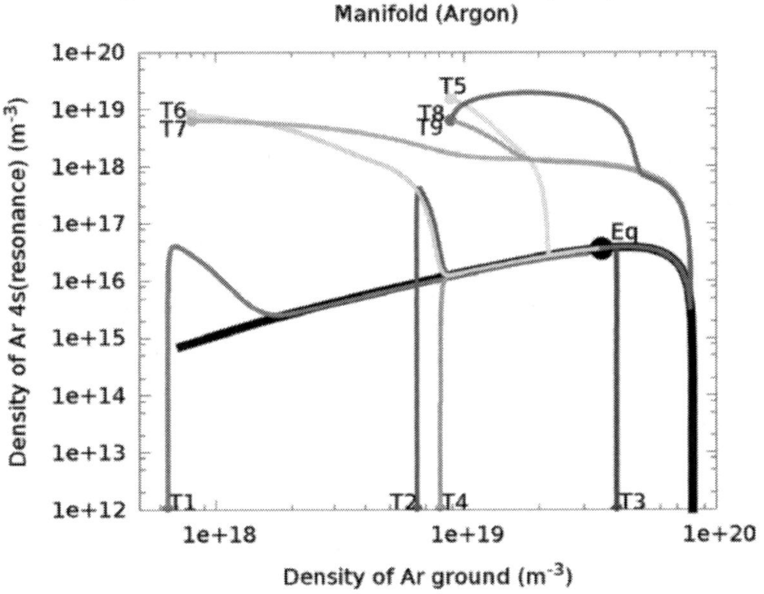

Figure 1. The time evolution of the state in the composition space for an argon plasma simulated with 5 distinct species. Each line (T1 to T9) represent a trajectory with distinct initial condition. The trajectories do not cross each other. It appears as though trajectories are crossing because it is a projection of a 6D composition space (5 species and electron) on a 2D plane. The trajectories first move to a 1D manifold (shown in black) and then along the manifold to the equilibrium point (labelled as Eq).

The fast time scale reactions can be considered in the steady state and the full description of the system can be given by the slow time scales without any significant loss in chemical kinetics description. Taking into account the difference in time scales, the ILDM method finds a lower dimensional space inside the complete composition space. After a short interval of time the fast time scale processes will quickly move towards this lower dimensional manifold and the slow time scale processes will move tangential or along the manifold to finally reach the equilibrium point as shown in figure 1. All the trajectories quickly move towards a line (shown in black) and then along that line to the equilibrium point (labeled as Eq in the figure 1). The line on which the trajectories bundle together is the one-dimensional manifold. Identification of the low dimensional manifold allows the decoupling of fast and slow time scales. Once the fast time scales have attained partial equilibrium, the full system description is given accurately by the lower dimensional manifold.

The advantage of ILDM over other conventional techniques like QSSA (Quasi steady state assumption) is that the ILDM technique automatically extracts the required information from the full system description. The only input required for ILDM is the set of user defined parameters and the dimension of the manifold. It is not dependent on the so called *experience* or *intuition* of the modeler to distinguish the slow and fast time scales. By constructing a lower dimensional manifold the reaction space is defined in terms of a few parameters. It becomes possible to tabulate the results in terms of those few parameters. Once the data is tabulated in the form of a look up table, the continuity equations are solved for only a few parameters and the remaining values are given explicitly. In addition the ILDM method also reduces the stiffness present in the system. Once the stiffness is reduced, there is a speed up in the computation and longer times can be reached. In this paper the ILDM method is applied to generate a manifold for an argon model containing five levels. The effect of transport is not taken into account.

2. Mathematical description of ILDM

The mathematical model[7] for a system consists of a set of $N = n_s + 2$ equations, that is n_s species balances and 2 equations that define temperature and pressure. The species balances are given as

$$\frac{\partial \mathbf{n}}{\partial t} = \mathbf{S}(\boldsymbol{\psi}), \quad (1)$$

where $\boldsymbol{\psi} = (T, P, \mathbf{n})$ with T the temperature, P the pressure, **n** the set of densities for n_s species, and **S** the source term. For each point in the state space the eigenvalues of the Jacobian of the source term give the time scale of the movement in the space. The eigenvectors define the characteristic direction associated with the time scale. The Jacobian of the source term is

$$J_{ij} = \frac{\partial \mathbf{S}_i}{\partial \boldsymbol{\psi}_j}. \quad (2)$$

The eigenvalues and eigenvectors are obtained from diagonalization of the Jacobian matrix J.

$$J = V\Lambda V^{-1}, \quad (3)$$

where V is the eigenvector matrix and Λ is the diagonal matrix with eigenvalues as the diagonal element. Eigenvalues are related to the time scale of the process given as

$$t_i = \frac{1}{\lambda_i}. \quad (4)$$

If any chemical reaction system is given a small perturbation there are four possibilities of how that system will react depending on the nature of eigenvalue of the Jacobian matrix obtained from equation (2). For a real and positive eigenvalue, the perturbation increases and for a real and negative eigenvalue, the perturbation decreases. If the eigenvalue is 0 the perturbation does not change with time. Perturbation due to imaginary eigenvalue has an oscillating nature. The amplitude of oscillation increases, decreases or remains constant, depending on whether the real portion of the eigenvalue is positive, negative or 0 respectively.

Diagonalization is not always possible for a given Jacobian matrix. In many cases the Jacobian matrix will contain repeated eigenvalues. The presence of repeated eigenvalues can lead to linearly dependent eigenvectors, and inversion of the eigenvector matrix is not possible. The problem of diagonalization can be solved by using another basis namely, a basis of Schur vectors denoted as Q. The Schur decomposition of the Jacobian is given as

$$Q^T JQ = M, \quad (5)$$

where M is an upper triangular matrix with eigenvalues on the diagonal arranged in descending order of magnitude, and Q^T is the transpose of the Schur vector matrix. The Schur vectors give the characteristic direction associated with the particular time scale. The fast Schur vectors are perpendicular to the manifold. All the Schur vectors are orthogonal to each other in the N dimensional space, hence the slow Schur vectors are parallel to the manifold. The equation of the manifold is defined as the dot product of source vector with the fast Schur vectors.

$$Q_f^T . \boldsymbol{S}(\boldsymbol{\psi}) = \mathbf{0}, \quad (6)$$

where Q_f^T is the matrix consisting of fast Schur vectors. The fast Schur vector matrix is formed by neglecting the $n_C + n_P$ rows of the Schur vector matrix from the top. n_C is the number of conserved quantities like elements, total number of particles, and charge. n_P is the number of user defined parameters. The numbers of user defined parameters, defines the number of dimensions of the manifold. So we need additional $2 + n_C + n_P$ equations to complete the set of N equations.

The remaining equations are given as

1. Temperature

$$T - T_{ref} = 0. \quad (7)$$

Journal of Physics: Conference Series **682** (2016) 012035 doi:10.1088/1742-6596/682/1/012035

2. Pressure

$$P - \sum n_i k_B T = 0. \quad (8)$$

3. Conservation equations (n_C) for elements

$$f(\chi_i) - \tau_i = 0, \quad (9)$$

where $f(\chi_i)$ is the function for calculating the total number of element χ in the system from all the species present in the system. If the electrons and the charged species are present an additional equation for quasineutrality is added to the elemental conservation equations.

4. Parameter equation (n_P)

$$\varphi_i - p_i = 0, \quad (10)$$

where φ refers to the parameter and p is the value of the parameter. The parameter is the density of the species. The choice of the parameter and the number of parameters are user defined.

3. Application of ILDM to an argon plasma

The ILDM method is applied to generate the manifold for an argon system[11]. The argon plasma that we study contains 5 levels: the ground level (Ar), 4s metastable ($Ar(4s)^m$), 4s resonance ($Ar(4s)^r$), 4p (Ar(4p)) and ion level(Ar^+). The temperature of the electrons is 3 eV and the temperature of the heavy particles is 600 K. The pressure of the background gas is 5 mTorr. The density of electrons is calculated from quasineutrality. The reactions are mainly excitation, de-excitation and ionization as given in table 1.

Table 1. Table showing the rate coefficient for argon system as taken from [11]

Reactions	Rate Coefficient
$Ar + e \rightarrow Ar(4s)^r + e$	$5.00 \cdot 10^{-15}\, Te^{0.74} \exp(-11.56/Te)$
$Ar + e \rightarrow Ar(4s)^m + e$	$5.00 \cdot 10^{-15}\, Te^{0.74} \exp(-11.56/Te)$
$Ar(4s)^r + e \rightarrow Ar + e$	$4.30 \cdot 10^{-16}\, Te^{0.74}$
$Ar(4s)^m + e \rightarrow Ar + e$	$4.30 \cdot 10^{-16}\, Te^{0.74}$
$Ar + e \rightarrow Ar(4p) + e$	$1.40 \cdot 10^{-14}\, Te^{0.71} \exp(-13.20/Te)$
$Ar(4p) + e \rightarrow Ar + e$	$3.90 \cdot 10^{-16}\, Te^{0.71}$
$Ar(4s)^r + e \rightarrow Ar(4p) + e$	$8.90 \cdot 10^{-13}\, Te^{0.51} \exp(-1.59/Te)$
$Ar(4s)^m + e \rightarrow Ar(4p) + e$	$8.90 \cdot 10^{-13}\, Te^{0.51} \exp(-1.59/Te)$
$Ar(4p) + e \rightarrow Ar(4s)^r + e$	$3.00 \cdot 10^{-13}\, Te^{0.51}$
$Ar(4p) + e \rightarrow Ar(4s)^m + e$	$3.00 \cdot 10^{-13}\, Te^{0.51}$
$Ar + e \rightarrow Ar^+ + e$	$2.30 \cdot 10^{-14}\, Te^{0.68} \exp(-15.76/Te)$
$Ar(4s)^r + e \rightarrow Ar^+ + 2e$	$6.80 \cdot 10^{-15}\, Te^{0.67} \exp(-4.20/Te)$
$Ar(4s)^m + e \rightarrow Ar^+ + 2e$	$6.80 \cdot 10^{-15}\, Te^{0.67} \exp(-4.20/Te)$
$Ar(4p) + e \rightarrow Ar^+ + 2e$	$1.80 \cdot 10^{-13}\, Te^{0.61} \exp(-2.61/Te)$
$Ar(4s)^m + e \rightarrow Ar(4s)^r + e$	$2.00 \cdot 10^{-13}$
$Ar(4s)^r \rightarrow Ar + h\upsilon$	$3.00 \cdot 10^{7}$
$Ar(4p) \rightarrow Ar + h\upsilon$	$3.20 \cdot 10^{7}$

Te is the electron temperature in eV and $h\upsilon$ is the energy released.

The equations used for generating a 1D manifold for the argon system are

1. Since the temperature and pressure are constant for the system, the total number of particles is constant. The total amount of argon in the system will remain constant.
2. The plasma is electrically neutral hence the charge conservation equation can be written as

$$\sum e\mu_i n_i - e n_e = 0, \quad (11)$$

where e is the electron charge, n_i is the density of the ion species i and μ_i is the charge number of the ion species i. (for electrons $\mu = -1$)

Journal of Physics: Conference Series **682** (2016) 012035 doi:10.1088/1742-6596/682/1/012035

3. Since we are generating a 1D manifold, there is one user defined parameter. The parameter can be any of the species densities as given in equation (10).
4. The remaining equations are given by the manifold equation (6) as described in section 2.

A 1D manifold can be constructed for an argon plasma as shown in figure 1.

4. Collisional Radiative Model for an argon plasma

A Collisional Radiative Model[12] takes into account the processes of electron collision and radiative transition for the change of state of an atom. All the species in the CRM are classified into two types: Transport Sensitive (TS) and Local Chemistry (LC) [13]. The levels in which the time scales of the chemical processes are comparable to the time scales of transport are termed as the transport sensitive levels. The levels in which the chemical processes are much faster than the transport are termed as the local chemistry levels. Since the chemical processes are very fast, the local chemistry levels can be assumed to be in quasi-steady state. Transport sensitive species are typically the ground and the ion species. The intermediate species are termed as local chemistry species. If the system is linear the source term can be written as

$$\mathbf{S} = \mathbf{F} \cdot \mathbf{n}, \tag{12}$$

where **S** is the source term, F is the Jacobian matrix and **n** is the density vector. The source term is split into TS and LC levels.

$$\begin{pmatrix} \mathbf{S_t} \\ \mathbf{S_l} \end{pmatrix} = \begin{pmatrix} \mathbf{F_{tt}} & \mathbf{F_{lt}} \\ \mathbf{F_{tl}} & \mathbf{F_{ll}} \end{pmatrix} \cdot \begin{pmatrix} \mathbf{n_t} \\ \mathbf{n_l} \end{pmatrix} \tag{13}$$

The local chemistry species are assumed to be in a quasi-steady state. Hence, the source term for LC species is 0 and the density of LC species can be expressed in terms of TS species as

$$\mathbf{n_l} = -\left(\mathbf{F_{ll}^{-1}} \cdot \mathbf{F_{tl}}\right) \cdot \mathbf{n_t} \tag{14}$$

Substituting the density of LC species from equation (14), the source term for TS species can be given as

$$\mathbf{S_t} = \left(\mathbf{F_{tt}} \cdot \mathbf{n_t}\right) - \left(\mathbf{F_{lt}} \cdot \mathbf{F_{ll}^{-1}} \cdot \mathbf{F_{tl}} \cdot \mathbf{n_t}\right) \tag{15}$$

The source for TS species in equation (15) has two terms. The first term in the equation denotes the direct transition between the ion and the ground level. The second term represents the step wise processes of excitation/de-excitation and ionization/de-ionization through the local chemistry levels.

5. Comparison of CRM and ILDM for an argon plasma

CRM is a commonly used reduction technique for plasma chemistry. In case of CRM there is no mass conservation. For comparison of the manifold of ILDM with CRM the mass conservation is not taken into account (unlike as in section 3). The only equations are charge conservation and parameter equations. The ground density is kept fixed to a constant value as it is high compared to the density of intermediate species and the ion density. The ion density is taken as the parameter. The ILDM manifold is found from equation (6). The manifold of CRM is given by equation(14). The result of the comparison of manifold between ILDM and CRM shows a good agreement for the argon system considered here. The ILDM method can be used to simplify the plasma chemistry.

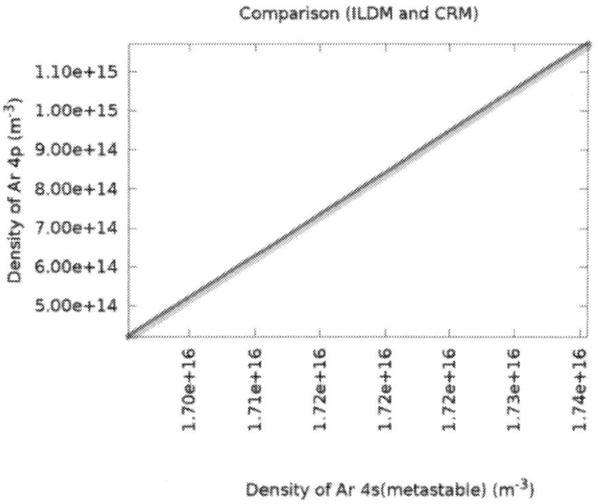

Figure 2. Comparison of the manifold for five species argon model as found from CRM (red) and the ILDM (green). The plot is a projection of 6D composition space on a 2D plane represented by the density of 4s metastable state and 4p state of argon. Similar projections can also be drawn on a different plane represented by different states of argon.

6. Conclusion

A manifold is generated using ILDM for an argon plasma containing two temperatures. The manifold was also compared with another reduction technique used in plasma modelling that is CRM which relies on quasi steady state assumption. The ILDM method has the advantage of being a mathematical technique that separates the slow from the fast reaction. The manifold is described by a look up table as discreet points. Once the look up table is generated the continuity equations are solved for fewer species and the values of the rest of the species are read directly from the table. Thus the computational costs are reduced by using the reduction techniques like ILDM.

7. References

[1] Stoffels E, Flikweert A J, Stoffels W W and Kroesen G M W 2002 *Plasma Sources Science and Technology* **11** 383-388
[2] Chang M B and Chang J S 2006 *Ind. Eng. Chem. Res.* **45** 4101-4109
[3] Waymoth J F and Bitter F 1956 *Journal of Applied Physics* **27** 122-131
[4] Roschek T, Repmann T, Muller J, Rech B and Wagner H 2001 *Journal of Vacuum Science & Technology A* **20(2)** 492-498
[5] Laroussi M and Lu X 2005 *Applied Physics Letters* **87** 113902(1)-113902(3)
[6] De Bie C, Verheyde B, Martens T, Dijk J v, Paulussen S and Bogaerts A 2011 *Plasma Processes and Polymers* **8** 1033-1058
[7] Maas U and Pope S B 1992 *Combustion and Flame* **88** 239-264
[8] Wang M, Huang J and Bushe W K 2007 *Proceedings of the Combustion Institute* **31** 1701-1709
[9] Oijen J A van and de Goey L P H 2002 *Combust. Theory and Modelling* **6** 463-478
[10] Sutherland J C and Parente A 2009 *Proceedings of the Combustion Institute* **32** 1563-1570
[11] Ashida S, Lee C and Lieberman M A 1995 *Journal of Vacuum Science & Technology A* **13(5)** 2498-2507
[12] Dijk J van, Hartgers B, Jonkers J and van der Mullen J 2000 *J. Phys D: Appl. Phys.* **33** 2798-2806
[13] Benoy D A, van der Mullen J A M, van der Sijde B and Schram D C 1991 *J. Quant. Spectrosc. Radiat. Transfer* **46(3)** 195-210

Journal of Physics: Conference Series **682** (2016) 012036 doi:10.1088/1742-6596/682/1/012036

Nonselective and polarization effects in time-resolved optogalvanic spectroscopy

D Zhechev[1] and V Steflekova

Acad. G. Nadjakov Institute of Solid State Physics, Bulgarian Academy of Sciences, 72 Tzarigradsko Chaussee, 1784 Sofia, Bulgaria

E-mail: spectron@issp.bas.bg

Abstract. Three interfering effects in optogalvanic (OG) spectroscopy are identified in a hollow cathode discharge (HCD) – OG detector. The laser beam is found to generate *two nonselective processes,* namely *photoelectron emission (PE)* from the cathode surface with a sub-breakdown bias applied, and *nonresonant space ionization.* The convolution of these galvanic contributions was determined experimentally as an instrumental function and a deconvolution procedure to determine the actual OG signal was developed. *Specific plasma conductance* is detected dependent on the *polarization* of the laser beam irradiating. *Linearly/circularly* polarized light beam is found to induce OG signals differ in amplitude (and their shape parameters in the time-resolved OG signals (TROGS)). The phenomena *coherence* and *specific conductance* are found to be in causal relationship. The additional conductance due to coherent states of atoms manifests itself as an intrinsic instrumental property of OG detector.

1. Introduction

The light portion absorbed on a selected optical transition $i \rightarrow k$ transforms itself into an additional conductance (OG effect) [1, 2] in the OG detector. The measured OG signal represents the galvanic effect due to both light induced small initial departure Δn_{ik} from/to the steady state population density n_i/n_k and its relaxation. When this effect is time-averaged an amplitude OG signal is measured; when a short enough laser pulse is absorbed the time-resolved OG signal (TROGS) is observable and the relaxing processes (spontaneous decay of *k*- state, ionization Δn_e and electron gas heating Δv_e) may be identified.

The hollow cathode discharge (HCD) is the preferred medium – OG detector due to its inherent low level of conductivity fluctuations or galvanic noise. On the other hand, due to the commensurability of the large contact surfaces: *negative glow (NG) – cathode dark space (CDS) – cathode surface*, the HCD is known to be a reservoir of sputtered atoms mostly residing in the ground state. Another feature of the HCD is the characteristic electron energy distribution function (EEDF), which contains three groups of electrons, including one group with energy of up to a few hundred *eV*. Thus, the HCD extends the application of OG spectroscopy to states of sputtered atoms/ions [3] and also to high lying states of buffer gases [4]. Apart from the usual OG technique, Alternating Current Electro-Luminescent (ACEL) structure light emission has also been studied by using another OG

[1] To whom any correspondence should be addressed.

scheme containing a HCD [5]. However, the above intrinsic HCD properties also introduce some specific complications to the measured OG signal in HCDs [6] including the important contribution of photon or γ- processes to the OG signal [4, 7].

Instrumental effects are a basic aspect of any measuring technique and a full understanding of their origin and magnitude is essential for the successful extraction of the real physical parameters from the measured parameter set. As for HCD the above spectral advantages are the reason for some interfering effects in HCD- OG detector. The laser light beam generates additional conductance due to its interaction with both cathode surface and highly excited states of buffer gas. To date the transfer function between *input light stimulus* and *the output galvanic signal* has been only partially discussed within the context of OG spectroscopy, based on the HCD, have already been reported [7].

In this study three light-induced instrumental effects in HCDs are discussed, namely: *i/* the photoelectron emission (PE) from the cathode surface in the absence of a discharge but with an applied bias; *ii/* nonresonant space ionization (SP) from the highly excited states of buffers; *iii/* the state of polarization of the irradiating light beam. These instrumental issues, and the need to explore, quantify and account for them, came to light during a systematic study of the coherent conductivity of aligned and oriented ensembles of atoms [8] in HCDs irradiated by linearly and circularly polarized laser light. We present the results of this study here.

2. Experimental set-up

Time-resolved measurements of the modulation in the voltage across the hollow cathode lamp in the absence (HC) and presence (HCD) of a discharge respectively were performed. Figure 1 show the experimental set-up used to observe time-resolved PE and OG signals. The trade marked HCD lamps Ne/Ca, Ne/Li and Ne/Cd ("Narva"), were used. The discharge was produced by applying a highly stable DC voltage and was operated in negative glow (NG) regime. A standard experimental scheme for OG detection was employed. Time-resolved PE- and measured OG (MOG) signals were induced by irradiating the HC (HCD) with an OPO system which delivered a pulse of duration 4 ns FWHM at a repetition rate of 10 Hz and an average output power 25 mW and were detected across R_m = 20 kΩ with a digital oscilloscope. The RC constant was chosen low enough to allow one to resolve the time-dependent signal.

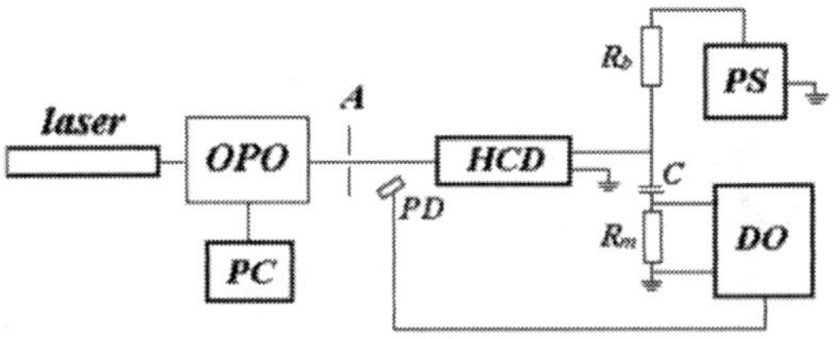

Figure 1. Schematic diagram of the experiment: *HCD* - hollow cathode discharge, *OPO* - optical parametric oscillator system, *P* - polarizer (linear or achromatic waveplate *λ/4,* 450-800 nm), *A*- aperture, *PD* - photo detector, *C* - condenser, R_b – ballast resistor, R_m =20 kΩ– measuring resistor, *PC* – personal computer, *PS* - power supply, *DO* – digital oscilloscope.

3. Experimental results and discussion

Three types of light-induced time-resolved galvanic responses were measured, i.e. the photoelectron emission $\Delta U_{PE}(t)$ from the cathode surface in the absence of a discharge but with an applied bias (figure 2a), the conductivity change $\Delta U_{PE+SP}(t)$ (figure 2b) due to the nonresonant irradiation of the HCD and the conductivity change of HCD due to resonant light pulse irradiation, i.e. the measured TROGS signal $\Delta U_{MOG}(t)$. TROGSs due to both circularly and plane polarized light irradiation are also compared in figure 4.

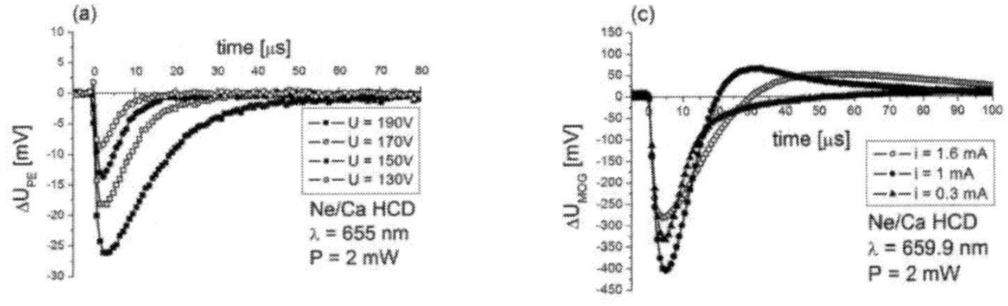

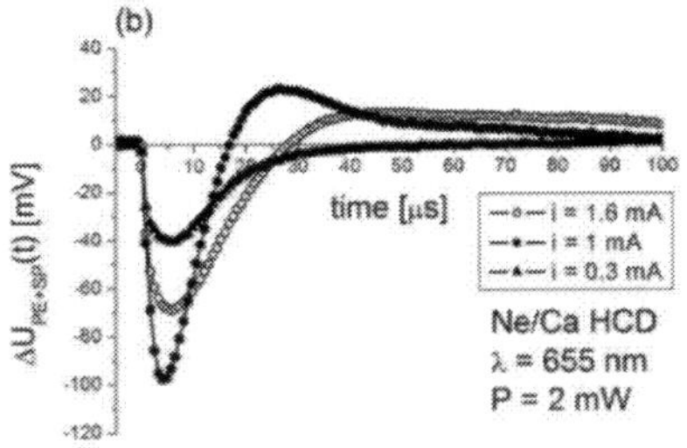

Figure 2. Light-induced time-resolved galvanic responses: (a) photoelectron emission (PE) from the cathode in absence of a HCD; (b) additive nonresonant TROG response $\Delta U_{PE+SP}(t)$ due to both PE- and space ionization effects; (c) measured resonant TROG signal $\Delta U_{MOG}(t)$.

3.1. Photoelectron emission (PE) from the cathode surface and nonresonant space ionization of buffers. Excluding of the background OG component

Figure 2a displays the light-induced PE time-resolved response $\Delta U_{PE}(t)$ of the cathode surface to the absorbed photons at various subthreshold voltage U, i.e. $0 < U < U_{thr}$ (U_{thr} , the threshold voltage of HCD on). The PE maximum ΔU_{PE}^{max} occurs at a time (t_{max}) up to 1 μs after the onset of the signal depending on the voltage applied. The observed time decay of the responses is (15÷80) μs which is typical for TROGS too. Ultimately, the data $\Delta U_{PE}(t)$ measured at values $U<U_{thr}$ characterize the PE properties of the cathode surface material but not the PE of the cathode surface in the presence of a discharge (HCD mode *on*). In HCD mode the voltage applied is confined to a thin layer of 1-2 mm thickness near to the cathode surface and forms the cathode dark space (CDS) region. Earlier we estimated the electric field in the CDS by measuring the peak shift of some He I spectral lines [9] where values in the range 3-5 *kV/cm* were obtained. This voltage (300 V ÷ 1000 V) exceeds the values used above of $U < U_{thr}$ and there is in fact no reason for the PE contribution to be neglected a priori in the measured galvanic signal.

Apart from this PE contribution, the laser light induces one more nonresonant galvanic process, i.e. space ionization. The latter takes place in HCD mode and refers to ionization from atoms residing in excited states close enough to the ionization threshold to be easily ionized. These ensembles of atoms, which reside in levels at high excitation energies, are characteristic of HCDs because they are intimately connected to its EEDF and the corresponding galvanic contribution $\Delta U_{SP}(t)$ should also be accounted for as they have an instrumental impact. The function $\Delta U_{SP}(t)$ manifests itself together with $\Delta U_{PE}(t)$ only in the cup-shaped HCD lamps studied. Figure 2b illustrates the additive effect of the two separate contributions as the function $\Delta U_{PE+SP}(t)$ measured at the nonresonant light irradiation by $\lambda =$ 655 nm.

The measured resonant TROGSs $\Delta U_{MOG}(t)$ (figure 2c) are induced on the transition NeI ($2s_2$-$1p_2$) irradiated by its own line NeI 659.9 nm. The values of $\Delta U_{MOG}(t)$ are affected by the above PE- and SP processes. The function $\Delta U_{PE+SP}(t)$ is measured at the same discharge current 1 mA and laser power 2mW of the nonresonat line $\lambda = 655$ nm. The closest spectral lines 659.9 nm and 653.3 nm are far enough away from the laser wavelength not to contribute to the resonant measurement. Thus $\Delta U_{PE+SP}(t)$ can be taken as an instrumental function.

Then the actual TROGS $\Delta U_{OG}(t)$ signal can be found by the deconvolution of $\Delta U_{MOG}(t)$:

$$\Delta U_{MOG}(t) = \int_{-\infty}^{t} \Delta U_{OG}(t_1) \frac{d}{dt} \Delta U_{PE+SP}(t - t_1) dt_1 \quad (1)$$

The deconvolution of $\Delta U_{MOG}(t)$ by using the instrumental function $\Delta U_{PE+SP}(t)$ gives the actual signal $\Delta U_{OG}(t)$ in figure 3. The algorithm of Petrov [10], which is based on the Tikhonov regularization method [11], was used to compute the deconvolution.

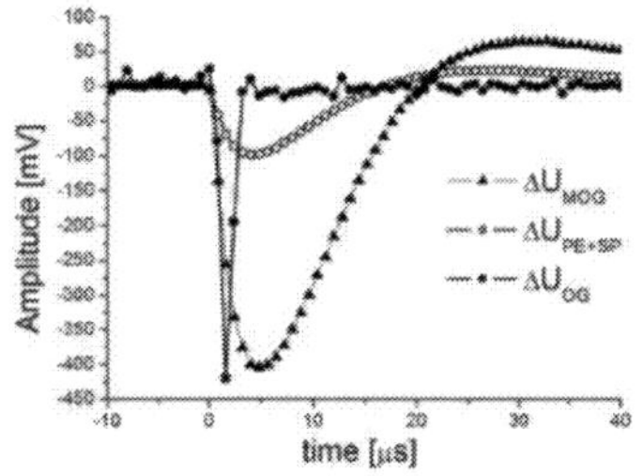

Figure 3. Deconvolution of the measured signal $\Delta U_{MOG}(t)$ by using the instrumental function $\Delta U_{PE+SP}(t)$. $\Delta U_{OG}(t)$ - the actual OG signal in Ne/Ca HCD at discharge current $I = 1$ mA, $P = 2$ mW.

3.2. TROG signal and polarization of light irradiating

Linearly/circularly polarized light is known to induce coherence, namely *alignment/orientation* of the atoms [8]. Within the framework of a comparative study of *aligned/oriented* ensemble of atoms, the TROGSs due to *linearly/circularly* polarized light at one and the same optical transition and at identical discharge current and laser power using an OPO system (figure 1) were detected. The measured TROGSs $\Delta U^{OG}(t)$ signals (figure 4 (a, b)) give evidence two general regularities of their *structure* (the evolutions $\Delta U^{OG}(t)$ are composed by a fast rising peak followed by an exponential decay to either base line or signal with an opposite sign, which returns to the base line) and *timeline duration/damping* (different durations dependent on the optical transition) but allows in the μs time scale).

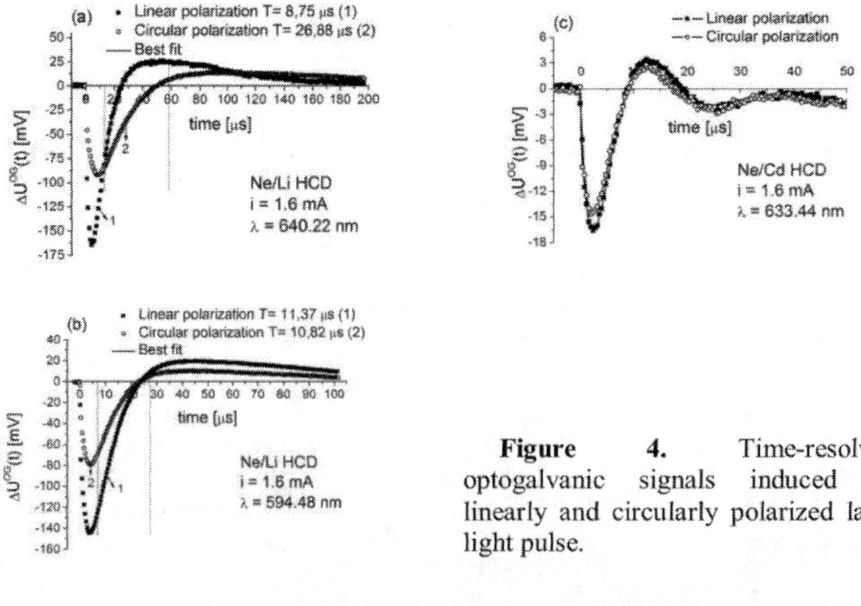

Figure 4. Time-resolved optogalvanic signals induced by linearly and circularly polarized laser light pulse.

In general, the corresponding evolutions $\Delta U^{OG}(t)^{linear}$ and $\Delta U^{OG}(t)^{circl}$ are not identical. They differ in waveform $\Delta U^{OG}(t)$ parameters:

i/ The peak $|\Delta U^{OG}(t)^{\text{linear max}}|$ of the aligned atoms dominates in amplitude that of oriented ones. It is more evident on the transition $1s_5 \rightarrow 2p_9$ (figure 4(a, b));

ii/ The peaks $\Delta U^{OG}(t)^{max}$ lie at *t*- coordinates different for aligned and oriented ensembles $\Delta U^{OG}(t)^{linear}$ and $\Delta U^{OG}(t)^{circl}$ ($\Delta U^{OG}(t)^{linear/circl}$ A clear shift of nearby 4 μs is observable (figure 4a);

iii/ Different decay slopes characterize the evolutions $\Delta U^{OG}(t)^{linear}$ and $\Delta U^{OG}(t)^{circl}$. The best fit of the exponential part of the evolution gives different time of relaxation T in the single-exponential approximation $y = A*exp(-t/T)$ applied.

Figure 4 (a,b) displays that the waveform parameters of TROGS depend on the polarization if inducing light.

The TROGS in figure 4c has to be analyzed separately from the rest. Earlier TROGSs of similar oscillating shape were observed near *I-V* inflection points [7] and taken as an extremal instrumental manifestation of the transient properties of the measuring circle. The oscillating TROGS characterizes the OG detector to stability for disturbance only and gives no information on the optical transition, including *alignment/orientation*. However the uniformity of the two oscillations in figure 4c gives further evidence of the normal (nonoscillating) TROGS sensitivity to the polarization observed in figure 4 (a, b).

Generally, one may write down for the OG signal:

$$\Delta U_{MOG} \propto \pm | \Delta U(\lambda_{ik}) \pm \Delta U(\Delta m = \pm 1 \text{ or } 0, \pm 2) | , \qquad (2)$$

where the dominating energy term $\Delta U(\lambda_{ik})$ describes the OG efficiency of population transfer $i \rightarrow k$. The second term $\Delta U = \Delta U(\Delta m)$ describes the OG contribution of the magnetic states m of the *aligned* ($\Delta m = 0; \pm 2$) or *oriented* ($\Delta m = \pm 1$) atoms [8]. This term includes also the spontaneous coherence, namely *self-alignment* arising in any gas discharge due to space anisotropy of the excitation processes [8]. So far, the term $\pm \Delta U(\Delta m)$ has not been taken into account. On the other hand both *(self-) aligned* and *oriented* states can be influenced by a magnetic field [8, 12], including the lab. Therefore, the term $\pm \Delta U(\Delta m)$ manifests itself as an instrumental factor vs any measured OG signal.

4. Conclusions

Two nonresonant light induced processes, i.e. photoelectron emission (PE) from the cathode surface and space ionization SP are specific to the cup-shaped HCD lamps – OG detector. The measurements in HCD *off* illustrate the significance of PE process. The observed time decay of the responses $\Delta U_{PE}(t)$ is (15÷80) μs, which is typical for TROGS too. In HCD mode for operation the two effects manifest themselves simultaneously as a galvanic background and may be excluded from the measured TROG signal by measuring their convolution $\Delta U_{PE+SP}(t)$ in vicinity of the selected optical transition.

The opto-galvanic signal depends also on the polarization of the illuminating light via the coherent state *(alignment/orientation)* of atoms. Therefore, an additional instrumental influence due to any change in either initial polarization or magnetic field around the OG detector may have to be accounted for in OG experiments.

References

[1] Penning F 1928 *Physica* **8** 137-140
[2] Barbieri B and Beverini N 1990 *Rev. Modern. Phys.* **62** 603-644
[3] Rao G N, Govindarajan J and Reddy M N 1987 *Hyperfine Interactions* **38** 539-552
[4] Zhechev D and Atanassova S 2001 *Opt. Commun.* **194** 359-365
[5] Atanassova S, Yourukova L, Zhechev D, Kolentsov K and Nedev N 1997 *Spectr. Lett.* **30** 1155-1163

[6] Zhechev D and Todorov G 1997 *Opt. Comm.* **136** 227-230
[7] Zhechev D and Atanassova S 1998 *Opt.Comm.* **156** 400-408
[8] Aleksandrov E B, Chaika M P and Khvostenko G I 1993 Interference of atomic states *Springer Series on Atoms and Plasmas* **7** (Springer Berlin Heidelberg New York)
[9] Zhechev D and Steflekova V 2011 *Spectr. Lett.* **44** 47-51
[10] Petrov G 2002 *JQSRT* **72** 281-287
[11] Tikhonov A 1963 *Sov. Phys-Dokl.* **151** 501-504 (in Russian)
[12] Zhechev D, Todorov G, Steflekova V, Slavov D, Polischuk V and Gorbenko A 2009 *J. Phys. B: At. Mol. Opt. Phys.* **42** 035701

New effects in Langmuir and Langmuir-Blodgett monolayers from fluorescently labelled phospholipids and their possible use for water quality control

G R Ivanov[1 *] and N I Geshev[2]

[1] Department of Physics, Faculty of Hydraulic Engineering, University of Architecture, Civil Engineering and Geodesy (UACEG), blvd. Hr. Smirnenski 1, 1046 Sofia, Bulgaria & Advanced Technologies Ltd.
[2] 2nd year student, Civil Engineering, UACEG

george@at-equipment.com

Abstract. Secondary water contamination poses significant challenges to the sensitivity and selectivity of sensors used for its detection and monitoring. Currently only lab tests can detect these contaminants and by the time this happens the contaminated water has entered the city water supply system. Fluorescent chromophore NitroBenzoxaDiazole (NBD) is very suitable and had been successfully used in biosensor applications due to its high sensitivity to close proximity polarity of the medium. Over the years we have discovered 3 new effects in NBD-labelled phospholipids which can significantly improve the performance of biosensors. The phospholipid matrix provides flexible biocompatible environment for immobilization of selectively reacting enzymes, microorganisms, DNA, immunoagents, whole cells. Use of single layer (3.1 nm thickness) films at the air-water interface (Langmuir films) or deposited on solid support as Langmuir-Blodgett (LB) film gives fast response times for real time monitoring (no slow diffusion processes) and precise molecule ordering and orientation. The first new effect was fluorescence self-quenching in Langmuir and LB films. In the liquid phase films exhibit normal fluorescence. Upon transition to solid phase fluorescence intensity is almost completely self-quenched and fluorescence lifetimes in the nanosecond region decrease 2 times. This is easily measured. Usually large heavy metal atoms quench fluorescence. We observed the opposite new effect when LB film is deposited in the solid phase from a subphase containing heavy metals. The third new effect is the obtaining of nanosized cylinders with bilayer thickness, which remain stable at least for months, when LB monolayer is deposited in the phase coexistence region at thermodynamic equilibrium. This greatly increases reacting surface and sensitivity of possible sensors. Almost all possible optical experimental methods were used for this research. This includes polarized ATR FTIR and polarized UV-VIS spectroscopies, Fluorescence Lifetime Imaging Microscopy (FLIM), Scanning Near-field Optical Microscopy (SNOM).

1. Introduction

Reliable environmental monitoring strongly depends on the quality of chemical and biochemical sensors. There are still some unsolved problems especially when higher selectivity is required. Special

* To whom any correspondence should be addressed.

Journal of Physics: Conference Series **682** (2016) 012037 doi:10.1088/1742-6596/682/1/012037

problems pose secondary water contamination e.g. medicine disposal into drains [1, 2]. In this paper we propose a new class of materials – fluorescently labeled phospholipids, which can be used as chemical and biological sensors. We focus our attention on head labeled with nitrobenzoxadiazole (NBD) phosphatidyl ethanolamine. NBD molecule is considered one of the most suitable chromophores for sensor applications because of easily measurable fluorescence change on local changes of the polarity of the medium that surrounds the chromophore. In the nineties this effect was used for the fabrication of a number of biosensors by the Krull's group in Toronto [see e.g. 3]. More recently NBD labelled molecules were used as sensors for phosphatidylserine containing membranes [4], cysteine/homocysteine [5] and for selective chemical sensing of hazardous compounds and drugs of abuse [6]. In several hundred papers small percent (0.5 to 1 % typically) NBD labelled lipids were used in a mixture with other lipids to visualize domain structure in Langmuir films with fluorescence microscopy or in FRAP measurements. However we were the first to study these compounds as single component layers. Three new phenomena were discovered for this material that can be used in sensor applications. In our research we use the Langmuir and Blodgett (LB) method for investigation of organic monolayers at the air-water interface and for thin film deposition. The Nobel prize in mid-thirties winning LB method has renewed interest in it because this is the best method for supramolecular architecture offering discrete layer-by-layer film deposition, precise control of thickness and molecular orientation, considerably smaller number of defects compared to alternative methods. A small review of the method is given in our book chapter [7]. More recently Langmuir film from water insoluble molecules was used as a carried to deposit e.g. adsorbed proteins from the water subphase or as matrix for deposition of inorganic nano particles [8].

2. New effects discovered by us

In the past 25 years we have focused on investigations of Langmuir and Langmuir-Blodgett (LB) films from fluorescently labelled phospholipids mainly Dipalmitoyl Phospahatydyl Ethanolamine head labelled with NitroBenzoxaDiazole (DP-NBD-PE). Such long lasting focus allowed us to discover 3 new effects. First one is fluorescence self-quenching in Langmuir [9] and LB films [10]. Isotherms on Langmuir films from DP-NBD-PE on pure water subphase were measured on the 5th generation LB film system from Advanced Technologies Ltd. (Sofia, Bulgaria) at different temperatures (figure 1).

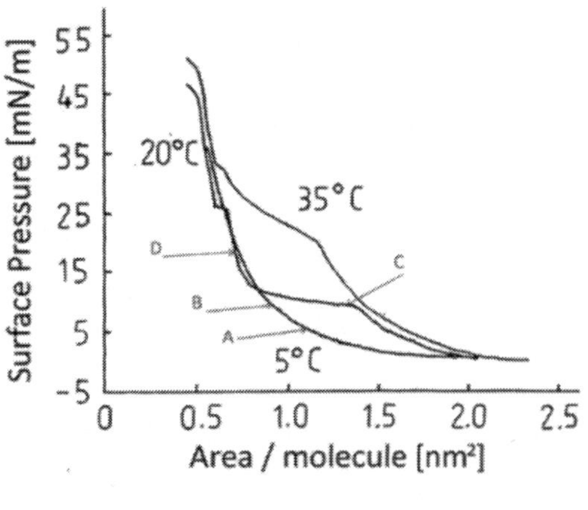

Figure 1. Isotherms of Langmuir films from DP-NBD-PE on pure water subphase at different temperatures (modified from [7]).

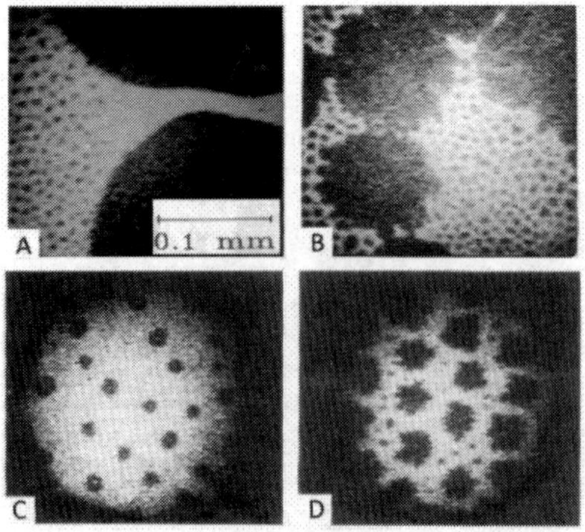

Figure 2. Fluorescence microscopy of a Langmuir film from DP-NBD-PE at different surface pressures indicated with red arrows on figure 1 and at different temperatures: 5° C – A) and B); 20° C – C) and D) (modified from [7, 9]).

Journal of Physics: Conference Series **682** (2016) 012037 doi:10.1088/1742-6596/682/1/012037

During compression optical epi Fluorescence (FL) microscopy with high sensitivity camera is used to look at the monolayer. Some image examples are shown on figure 2. At 5° C there is no 1st order phase transition from liquid-expanded to liquid condensed phase as seen from both figures. In the solid phase the fluorescence is strongly self-quenched (dark domains) due to radiationless energy transfer between the NBD groups. In the liquid phase FL remains strong, only partly quenched, due to interaction with water. This phenomenon remains upon transfer of the monolayer to solid support as LB film (figure 3). During the transfer process the structure of the film is preserved as observed by FL microscopy.

Another new phenomenon is shown and described on figure 4. Large paramagnetic metal atoms are known as good FL quenchers. However we observe the opposite effect. If large metal ions in the water subphase interact with the FL self-quenched DP-NBD-PE molecules in the solid phase then they restore the FL. If the monolayer is in the solid phase the fluorescent is self-quenched. If large bivalent metal ions interact with DP-NBD-PE molecules they attach to the negatively charged head group and separate the molecules in the solid phase. If this separation is larger than the critical distance R_c FL is de-quenched and emission is restored. This is what we observe for Cadmium (Cd) ions. Cd is very health threatening contaminant. This effect can be used for express monitoring of heavy metal contamination in water.

The 3rd new effect – obtaining a highly developed surface from DP-NBD-PE monolayers goes beyond the scope of the current paper.

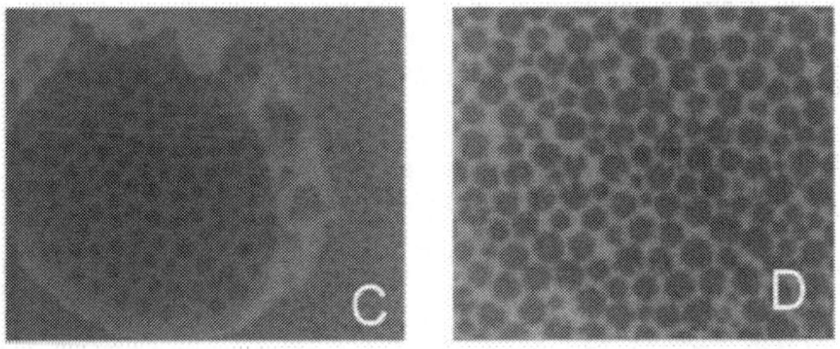

Figure 3. FL microscopy of an LB film from DP-NBD-PE deposited from pure water at room temperature. Some defects are seen in the left picture.

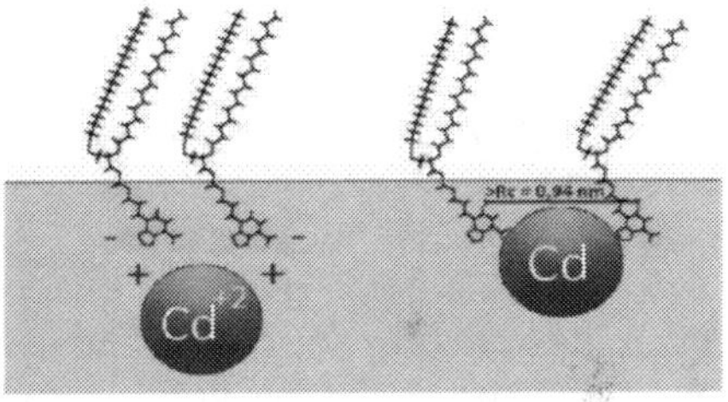

Figure 4. Second new effect. Large bivalent metal ions de-quench FL though typically they quench it.

Fluorescence spectroscopy of an LB film monolayers from DP-NBD-PE was recorded. When deposition is carried in the liquid phase below the main phase transition large well pronounced maximum at 530 nm is observed. If the deposition is carried in the solid phase above 26 mN/m then the fluorescence is almost completely quenched with just a slightly visible shoulder on the curve. This 95% quenchiung of the fluorescence is easily measurable. A model for the fluorescence self-quenching was developed [7]. It considers the probability for static quenching by the formation of emissionless traps consisting of pairs of statistical DP-NBD-PE molecules which are below the critical distance R_c. The model also considers the dynamic quenching due to Förster transfer of energy from DP-NBD-PE monomers to the traps. Assumptions in this model are: 1) statistical traps are formed according to two-dimensional equation of Perrin; 2) all DP-NBD-PE molecules that do not participate in the traps are uniformly distributed throughout the monolayer; 3) there is no diffusion during the lifetime of the excited state, 4) energy can move between and among fluorophores and traps, but once traps are reached energy immediately and without emission decreases; 5) passing of energy in more than one DP-NBD-PE molecule before reaching the trap is negligible. It is estimated that the distance at which the efficiency of Förster transfer of energy becomes 50% R_0 = 2,55 nm and that R_c = 0,94 nm. The insertion of the large metal ion (figure 4) between two DP-NBD-PE molecules increases the distance between the NBD chromophores above R_c. As a result de-quenching is observed and this effect can directly be used for express monitoring of heavy bivalent or trivalent metal water contaminations as discussed below. Currently there are no express methods to do so.

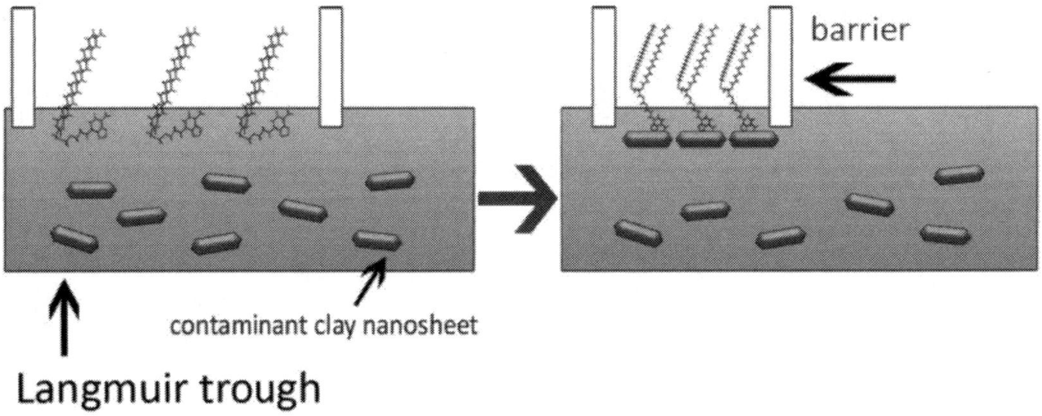

Figure 5. Langmuir trough method for monitoring of water purity.

3. Use of these new effects for measuring the quality of water in natural basins

The development of rapid, economical and sensitive techniques for characterization of the purity of natural and drinking water represents leading ecological problem. The surface properties of natural waters (sampled from rivers, lakes and gulfs) are already successfully used for evaluation of the ecology standard and purity of water basins. Recently it is proposed a Langmuir monolayer based technique which by measuring the surface pressure - area isotherm of the samples collected from a range of natural water basins, yields the so called "structural signatures" of water, which adequately predicted the quality and the purity of the basin [11]. Major advantage of the Langmuir monolayer technique is that it combines ease of use, high sensitivity and possibility for rapid field application with much lower price in comparison with the most commonly used chromatography techniques.

The "structural signatures" of samples of natural water result from the generalized scaling procedures applied to the surface pressure-area isotherms of the natural films. They appear to reflect in a quantitative and sensitive way the film composition, film solubility and the miscibility of its components, the kinetic mobility of surfactant molecules, and the compound's surface concentration. It is suggested that certain classes of film-forming components or "end-members" may dominate the static and dynamic surface properties. Variation in the surface rheological parameters of source-specific surfactants is postulated to reflect organic matter dynamics in natural waters. The reported results demonstrate that natural films are complex mixtures of biopolymeric molecules covering a wide range of solubility, surface activities and molecular masses with a complex interfacial architecture.

An improvement of the method would be to use a Langmuir film from DP-NBD-PE molecules (figure 5). If some contaminants attach to the negatively charged NBD chromophore this will lead to changes in local polarity of the medium and subsequently change in the NBD fluorescence. Thus in addition to isotherm data also FL spectroscopy data will be available which can help with selectivity and sensitivity of the sensor. The described above effect in figure 4 can be used for monitoring heavy metal contamination.

Further extension of this method would be to mix selectively sensing molecules e.g. enzymes (figure 6) in the Langmuir film and use the FL self-quenching effect if some conformational changes in the enzyme when the analyte attaches leads to change in surface pressure, self-quenching and subsequent easily measurable optical signal. FL quenching can be measured not only by measuring intensity of the signal but in addition FL maximum peak position and FL lifetime. Quenched FL has roughly 2 times shorter lifetimes of around 2 ns compared to unquenched signal (not shown).

Figure 6. The use of FL self-quenching for biosensor applications. If a selectively sensing molecule is mixed with DP-NBD-PE the interaction with analyte can cause increase of lateral pressure and transition of part of the molecules from liquid to solid phase and FL self-quenching with easily measurable signal.

4. Conclusion

Over the years we have discovered 3 new effects in NBD-labelled phospholipids which can significantly improve the performance of biosensors. The phospholipid matrix provides flexible biocompatible environment for immobilization of selectively reacting enzymes, microorganisms, DNA, immunoagents, whole cells. Use of single layer (3.1 nm thickness) films at the air-water interface (Langmuir films) or deposited on solid support as Langmuir-Blodgett (LB) film gives fast response times for real time monitoring (no slow diffusion processes) and precise molecule ordering and orientation. The first new effect discovered was fluorescence self-quenching in Langmuir and LB films. In the liquid phase films exhibit normal fluorescence. Upon transition to solid phase fluorescence intensity is almost completely self-quenched and fluorescence lifetimes decrease 2 times. This is easily measured. Usually large heavy metal atoms quench fluorescence. We observed the opposite new effect when LB film is deposited in the solid phase from a subphase containing heavy metals. Possible use of these effects for monitoring water quality was discussed here.

Acknowledgements

This work is partly supported under contract BN-176/15 with Center for Research and Design of UACEG and a donation from Advanced Technologies Ltd.

References

[1] Richardson S D and Ternes T A, 2014 "Water Analysis: Emerging Contaminants and Current Issues", Review, Anal. Chem., 86 (6), pp 2813–2848

[2] Rivera-Utrilla J, Sánchez-Polo M, Ferro-García M A, Prados-Joya G, Ocampo-Pérez R, 2013 "Pharmaceuticals as emerging contaminants and their removal from water. A review", Chemosphere **93**, (7), pp 1268–1287

[3] Brennan J D, Kallury K M R and Krull U J, 1994 Thin Solid Films, **244**, p. 898

[4] Lakshmi C, Hanshaw R G, Smith B D, 2004, "Fluorophore-linked zinc(II)dipicolylamine coordination complexes as sensors for phosphatidylserine-containing membranes", Tetrahedron 60 (49) pp 11307-11315

[5] Chen Y-H, Tsai J-C, Cheng T-H, Yuan S-S, Wang Y-M, 2014 "Sensitivity evaluation of NBD-SCN towards cysteine/homocysteine and its bioimaging applications", Biosensors and Bioelectronics, **56**, pp 117-123

[6] Sharma P S, D'Souza F and Kutner W, 2012 "Molecular imprinting for selective chemical sensing of hazardous compounds and drugs of abuse", Trends in Analytical Chemistry **34**, pp 59-77

[7] Ivanov G R, Georgiev G and Lalchev Z 2011 "Fluorescently Labeled Phospholipids - New Class of Materials for Chemical Sensors for Environmental Monitoring", in Relevant Perspectives in Global Environmental Change, ed J I Agboola, ISBN: 978-953-307-709-3, Rijeka: (InTech) chapter 7 pp 89-112

[8] Angelov T, Radev D, Ivanov G R, Antonov D, Petrov A G, 2007, "Hydrophobic magnetic nanoparticles: synthesis and LB film preparation", J. Optoelectronics and Advanced Materials, **9** (2) pp 424 – 426

[9] Ivanov G R, 1992 "First observation of fluorescence self-quenching in Langmuir films", Chem. Phys. Lett., **193** p 323

[10] Ivanov G R, J. Nanoscience and Nanoengineering, submitted

[11] Pogorzelski S J, 2001 "Structural and thermodynamic characteristics of natural marine films derived from force-area studies", Colloids Surfaces A: Physicochem. Eng. Aspects., **189** pp 163–176

Institute of Physics
Dirac House, Temple Back
Bristol BS1 6BE UK

ISSN: 1742-6588
ISBN 978-1-5108-2092-0

9 781510 820920